90 0639581 X

Groundwater in the Celtic Regions:
Studies in Hard Rock and Quaternary Hydrogeology

Special Publication reviewing procedures

The Society makes every effort to ensure that the scientific and production quality of its books matches that of its journals. Since 1997, all book proposals have been refereed by specialist reviewers as well as by the Society's Publications Committee. If the referees identify weaknesses in the proposal, these must be addressed before the proposal is accepted.

Once the book is accepted, the Society has a team of series editors (listed above) who ensure that the volume editors follow strict guidelines on refereeing and quality control. We insist that individual papers can only be accepted after satisfactory review by two independent referees. The questions on the review forms are similar to those for *Journal of the Geological Society*. The referees' forms and comments must be available to the Society's series editors on request.

Although many of the books result from meetings, the editors are expected to commission papers that were not presented at the meeting to ensure that the book provides a balanced coverage of the subject. Being accepted for presentation at the meeting does not guarantee inclusion in the book.

Geological Society Special Publications are included in the ISI Science Citation Index, but they do not have an impact factor, the latter being applicable only to journals.

More information about submitting a proposal and producing a Special Publication can be found on the Society's web site: www.geolsoc.org.uk.

GEOLOGICAL SOCIETY SPECIAL PUBLICATION NO. 182

Groundwater in the Celtic Regions: Studies in Hard Rock and Quaternary Hydrogeology

EDITED BY

N. S. ROBINS
British Geological Survey, Wallingford, UK

B. D. R. MISSTEAR
University of Dublin, Trinity College, Ireland

2000

Published by

The Geological Society

London

THE GEOLOGICAL SOCIETY

The Geological Society of London was founded in 1807 and is the oldest geological society in the world. It received its Royal Charter in 1825 for the purpose of 'investigating the mineral structure of the Earth' and is now Britain's national society for geology.

Both a learned society and a professional body, the Geological Society is recognized by the Department of Trade and Industry (DTI) as the chartering authority for geoscience, able to award Chartered Geologist status upon appropriately qualified Fellows. The Society has a membership of 9099, of whom about 1500 live outside the UK.

Fellowship of the Society is open to those holding a recognized honours degree in geology or cognate subject, or not less than six years' relevant experience in geology or a cognate subject. A Fellow with a minimum of five years' relevant postgraduate experience in the practice of geology may apply for chartered status. Successful applicants are entitled to use the designatory postnominal CGeol (Chartered Geologist). Fellows of the Society may use the letters FGS. Other grades of membership are available to members not yet qualifying for Fellowship.

The Society has its own publishing house based in Bath, UK. It produces the Society's international journals, books and maps, and is the European distributor for publications of the American Association of Petroleum Geologists, (AAPG), the Society for Sedimentary Geology (SEPM) and the Geological Society of America (GSA). Members of the Society can buy books at considerable discounts. The publishing House has an online bookshop (*http://bookshop.geolsoc.org.uk*).

Further information on Society membership may be obtained from the Membership Services Manager, The Geological Society, Burlington House, Piccadilly, London W1V 0JU, UK. (E-mail: *enquiries@geolsoc.org.uk*: tel: +44 (0)207 434 9944).

The Society's Web Site can be found at *http://www.geolsoc.org.uk/*. The Society is a Registered Charity, number 210161.

Published by The Geological Society from:
The Geological Society Publishing House
Unit 7, Brassmill Enterprise Centre
Brassmill Lane
Bath BA1 3JN, UK

(*Orders*: Tel. +44 (0)1225 445046
Fax +44 (0)1225 442836)
Online bookshop: *http://bookshop.geolsoc.org.uk*

British Library Cataloguing in Publication Data
A catalogue record for this book is available from the British Library.

ISBN 1-86239-007-0
ISSN 0305-8719

Typeset by Aarontype Ltd, Bristol, UK

Printed by Alden Press, Oxford, UK

Distributors
USA
AAPG Bookstore
PO Box 979
Tulsa
OK 74101-0979
USA
Orders: Tel. +1 918 584-2555
Fax +1 918 560-2652
E-mail: *bookstore@aapg.org*

Australia
Australian Mineral Foundation Bookshop
63 Conyngham Street
Glenside
South Australia 5065
Australia
Orders: Tel. +61 88 379-0444
Fax +61 88 379-4634
E-mail: *bookshop@amf.com.au*

India
Affiliated East-West Press PVT Ltd
G-1/16 Ansari Road, Daryaganj,
New Delhi 110 002
India
Orders: Tel. +91 11 327-9113
Fax +91 11 326-0538
E-mail: *affiliat@nda.vsnl.net.in*

Japan
Kanda Book Trading Co.
Cityhouse Tama 204
Tsurumaki 1-3-10
Tama-shi
Tokyo 206-0034
Japan
Orders: Tel. +81 (0)423 57-7650
Fax +81 (0)423 57-7651

Contents

It is recommended that reference to all or part of this book should be made in one of the following ways:

ROBINS, N. S. & MISSTEAR, B. D. R. (eds) 2000. *Groundwater in the Celtic Regions: Studies in Hard Rock and Quaternary Hydrogeology*. Geological Society, London, Special Publications, **182**.

FOX, I. A. 2000. Groundwater protection in Scotland. *In*: ROBINS, N. S. & MISSTEAR, B. D. R. (eds) *Groundwater in the Celtic Regions: Studies in Hard Rock and Quaternary Hydrogeology*. Geological Society, London, Special Publications, **182**, 67–70.

Referees

The Editors are grateful to the following people for their assistance with the reviewing of papers submitted to this Special Publication.

Mr B. Adams	British Geological Survey, Wallingford
Mr J. Anderson	Scottish Environment Protection Agency, Perth
Dr R. P. Ashley	Mott MacDonald, Cambridge
Mr D. Banks	Holymoor, Chesterfield
Dr R. Barker	School of Earth Sciences, University of Birmingham
Mr K. Beesley	European Geophysical Services Ltd., Shrewsbury
Dr S. Beeson	Mott MacDonald, Cambridge
Mr J. R. P. Bennett	Hydrogeological & Environmental Services Ltd., Belfast
Mr F. C. Brassington	Culcheth, Warrington
Dr W. G. Burgess	Department of Geological Sciences, University College London
Dr R. Carter	Institute of Water and Environment, Cranfield University, Silsoe
Mr R. S. Culverwell	Public Services Department, States of Jersey
Mr D. Daly	Geological Survey of Ireland, Dublin
Mr E. P. Daly	E. P. Daly Associates, Delgany
Mrs N. J. Dottridge	Komex, London
Dr D. Drew	Department of Geography, University of Dublin, Trinity College
Mr S. Dumpleton	British Geological Survey, Keyworth
Mr V. Fitzsimons	Geological Survey of Ireland, Dublin
Dr M. Forde	Dames & Moore, Manchester
Professor S. S. D. Foster	British Geological Survey, Wallingford
Dr T. Haines	Entec UK Ltd., Shrewsbury
Ms T. Hayes	K. T. Cullen & Company, Dublin
Dr K. M. Hiscock	School of Environmental Sciences, University of East Anglia
Dr P. Howsam	Institute of Water and Environment, Cranfield University, Silsoe
Dr A. G. Hughes	British Geological Survey, Wallingford
Ms P. Johnston	Environment Agency, Exeter
Mr C. R. C. Jones	Mott MacDonald, Cambridge
Professor G. P. Jones	Henley-on-Thames, London
Dr B. A. Klinck	British Geological Survey, Keyworth
Mr A. R. Lawrence	British Geological Survey, Wallingford
Professor J. W. Lloyd	School of Earth Sciences, University of Birmingham
Professor J. D. Mather	Department of Geology, Royal Holloway, University of London
Mr P. McConvey	Geological Survey of Northern Ireland, Belfast
Ms H. Nash	Wardell Armstrong, Newcastle-under-Lyme
Dr C. P. Nathanail	School of Chemical Environmental and Mining Engineering, University of Nottingham
Dr C. Neal	Centre for Ecology and Hydrology, Wallingford
Mr S. Neale	Environment Agency Wales, Cardiff
Dr M. Packman	Southern Water, Crawley
Mr R. C. Palmer	Soil Survey & Land Research Centre, Cranfield University, York
Dr A. Papaioannu	Entec UK Ltd., Shrewsbury
Mr M. Price	Postgraduate Research Institute for Sedimentology, University of Reading
Mr C. Shine	Minerex Environmental Ltd., Dublin
Dr J. H. Tellam	School of Earth Sciences, University of Birmingham
Dr P. Wood	School of Applied Sciences, University of Huddersfield
Mr G. Wright	Geological Survey of Ireland, Dublin

Introduction

Our perception of our environment and the ways in which it varies and evolves through time depends very much on our daily priorities. Geologists are happy to think about eras of millions of years, hydrogeologists routinely consider thousand year periods, historians are concerned with centuries, and human consciousness exists over a period of between six months and a hundred years: three generations generally mark the limit of our collective memory and understanding of our lives. Interaction between man and the environment may lead to conflict, precisely because of the different time scales over which natural phenomena and human society exist. In hydrogeological terms, groundwater reserves which have taken centuries or millennia to accumulate can all but disappear within a few years due to over-abstraction, while groundwater pollution which can occur in minutes or hours (such as an accidental spillage of petrol or industrial chemicals on permeable ground) can take many years or even decades to remediate.

Our use of groundwater is not only a modern phenomenon. There is a wealth of evidence that groundwater has been important in the Celtic regions for many thousands of years. Various Celtic societies have occupied large, though decreasing, areas of Western Europe since at least the eighth century BC: the few groups existing today on the edge of the Atlantic Ocean are only the remnants of a much larger prehistorical culture. The long history of European Celticity has left a legacy of place names and words associated with the earlier Celtic culture. Many of the myths and religious beliefs of this culture, particularly those associated with the supernatural or 'other world', focused on water, both surface and groundwater. Clues to these beliefs can be read in the remnants of Celtic languages which survive into modern times as names of places and environmental features.

Analysing words and how their meaning changes over time is one of the mainstays of historical linguistic studies. Part of the fascination of such studies is that we can use language explorations as echo-sounders into a hidden past to try and understand the mind-set of vanished peoples. By trying to interpret the cultural associations attached to words which survive from ancient languages, we attempt to gain insights into the social reality of past societies.

One important aspect of the history of societies has been that of cultural and linguistic over-writing, as earlier languages gradually gave way to replacement tongues. For the European Celtic languages this process started in North Italy around the third century BC, to be followed by Switzerland and South Germany (replaced by German or Romance tongues), Gaul and Spain (replaced by Latin, itself later changing to the modern languages of French and Spanish). It began in England in the sixth century AD as native Celtic began to be replaced by English, a process completed within a few hundred years, and has been observable in the remaining so-called Celtic fringes of the British islands from around the fourteenth century. These changes normally occur across at least several generations of bilingual speakers, giving rise to a replacement pattern which ranges from monoglots through several generations of bilingual speakers, to finish with monoglot replacement speakers. The vital linguistic link in this chain is the duration and extent of the period of bilingualism, which allows the names of objects and ideas important to a society to be filtered across the barrier of language shift. In a gradual process, which may extend over several centuries, culturally important words from the original languages, such as place names, can retain traces of their original meanings in the new overwriting culture. The linguistic identification of the roots and origins of particular words and names allows us to make the suggestions we do about their earlier cultural importance.

The Celtic supernatural world, peopled with gods and goddesses, acted as a link between their physical surroundings and those of their religious beliefs. Fundamentally, Celtic religion had a chthonic basis, with an underground otherworld, unlike the religions of the Mediterranean peoples where the sky was the abode of gods. Archaeologists tell us that another difference between early Celtic cultic practice and that of their Mediterranean neighbours was the paucity (or perhaps historical loss) of large numbers of built temples as sacred places for worship. Celtic religion appears to have used parts of their natural environment as sacred sites, particularly forest clearings. In addition, sunk shafts (some more than 36 m deep) appear to have been associated with religion, and although their

From: Robins, N. S. & Misstear, B. D. R. (eds) *Groundwater in the Celtic Regions: Studies in Hard Rock and Quaternary Hydrogeology*. Geological Society, London, Special Publications, **182**, 1–3. 1-86239-077-0/00/$15.00

precise cultic purpose is uncertain, they do show evidence of ritual deposits, including the remains of sacrifices to the gods.

It is a short enough cultural transfer from this ritual use of shafts, which must often have been prone to filling with water, to a belief in a relationship between gods and underground water. From here, an extension of belief to the more general notion of the water-surface as representing an interface between the world of everyday reality and the supernatural world seems to follow easily. Later in history, from the maritime as opposed to the continental areas of the Celtic world, we find in Ireland ideas such as *Tír na nÓg*, where the otherworld of unchanging youth is located in the western ocean. This must have been a way for the Celtic peoples of Ireland to come to terms with the terminal edge of their physical world. The inhabitants of Celtic Britain and Ireland appear in general to have believed that large areas of water – the sea or inland lakes – contained in their depths representatives of an otherworld (*Tír fó Thuinn*), whether in human form or in monster shape. The Celtic Sea was interpreted this way in Welsh tradition, and even today myths surrounding both Lough Neagh and Loch Ness attest to such beliefs.

It is easy to accept that the presence of underground water, and its emergence through wells or springs, represented a boon from the otherworld, and there is clear evidence for a Celtic interpretation of rivers as representatives of this chthonic world. Thus we find numerous examples of the names of female deities being given to rivers across the wider Celtic world. River names are a regular carry-over between one culture and its replacement: names such as Seine (*Sequana, goddess of healing*) and Boyne (*Boann*) are indications of this connection with goddesses. The spring sources of rivers were especially revered, and at the spring site of *Sequana* on the Châtillon Plateau in Burgundy many votive deposits have been found, including bronze and silver models of parts of human bodies deposited as requests for healing. Even without a direct connection with a river, wells and springs have been recognized for centuries as objects of veneration within Celtic countries (or in their replacement cultures), particularly for their reputed healing properties, as indicated by the nineteenth century example of Lourdes and other such religiously over-written sites. The words *tobar* (Irish, Scottish Gaelic) and *ffynon* or *pudew* (Welsh) are used for well names in these areas of the British Isles. Although hydrogeologists have an interest in wells or spring sites for their own purposes, any survey of such sources should include not only the necessary scientific analysis, but also record any local names given to wells or springs. Such language data would potentially serve to enhance relationships between the two disciplines, and perhaps allow us to extend our understanding of the historical continuity of relationships between the Celts and their water sources.

Above ground, the lake (*loch, linn, llyn*) acts as a natural reservoir, both physical and cultural, for its surrounding population, and here also we find evidence for the ritual use of lake sites across the Celtic world. In Continental and Insular Europe they were considered as important repositories for votive offerings, including human sacrifices. In the main such deposits consist of swords and other weapons, with their blades usually bent almost in two before deposition, to demonstrate that they were no longer functional in this world and that they took on the rôle of an offering to the otherworld. Lake Neuchâtel in Switzerland is an excellent type site for this practice, as projecting into this lake in the vicinity of La Tène a wooden platform was erected to act as a site of votive deposition. Several hundred brooches, spears, shields and swords have been recovered from its vicinity on the lake bottom.

In 106 BC the Roman general Caepio was sent with an army into South West France (*Gallia Narbonensis*) to put down a revolt. Like most other such generals in history, he took advantage of the opportunity for plunder after sacking the town of Tolosa (*Toulouse*), by seizing its sacred lakes and instructing his engineers to drain these and remove the votive treasures – mostly gold – deposited there. These materials had remained untouched by the Celtic inhabitants, being considered in much the same way as we would envisage our church or civic treasures nowadays. This is probably the earliest recorded instance in Europe of pollution of a water site, with a clear indication of the original meaning of pollution with its ritual significance. It was also, as is usual in such cases, carried out for financial gain, though no doubt there would have been Roman spin doctors to argue that the removal of such heavy metal pollution from the lake was of considerable benefit to its biological population, whether flora or fauna.

The papers in this volume indicate a concern with water abstraction and its use in what are essentially maritime environments, on islands with a broadly common geographical situation far removed from the original European homelands of the Celtic peoples. It is clear that on these islands we share a vulnerable hydrogeological situation, as well as an interrelated social and cultural history, and a shared understanding between scientists and humanities specialists can only be to our mutual benefit. In the 1960s, a local resident of West Town on Tory, a small granite island off the north-west coast of Donegal, revealed how proud the inhabitants were of the only well in the village, 'which had never run dry'. Boons from the ground are as welcome and necessary today as they were 2500 years ago to our ancestors. With modern exploration and drilling techniques and the construction of deeper boreholes we can reach much further into that Celtic otherworld to seek these boons.

As well as their common history, this volume emphasizes the shared environmental heritage of the remaining Celtic regions. It also illustrates the hydrogeological issues shared with the wider world. Today we are more likely to find elevated nitrates or chlorinated solvents in Celtic wells than votive offerings, and the impacts of global climatic change are also likely to be felt in the maritime Celtic environment. The preservation of useable water resources, both above and below the ground surface, is not a short term problem, and we all – in the humanities and the sciences – must attempt to develop and consolidate our understanding of a world which we currently tend to view exclusively through the separate lenses of individual subject perceptions. It is on this shared stage that we all must live, as a potential future without common access to useable water is untenable for all of us.

Cathair Ó Dochartaigh
Department of Celtic/Roinn na Ceiltis
University of Glasgow

Groundwater in the Celtic regions

N. S. ROBINS[1] & B. D. R. MISSTEAR[2]

[1] *British Geological Survey, Maclean Building, Wallingford, Oxfordshire OX10 8BB, UK*
[2] *Department of Civil, Structural & Environmental Engineering, University of Dublin, Trinity College, Dublin 2, Ireland*

Abstract: The Celtic regions of Britain and Ireland have a complex and diverse geology which supports a range of regionally and locally important bedrock aquifers and unconsolidated Quaternary aquifers. In bedrock, aquifer units are often small and groundwater flow paths short and largely reliant on fracture flow. Groundwater has fulfilled an important social role throughout history, and is now enjoying renewed interest. Groundwater quality is generally favourable and suitable for drinking with minimal treatment. However, many wells are vulnerable to microbiological and chemical pollutants from point sources such as farmyards and septic tank systems, and nitrate concentrations from diffuse agricultural sources are causing concern in certain areas. Contamination by rising minewaters in abandoned coalfields and in the vicinity of abandoned metal mines is also a problem in some of the Celtic lands.

At the close of the Hallstatt period the Celts occupied much of central and western Europe, and Britain and Ireland were home to the Gaels, the Picts and the Britons. Nowadays, the land of the Britons is divided between England and Wales, and the Celtic regions are considered principally as Wales, south-west England (the Cornish peninsula), Northumbria, Scotland and Ireland. Celtic regions have become those areas farthest from continental Europe and the English capital. History has repeatedly demonstrated that the people of the Celtic regions share an independence of outlook which in Britain has recently been reflected in political devolution. But as well as similarities in the aspirations, history and culture of the different Celtic peoples, there is a common thread through many of the hydrogeological and associated environmental issues affecting the Celtic regions.

The Celtic regions enjoy a maritime temperate climate with orographic rainfall over upland or hilly terrain on the western margins. Apart from these characteristic upland areas there are a number of broad tracts of lowland. The geology of these lands is diverse and complex, with ages and lithologies ranging from Precambrian granite and metasediments, through Palaeozoic limestone, Permo-Triassic sandstone and Cretaceous Chalk, to Palaeogene volcanic rocks (Fig. 1). Drift geology includes a widespread blanket mantle of till as well as a range of coarser granular Quaternary deposits. Above this the variety of soil types include peat, gley and sandy soils. Land use includes grassland characteristic of much of lowland Ireland, upland pastoral farming typical of much of central Wales and arable cultivation in eastern Scotland. Forestry is also extensive. Many lowland coastal areas support the highest population density as well as traditional heavy industry, manufacturing and service industry, and mining has left its impact over the Midland Valley of Scotland, South Wales, south-west England and parts of central and eastern Ireland.

Up and until now the main focus in hydrogeology has been groundwater development and resources evaluation. These activities preoccupied hydrogeologists throughout the 1980s and early 1990s, but since then the emphasis has moved towards environmental aspects of groundwater. Of particular importance are groundwater protection against pollution and the interrelationship between groundwater and surface water, especially in sensitive wetland habitats.

A number of specific technical issues are pertinent to the Celtic regions. These include:

(a) recharge is difficult to quantify, more so where Quaternary deposits are present;
(b) the hydrogeological characteristics of glacial tills require detailed research, especially the role of bypass flow;
(c) hydrogeological processes in peat are little understood (e.g. 14% of Scotland is covered by 1 m or more thickness of peat);

From: ROBINS, N. S. & MISSTEAR, B. D. R. (eds) *Groundwater in the Celtic Regions: Studies in Hard Rock and Quaternary Hydrogeology*. Geological Society, London, Special Publications, **182**, 5–17. 1-86239-077-0/00/$15.00

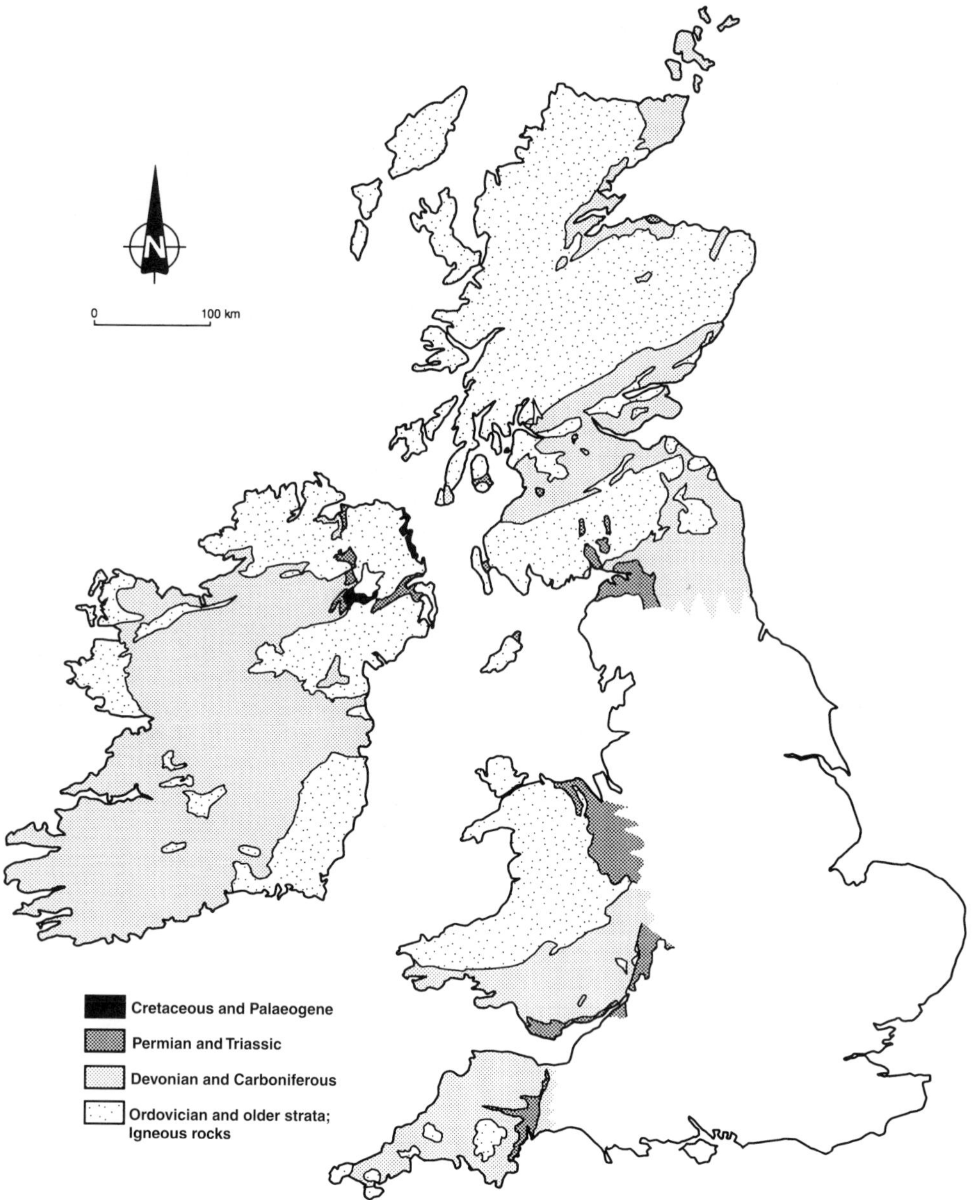

Fig. 1. Broad geological/hydrogeological groups in the Celtic regions.

(d) fracture flow of groundwater predominates, and is difficult to replicate by modelling;

(e) karst hydrogeology involves complex interactions with surface water, and requires intensive study;

(f) minewater discharges in the shallow and surface environment require evaluation and remediation;

(g) valley bottom hydrology/hydrogeology is poorly understood, particularly in upland catchments;

(h) local groundwater quality is often affected by point sources of pollution such as septic tank systems, and diffuse pollution from agriculture is also a threat to water quality in particular regions;
(i) development of brownfield sites and contaminated land has implications for groundwater quality.

Groundwater and people

Groundwater fulfils an important role throughout the Celtic lands in satisfying demand for water, both on a local and a national level. Groundwater is not generally considered as the mainstay of bulk public supplies in many areas, although its contribution to public supply is significant (Table 1). Exploitation for generally small volume private consumption, principally for domestic or farm use, is also very important, and larger volume uses for irrigation and industry are also surprisingly widespread.

There are a number of reasons why groundwater supply is important even in areas facing the prevailing moisture-laden westerly winds, and blessed with abundant surface water resources. The reasons for this include:

(a) the consistent and generally favourable quality of groundwater;
(b) ready access to groundwater, at least in small volumes;
(c) economics of development and supply;
(d) mainly licence-exempt development.

These factors conspire to increase the importance of groundwater for rural and island community supplies. Many peat-coloured village supplies in upland regions, and other surface water supplies of variable quality and associated treatment difficulties, are being replaced by small groundwater schemes targeted at village level. Groundwater is also valuable for island community supply where insufficient catchment area and space for surface water gathering make groundwater a preferred option (Fig. 2). Perhaps the greatest advantage of groundwater is that it is available at many, diverse locations, and it does not need long and expensive pipe runs connecting source and consumer.

Fig. 2. Exploratory drilling for public water supply on Arran.

Poor aquifer characteristics in many areas, especially in the older bedrock formations, mean that groundwater supplies are often relatively small by comparison to groundwater schemes in

Table 1. *Estimates of groundwater in supply*

	Groundwater as a % of total public supply	Volume of groundwater in public supply (Ml d^{-1})	Estimated number of private sources	Volume of water in private supply (Ml d^{-1})
Scotland	5	110	20 000	110
N Ireland	10	77	1 500	31
Ireland	15	270	>100 000	80
Wales	8	250	5 000	95
Northumbria	10	110	3 000	30
Cornwall	3	12	10 000	62

the Sherwood Sandstone and Chalk aquifers of central and southern England. However, there are a very large number of wells in the Celtic lands. Although it is not possible to give an exact number, one estimate is that there are more than 200 000 wells in Ireland alone (Wright 1999). In addition to wells, groundwater supplies are obtained from springs, infiltration galleries and Ranney wells (Jones & Singleton this volume). Some of the springs have very large flows e.g. the Schwyll spring in South Wales has an average flow of $390 l s^{-1}$ (Hobbs this volume) and the Pouladower spring in County Clare in western Ireland has mean flows of between 115 and $720 l s^{-1}$ (Deakin this volume).

The traditional social importance of groundwater sources is illustrated by the naming of certain springs and wells after saints. This practice conferred some measure of protection on the source from elements of evil, and the water could be drunk for cures and grace. Although pilgrimages to Holy wells are less common nowadays, many wells are still visited, especially on the Saints' day. Selected Clootie wells (Devil or Witch) in the Black Isle of northern Scotland are annually bedecked with a variety of torn coloured rags (clooties) to attract a cure from disease. Elsewhere, Holy wells – there are believed to be about 3000 in Ireland alone (Logan 1980) – are adorned with plaques bearing the scriptures and may be blessed periodically by the priest.

Nowadays, groundwater also contributes to bulk public supply. In Ireland about 15% of the water supplied by local authorities comes from groundwater, and groundwater accounts for about 25% of all water supplies in the country. In some counties the proportion of total water supply from groundwater is much greater, e.g. Roscommon 86%, Offaly 56% and Laois 52% (Department of the Environment and Local Government *et al.* 1999). There are also important wellfields for public supply in the Lagan Valley in Northern Ireland, in Dumfriesshire, Morayshire and Fife in Scotland, and in alluvial valley fill material in west Wales.

Groundwater has fostered a number of high profile industries. The major brewing centres all at one time relied on groundwater for make-up water, the mineralization of the water offsetting the need to add brewing salts and providing each centre with a unique product flavour. Distilleries, both in Scotland and Ireland, rely heavily on groundwater and jealously guard the chemistry of their many spring and borehole sources from competitors. Nowadays, Celtic groundwater is also packaged and sold neat, not only through the home market but also worldwide. This burgeoning industry plays on the image of green and pleasant lands to sell untreated bottled groundwater at prices that currently exceed those of petrol, volume for volume. However, not all bottlers are situated in such pristine environments as consumers might imagine. Some draw from shallow alluvial sources at risk to pollution by local farming activity, whilst others draw from sandstone aquifers adjacent to rivers of periodically indifferent quality and of undefined hydraulic connectivity with the groundwater.

Groundwater and the environment

Groundwater is a key to many environmental issues in the Celtic regions. Its role in maintaining low surface water flow is as fundamental to the maintenance of wetland areas, including groundwater fed peatlands, as it is to the maintenance of the integrity of stream flow (Soulsby *et al.* this volume). It also plays a part in moderating acid surface water environments, as baseflow is commonly alkaline even in areas apparently devoid of calcite, so that otherwise acid upland streams may temporally become alkaline.

Groundwater, although seemingly innocent, is not all good news. Acid and iron-rich minewaters accumulate in abandoned mineworkings and rebound to contaminate the shallow environment and ultimately surface waters (Younger & La Pierre this volume). Contamination from point sources such as landfill, industry or septic tank systems coupled with diffuse pollution from agricultural and other activities requires careful management of groundwater resources (Lewis *et al.* this volume; Misstear & Daly this volume). Source protection can be undertaken in many fractured hard rock aquifers according to the prevailing groundwater setting rather than by means of detailed data gathering and analytical modelling (Robins 1999). In addition, management activities need to be supported by enabling legislation in order that resource derogation is minimized (Fox this volume).

Groundwater occurrence

Classes of aquifer

Three classes of aquifer occur within the Celtic regions (Table 2). The first incorporates bedrock aquifers of regional importance such as the sandstone basins of North Wales, south-west Scotland and Northern Ireland; the limestone

Table 2. *Hydraulic characteristics of the three aquifer classes in the Celtic regions*

Class	Geology	Location	Properties	Transmissivity ($m^2 d^{-1}$)	Storativity	Borehole yield ($l s^{-1}$)
Regionally important bedrock	Permian and Triassic basins; some Devonian sandstones; Carboniferous Limestone; some volcanics	Generally in lowland areas; Karst limestone in Ireland & South Wales	Anisotropic, fracture flow dominant, regional and local flow paths	100 to 4000	0.01–0.20	5 to 40
Locally important bedrock	Precambrian and Lower Palaeozoic; some Upper Palaeozoic; some volcanics	Widespread, in lowland and upland areas; Lough Neagh basin	Anisotropic, secondary porosity dominant, local flow paths	20 to 100	<0.05	1 to 5
Superficial	Alluvium; granular glacial deposits; raised beach; and wind-blown sand	Widespread, best developed in valley bottoms and coastal areas	Primary porosity only, potential is limited by geometry	50 to 5000	Variable	3 to 40

aquifers of Ireland; and the Devonian and Carboniferous aquifers of South Wales and the Midland Valley of Scotland (Fig. 1). The second class comprises the bedrock aquifers of local importance, which are widespread, and include Lower Palaeozoic and Precambrian strata as well as volcanic aquifers, and some Devonian and Carboniferous rocks. The aquifers that occur within the Quaternary and younger superficial strata form the third class. These are also widely distributed, and in some areas can form aquifers of regional importance (e.g. the Curragh glacial gravel aquifer in eastern Ireland).

The regionally important bedrock aquifers are typically characterized by secondary permeability of considerably greater magnitude than the available intergranular permeability. The aquifers may be anisotropic with the prevailing geological structure influencing the preferred horizontal flow paths, and with vertical permeability controlled by bedding and the availability of vertical joints. The degree of preferential flow varies from the weakly fractured Devonian aeolian sandstones of the Midland Valley of Scotland (Gaus & Ó Dochartaigh this volume) to the conduit flow dominated karstic limestones of Ireland (Coxon & Drew this volume) and South Wales (Hobbs this volume). Although these aquifers are generally unconfined they may locally be confined by till, especially in low-lying areas. Whether physically confined or not, a confined response is not uncommon on pumping these aquifers, and this reflects the small storativity of a fracture that is locally remote from the 'water table'. Groundwater flow paths may be several kilometres in length, and may pass across a number of surface water catchments between principal recharge area and discharge area. The majority of the groundwater transport is shallow except where deeper intergranular porosity and dilated fractures are present as, for example, in parts of the Permo-Triassic sandstones in Scotland and Northern Ireland. Good data on aquifer properties, such as transmissivity and specific yield, are relatively scarce in the Celtic regions, and aquifer performance is often expressed in terms of potential well yield and/or specific capacity (Wright this volume). Sustainable borehole yields are typically in the range 5 to $20 l s^{-1}$, but yields up to $40 l s^{-1}$ have been obtained in some areas.

Locally important bedrock aquifers generally have little intergranular permeability except in isolated horizons, so that nearly all groundwater storage and transport takes place within available secondary porosity. Enhanced storage may be available if the bedrock aquifer is in hydraulic contact with saturated granular superficial deposits. These aquifers are strongly anisotropic and may react to pumping as if they were confined, for the reason given above. Borehole yields are typically in the range 1 to $5 l s^{-1}$ although some higher yielding sources, notably spring flows, have been recorded. Groundwater flowpaths are generally short and usually correspond to groundwater catchment scale (Robins *et al.* this volume).

Groundwater in unconsolidated superficial deposits offers intergranular porosity for storage and transport of groundwater. The geometry of the deposit may constrain the storage potential unless the aquifer is in hydraulic contact with a nearby stream or river. Different lithologies may be interbedded within one deposit and horizontal permeability may exceed vertical permeability with, for example, a gravel layer offering greater potential than a silty sand. Peat and clay layers may act as local aquicludes within superficial sequences. Sustainable yields from these deposits are variable, but range up to $40\,l\,s^{-1}$ in selected locations which have both enhanced storage from a surface water body and suitably high transmissivity (Hiscock & Paci this volume). In this respect, two major bankside wellfield schemes have been developed in alluvial gravels in Scotland in recent years (Jones & Singleton this volume). Thin aquifers are self-limiting wherever the pumping drawdown from a groundwater source reduces the effective aquifer thickness, and hence also transmissivity, leading to declining yields. As well as forming aquifers in their own right, permeable superficial deposits are often significant in terms of the additional storage they provide for underlying bedrock aquifers. They may also act as pathways for groundwater discharge. For example, there are many springs in Ireland fed by limestone aquifers that issue via sand and gravel horizons (Daly 1995).

There are few non-aquifers although many unweathered rocks have low permeabilities. A non-aquifer comprises massive rock with no transmissive discontinuities and such conditions do not generally exist in the relatively shallow weathered zone other than on a limited scale. Most of the geological formations in the Celtic regions are capable of providing at least a small water supply, and so cannot be regarded as non-aquifers.

Role of secondary permeability

Deformation and weathering of consolidated rock creates fractures that promote ingress of groundwater and further potential for weathering. The blocks of unweathered material may break down to form a granular weathering product, and subsequently a clay grade material. In northern Britain these products have been removed or reworked by glacial action, but the underlying fractured bedrock remains in place. The fractures provide secondary porosity which enables groundwater transport in rocks with little if any primary or intergranular porosity (Fig. 3). This is illustrated by comparing laboratory determined permeability and porosity values with hydraulic conductivity and storativity derived from field testing in Permian and Triassic aquifers (Table 3). The field conductivity

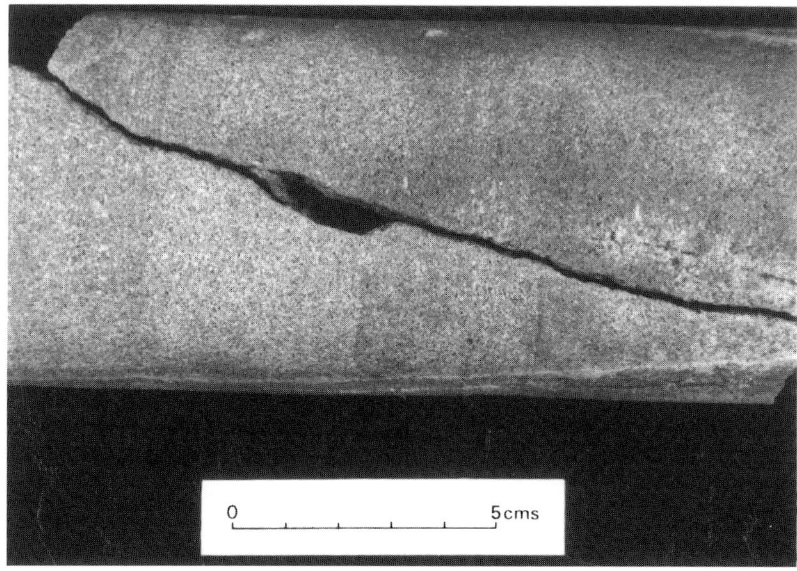

Fig. 3. Open fracture in Devonian sandstone from Morayshire.

Table 3. *Aquifer evaluation data for boreholes in the Permian and Triassic aquifers in south-west Scotland and Northern Ireland*

Area	Yield ($l\,s^{-1}$)	Hydraulic conductivity ($m\,d^{-1}$)	Laboratory permeability ($m\,d^{-1}$)	Storativity	Laboratory porosity
South-west Scotland					
Dumfries	53	1	0.001	0.01	0.15
Dumfries	17	5	–	0.1	–
Dumfries	8	20	0.01	–	0.19
Mauchline	41	7	–	0.0001	–
Mauchline	19	9	1	0.01	0.26
Mauchline	46	10	1	0.01	0.27
Thornhill	24	3	1	–	–
Northern Ireland					
Lagan Valley	12	0.3	0.001 to 0.01	–	0.15 to 0.27
Haw Hill	20	3	0.1	–	–

values are between 3 and 1 000 times greater than the laboratory derived values for permeability, depending on the degree of fracturing in the vicinity of the test borehole, whilst storativity often indicates an elastic, rather than an unconfined, response to pumping (Lovelock 1977; Robins 1990).

Hydraulic enhancement of secondary permeability is common to all fractured aquifers and it is critical to the occurrence of groundwater in many of the Celtic regions. There are a number of special cases in Ireland. One is the widespread karstification in the Carboniferous Limestone and the Ulster White Limestone, whereby fracturing has developed through solution and attrition into extensive cavern and conduit systems (Barnes this volume; Coxon & Drew this volume; Deakin this volume). These act as preferred pathways for rapid groundwater transport, often between sink-holes and emergences. A second case is the important influence dolomitization has had on secondary permeability in carbonate aquifers, especially in the south Midlands (Daly 1995). Another special case is the Palaeogene Antrim Plateau Basalt which contains relatively conductive fossil soil horizons, or 'boles', which developed between lava flows, and combined with a sub-vertical conjugate fracture system provide an otherwise weakly permeable volcanic rock with significant potential for groundwater transport and storage. On a smaller scale the Durness Limestone and the Loch Tay Limestone in Scotland also show karst features.

Little quantitative analysis of the role of fracture flow has been carried out in basement rocks. However, investigation into the Strath Halladale Granite and Precambrian metasediments in Caithness, as part of the UK radioactive waste research programme, demonstrated that groundwater transport readily takes place to considerable depths even though the intergranular porosity of these rocks is less than 0.01 (Mather 1997). Packer testing of exploratory boreholes demonstrated that the hydraulic conductivity of these strata ranges from $0.1\,m\,d^{-1}$ down to very low, at depths from just beneath the water table to about 150 m. Values fall rapidly at greater depths. Hydrochemical evidence demonstrated both shallow and relatively short groundwater flowpaths which are up to 2 km in length, and longer groundwater flowpaths with older water circulating to depths of over 200 m along a lateral distance of up to 5 km (Kay *et al.* 1983). Total infiltration to the groundwater system was estimated with the aid of thermal imaging of springs and seepages to be about 100 mm a^{-1} (Brereton & Lee 1987).

Similar flow path systems have been identified in other basement rock types. The majority of shallow flow takes place in the upper 25 m of saturated aquifer over catchment-scale flow paths.

Deep, but small volume, groundwater circulation has been identified at a number of saline discharges at former spa sites in Silurian and Ordovician rocks. These include Melrose, Moffat and Hartfell in the Southern Uplands of Scotland, and the small discharges in mid-Wales: $0.11\,l\,s^{-1}$ at Llandrindod, and only $0.05\,l\,s^{-1}$ at Builth Wells (Edmunds *et al.* 1998). In Ireland the former spa at Mallow is one of a group of warm water discharges in Munster indicative of deep circulation in Carboniferous strata. A similar group occurs in Leinster to the west of Dublin (Burdon *et al.* 1983).

Significance of till

With the exception of south-west England, the Celtic regions were covered by ice through much of the Quaternary Period. Glacial deposits of all kinds are, therefore, characteristic of many areas. These include permeable deposits such as outwash fans, eskers and kames, lacustrine deposits and associated alluvial terrace material, as well as low permeability lodgement till. The extent of these deposits is such that 90% of the land area of Ireland is covered by drift. Much of this is till or ground moraine, which is an irregular layer of variable lithology, typically containing poorly sorted silt- or sand-grade material in a matrix of clay, with cobbles and boulders and occasional sand and gravel horizons. Till may vary in thickness up to 100 m but is generally less than 20 m thick.

The overall hydraulic conductivity of till is generally low and it will not readily conduct infiltrating water. It therefore has a significant role as an inhibitor to recharge as well as a protective shield to bedrock aquifers otherwise vulnerable to surface pollution. The integrity of the deposit is usually unknown and small scale variations in lithology and the presence of fractures or deep root zones offer the potential for rapid by-pass transport to the base of the till. As a general rule, till reduces the recharge potential to an underlying aquifer by about 30%, but actual determination of recharge is difficult to determine and expensive to attain. In addition, available drift mapping is inadequate even for zoning areas of greater and lesser recharge potential although modern mapping techniques, such as drift domain mapping, are beginning to address this difficulty (McMillan *et al.* 2000).

Groundwater flowpaths

The first attempt at delineating groundwater flowpaths in a Celtic aquifer was made by Hartley (1935) in an analysis of the hydraulics of the Triassic aquifer in the Lagan and Dundonald valleys of Northern Ireland. This surprisingly robust analysis has stood the test of time and Hartley's piezometric contour map and groundwater flow paths are reproduced in the marginalia of the Hydrogeological Map of Northern Ireland (BGS 1994). The map shows a groundwater flow system with recharge areas in the higher ground with least cover of till, and discharge areas in the lower parts of the valleys towards the coast. It clearly defines the groundwater divide in the Dundonald Valley, and lateral rather than longitudinal flow in the upper Lagan Valley.

Similar analyses are now available for most of the significant bedrock aquifers in the Celtic lands. For example, owing to extensive structural deformation, the main Palaeozoic bedrock aquifers in the Irish Republic form relatively small discrete aquifer units, frequently with complex boundaries. Faults may occur at separations of 0.5 to 1 km and have a major impact on the flow regime in these aquifers (Daly 1995). Groundwater flow paths are constrained by the geometry of the aquifer unit and the head differences across it. A detailed study of the groundwater flow system of the Castlecomer Plateau showed that the Wesphalian sandstone aquifer system is divided by faulting into three main blocks, with little or no hydraulic connection between them (Misstear *et al.* 1980).

Groundwater flowpaths in granular drift deposits are also limited by the geometry of the aquifer. In valleys, where these deposits are most common, hydraulic contact with surface waters may induce local flow systems between gaining and losing reaches of a river. Where there is significant baseflow, the dominant groundwater flow direction may be laterally towards the river rather than longitudinally down the length of the valley.

Karst

About 50% of the Irish Republic is underlain by Lower Carboniferous Limestone (Coxon & Drew this volume), which forms the most important aquifer in the country. Carboniferous Limestone is also an aquifer of regional significance in South Wales (Hobbs this volume). The degree and extent of karstification is varied. In areas with well developed karst, groundwater flow occurs via large conduits – with recorded flow rates of up to $200\,\mathrm{m\,h^{-1}}$, comparable to surface water velocities (Drew & Daly 1993) – and via a diffuse network of smaller fractures, in which flow rates are much lower.

The uneven distribution of permeability is reflected by small numbers of high yielding wells in areas otherwise characterized by low yielding wells. Drew and Daly (1993) analysed the performance characteristics of 225 wells in limestone aquifers in the west of Ireland, and some of the results are reproduced in Table 4. Yields vary from zero up to $76\,\mathrm{l\,s^{-1}}$, with a relatively low median value of $1.2\,\mathrm{l\,s^{-1}}$. More recently, as part of a national study of well yields and specific capacities, Wright (this volume) found that wells in the Lower Carboniferous Limestone

Table 4. *Summary of performance characteristics of 225 Carboniferous Limestone wells in the west of Ireland (after Drew & Daly 1993)*

Variable	Maximum	Minimum	Mean	Median	Standard deviation
Depth (m)	177	3	57	53	27
Yield ($l\,s^{-1}$)	76	0	2.4	1.2	6
Specific capacity ($l\,s^{-1}\,m$)	7.6	0	0.8	0.2	1.4

aquifers in the west of Ireland covered almost the whole range of yields.

A key feature of karst hydrogeology is the close interrelationship between groundwater and surface water. This is especially the case in lowland areas, since surface water drainage is rare in upland karst (Coxon & Drew this volume). Groundwater and surface water catchments rarely coincide, and it is difficult to define an exact zone of contribution to a well or spring (Deakin this volume; Hobbs this volume). This has implications for groundwater protection and for managing turloughs, the temporary groundwater-fed lakes that are sites of ecological importance (Coxon & Drew this volume). The combination of rapid conduit flow, low storativity and close relationship with surface water means that karst areas can be prone to extreme flooding events. In the Gort area in County Galway, major flooding problems occurred in response to heavy winter rainfall in both 1990 and 1995 (Drew & Daly 1993, Coxon & Drew this volume). Conversely, spring flows and borehole water levels in karst aquifers may decline rapidly during prolonged periods of drought.

Recharge, throughput and baseflow

In contrast to south-east Britain and continental Europe, the Celtic regions possess a relatively humid temperate climate. Rainfall increases from east to west, and with elevation. In Ireland, mean annual precipitation varies from about 700 mm around Dublin to over 2000 mm in upland areas in the west and south-west (Collins & Cummins 1996), where soil is retained at field moisture capacity for a large part of the year. Potential recharge – i.e. the excess moisture after evaporation is deducted from precipitation – varies from 400 to 1000 mm (Daly 1995). Despite the relatively high potential recharge, actual recharge is often much less because of low aquifer storativity. Evidence of 'rejected recharge' includes high level springs, flooding and water logging of land. In Ireland, annual groundwater abstractions are estimated at only about 3% of the estimated annual recharge (Daly 1993), and there is an estimated surplus of resource over abstractions equivalent to 50 mm of rainfall over the area of the whole country (Wright *et al.* 1982).

The main mechanisms for recharge are:

(a) direct recharge from rainfall infiltrating through outcrop, soil and drift;
(b) secondary recharge from losing streams, additional runoff from nearby low permeability areas and, in karst, via swallow holes.

Several of the mechanisms may be important in any one area. The large Schwyll spring near Bridgend in South Wales is recharged from infiltrating rainfall over the limestone aquifer outcrop, from stream sinks and by influent sections of river (Hobbs this volume). Quantification of recharge is usually by relatively simple methods such as soil moisture balance and analysis of baseflow. More sophisticated analyses are rare owing to a lack of knowledge on recharge processes such as infiltration rates in low permeability tills.

In the Dumfries basin in Scotland recharge is estimated to be 436 mm a^{-1} or about 31% of the mean annual rainfall (Gaus & Ó Dochartaigh this volume). In Fife, on the east side of Scotland, recharge is only between 115 and 145 mm a^{-1} but this represents between 18 and 28% of the long term rainfall. Similarly in west Wales, potential infiltration of 535 mm a^{-1} represents 38% of the long term rainfall (Robins *et al.* this volume).

The groundwater contribution to rivers as baseflow is often significant. Analysis of river flow data from 1972 to 1981 in the mainly Carboniferous Limestone Nore River Basin in southern Ireland indicated the groundwater component to be 50% of the total flow, equivalent to 26% of the total rainfall (Daly, E. 1994).

Groundwater quality

Most of the aquifers in the Celtic lands are unconfined, with shallow flow systems, and so the groundwaters are generally well-oxygenated and weakly mineralized, with calcium and bicarbonate forming the main ionic components (Daly

1989; Robins in press). More mineralized groundwaters, depleted in oxygen, occur in deeper aquifer systems such as Devonian sandstones in Scotland and Westphalian sandstones in Ireland. A high degree of mineralization, dominated by sodium and chloride ions, is found locally in coastal aquifers including the Carboniferous Limestone of south Galway, the Permo-Triassic sandstone near Belfast and the sandstone aquifers of East Lothian. The hydrochemistry of the recharge water is strongly influenced by the mineral composition of the drift deposits (Robins *et al.* this volume).

Natural groundwater quality problems are usually associated with excessive hardness in carbonate and some drift aquifers, where concentrations are often in the range 200–400 $mg\,l^{-1}$ $CaCO_3$ (Daly, D. 1994). Elevated iron and manganese concentrations are present in groundwaters in igneous and other basement rocks, muddy limestones, sandstones, peat and alluvial gravels (Jones & Singleton this volume; Robins *et al.* this volume). Manganese, in particular, can be a problem for water supply, as the maximum admissible concentration in drinking water (MAC) is only 0.05 $mg\,l^{-1}$.

There are a large number of anthropogenic impacts on groundwater quality. Pressures from point sources of pollution include farmyards, sheep dips, septic tank systems, solid waste disposal, industry and mining. Diffuse pollution sources from agriculture include the application of inorganic fertilizers and landspreading of organic wastes. Diffuse pollution sources associated with infrastructure developments such as roads and housing are a growing concern.

Many of the small private well abstractions are affected by local point source contaminants, especially microbiological. The problems are often compounded by a combination of factors: poor well construction, especially the absence of a proper sanitary seal to the upper casing; high aquifer vulnerability; the predominance of fracture flow of groundwater; the presence of a nearby contamination source such as a septic tank system, itself often badly constructed; and a low well abstraction rate and hence low potential for dilution (Fig. 4). Microbiological contamination is usually detected by the presence of faecal bacteria. However, there is growing concern that other microbiological pollutants such as viruses and protozoa (including *cryptosporidium*) could be present in wells that are susceptible to pollution from livestock activities.

Chemical contaminants associated with point sources in rural areas include chloride, sodium, potassium, iron, manganese, ammonia and nitrate. High nitrate concentrations also result from diffuse pollution by inorganic and organic fertilizers. In Scotland, high nitrate levels are found widely in areas of intensive cultivation or grassland in Fife and the Southern Uplands (Robins in press). In Ireland, groundwater contamination by nitrate is generally only widespread in the intensive arable areas in the east and south, including the alluvial aquifer of the Barrow valley (Daly, D. 1994). A national survey of groundwater quality in Ireland between 1995 and 1997 identified nitrate at above the MAC of 50 $mg\,l^{-1}$ NO_3 at only 2.5% of the sites sampled, and concluded that nitrate pollution was generally a localized problem (Environmental Protection Agency 1999).

Fig. 4. A typical farmyard well-head in the north of Ireland.

Another agricultural contaminant, phosphorus, has received much less attention than nitrate. This is mainly because phosphorus levels in groundwater are generally quite low compared to the MAC of 2.2 $mg\,l^{-1}$ P. However, phosphorus can give rise to eutrophication problems in lakes and rivers at concentrations of as little as 20 and 30 $\mu g\,l^{-1}$, respectively. Hence phosphorus at these low levels in groundwater may be of concern where the groundwater is discharging into a sensitive surface water body. In a recent study of phosphorus in groundwater in the west of Ireland (counties Limerick, Mayo and Clare), 38% of groundwater samples collected in the summer of 1998 had phosphorus concentrations above 30 $\mu g\,l^{-1}$ (Kilroy *et al.* 1999). The authors go on to highlight that more research is needed into the means by which phosphorus reaches surface waters.

Trace organics occur in groundwater throughout the Celtic regions as a response to human activities. The most prevalent compounds are petroleum hydrocarbons, chlorinated solvents and pesticides. The risks posed to the aquatic environment by the disposal of spent sheep dip is a particular Celtic concern. Sheep dip insecticides have traditionally been of the organophosphate type, but are now being increasingly replaced by synthetic pyrethroids which are neither toxic to sheep or humans. Whereas biodegradation of organophosphates can produce yet more toxic products, the effect of sunlight and hydrolysis on the synthetic pyrethroids reduces their harmful potential.

Elevated metal concentrations occur in groundwaters and surface waters in former mining districts. Important historical mining activities included: tin in Cornwall; copper in Ireland, Wales and Cornwall; and coal in South Wales, Northumbria and Scotland. Whereas modern mining developments such as the new lead/zinc mines at Galmoy and Lisheen in Ireland are subject to strict environmental controls, the abandonment of old metalliferous and coal mines has led to serious pollution problems. Perhaps the best known case occurred in Cornwall in 1992, when a large discharge of acid mine drainage from the former Wheal Jane tin mine contaminated the River Carnon, resulting in concentrations of cadmium and zinc in the river of up to 600 $\mu g\,l^{-1}$ and 450 $mg\,l^{-1}$, respectively (National Rivers Authority 1994). At the Avoca copper mines in County Wicklow, which were finally closed in 1982, mine drainage with a pH of about 3.5 is discharging to the surface rendering the Avoca river effectively biologically dead for a reach of several kilometres (Wright *et al.* 1999). In Wales, over 50 km of rivers are adversely affected by discharges from abandoned coal mines (National Rivers Authority 1994), and there are reported to be more than 80 adits from coal and oil shale mines in the Midland Valley of Scotland that are releasing polluted mine water to surface water sources.

Influences of change

Change is currently being effected from a number of different directions. These include legislative drivers intended to regulate anthropogenic activities as part of the resource management process, environmental influences such as climate change, and changing patterns of demand, and of supply to meet demand.

Legislative influences are ongoing. The main changes over the next few years will stem from the proposed European Water Framework Directive, which requires integrated management of both groundwater and surface water to be carried out within the geographical unit of the river basin. This will not be easy in the Celtic regions where aquifer units tend to be small and compartmentalized, and consideration of individual catchments within each river basin unit will be required to enable resource management. Implementation of the Water Framework Directive will also involve the introduction of abstraction licensing in Ireland, Northern Ireland and Scotland.

The implications of climate change include increased annual rainfall over much of the Celtic regions with a small increase also in potential evapotranspiration (Hulme & Jenkins 1998). This may influence the hydrology of wetland areas if the elevation of the water table changes, a particularly critical issue in the many coastal wetland and duneland areas of eastern Scotland (Malcolm & Soulsby this volume). The implications of rising sea level and increased river base levels, particularly in lowland areas, are also significant.

Changing patterns of demand and supply have had a profound influence on groundwater abstraction in the past, and continue to do so. The decline of water intensive industry in the 1960s and 1970s reduced groundwater consumption in the Lagan Valley of Northern Ireland, the lower Clyde area of the Midland Valley of Scotland, and the industrial areas of South Wales. More recently, increased costs of public supply have created renewed interest in groundwater supplies for industry, and extensive development programmes have been carried out with varying degrees of success.

Alternative sources for public supply in rural and island areas have also seen an upturn in interest in groundwater by many water undertakers. Welsh Water is actively developing bulk supplies from valley bottom Quaternary fill material in west Wales, so that the area south of Aberystwyth, for example, is largely groundwater dependent (Hiscock & Paci this volume). However, the Water Service in Northern Ireland has pursued a policy of discarding smaller sources and this has reduced its dependence on groundwater.

Conclusions

Groundwater occurrence and its role within the environment have a distinct character in the Celtic regions. Hydrogeological processes are dominated by small aquifer units, short and shallow groundwater flow paths, fracture flow in otherwise poorly permeable strata, and complex interrelationships between bedrock and drift groundwater and surface water. Prevailing westerly air streams ensure that soils are kept at field moisture capacity for much of the year, although infiltration is inhibited by a widespread blanket cover of till.

The social and historical importance of groundwater can be traced from the Celtic peoples, through medieval times to the present day. Abundant surface water resources have not prevented the use and development of groundwater owing to its wide geographical availability, low cost of development and its generally favourable and constant quality. Renewed interest in groundwater for both public and private supply has greatly increased the number of sources and the volume of groundwater in supply in recent years. Other influences include changing legislative requirements for the management and protection of groundwater and the possible impact from climate change.

A number of the papers in this volume derive either from presentations at a joint meeting of the Geological Society Hydrogeological Group and the Scottish Hydrological Group of the Institution of Civil Engineers, which was held at Glasgow in March 1999, or from the annual meeting of the International Association of Hydrogeologists Irish Group, held at Portlaoise in April 1999.

References

BARNES, S. 2000. Groundwater resources and vulnerability in the Cretaceous Chalk of Northern Ireland. *This volume.*

BGS. 1994. Hydrogeological Map of Northern Ireland, 1:250 000 scale. British Geological Survey, Nottingham.

BRERETON, N. R. & LEE, M. K. 1987. Fluid flow in crystalline rocks: relationships between groundwater spring alignments and other surface lineations at Altnabreac, United Kingdom. *Journal of Geophysical Research*, **92**, 7797–7806.

BURDON, D. J., BURNS, D. J. & PEEL, S. 1983. Geothermal energy investigations in Ireland. *Proceedings of the Third International Seminar on the Results of EC Geothermal Energy Research*, Munich 1983, 350–360.

COLLINS, J. F. & CUMMINS, T. (eds) 1996. *Agroclimatic atlas of Ireland.* AGMET, Dublin.

COXON, C. & DREW, D. 2000. Interdependence of groundwater and surface water in lowland areas of western Ireland: management issues arising from water and contaminant transfers. *This volume.*

DALY, D. 1993. Groundwater resources in Ireland. *In*: MOLLAN, C. (ed.) *Water of Life*. Royal Dublin Society Seminar Proceedings, Number 5.

——1994. Chemical pollutants in groundwater: a review of the situation in Ireland. *In*: *Chemicals – A Cause for Concern?* Cork 3–4 November 1994, Sherkin Island Marine Station.

DALY, E. 1989. Natural chemistry of groundwater. *In*: *Proceedings of the 9th Annual Groundwater Seminar, International Association of Hydrogeologists (Irish Group)*, Portlaoise, 25–26 April 1989.

——1994. *Groundwater resources of the Nore River Basin.* Geological Survey of Ireland Report **RS94/1**.

——1995. The principal characteristics of the flow regime in Irish aquifers. *In*: *Proceedings of the 15th Annual Groundwater Seminar, International Association of Hydrogeologists (Irish Group)*, Portlaoise, 25–26 April 1995.

DEAKIN, J. 2000. Groundwater protection zone delineation at a large karst spring in western Ireland. *This volume.*

DEPARTMENT OF THE ENVIRONMENT AND LOCAL GOVERNMENT, ENVIRONMENTAL PROTECTION AGENCY & GEOLOGICAL SURVEY OF IRELAND. 1999. *A scheme for the protection of groundwater.* Geological Survey of Ireland, Dublin.

DREW, D. & DALY, D. 1993. *Groundwater and karstification in mid-Galway, south Mayo and north Clare.* Geological Survey of Ireland Report **RS 93/3**.

EDMUNDS, W. M., ROBINS, N. S. & SHAND, P. 1998. The saline waters of Llandrindod and Builth, Central Wales. *Journal of the Geological Society, London*, **155**, 627–637.

ENVIRONMENTAL PROTECTION AGENCY. 1999. *Water quality in Ireland 1995–1997.* Environmental Protection Agency, Wexford.

FOX, I. A. 2000. Groundwater protection in Scotland. *This volume.*

GAUS, I. & Ó DOCHARTAIGH, B. É. 2000. Conceptual modelling of data-scarce aquifers in Scotland: the sandstone aquifers of Fife and Dumfries. *This volume.*

HARTLEY, J. J. 1935. The underground water resources of Northern Ireland. *The Institution of Civil Engineers, Belfast and District Association Pamphlet.*

HISCOCK, K. & PACI, A. 2000. Groundwater resources in the Quaternary deposits and Lower Palaeozoic bedrock of the Rheidol catchment, west Wales. *This volume.*

HOBBS, S. 2000. Influent rivers – a pollution threat to Schwyll Spring, South Wales? *This volume.*

HULME, M. & JENKINS, G. 1998. *Climate change scenarios for the United Kingdom: UKCIP. Technical Report 1.* UK Climate Impacts Programme, Oxford.

JONES, C. & SINGLETON, A. 2000. Public water supplies from alluvial and glacial deposits in northern Scotland. *This volume.*

KAY, R. L. F., ANDREWS, J. H., BATH, A. H. & IVANOVITCH, M. 1983. Groundwater flow profile and residence times in crystalline rocks at Altnabreac, Caithness, UK. *Isotope Hydrology*, 231–248.

KILROY, G., COXON, C., ALLOTT, N. & RYBACZUK, R. 1999. The contribution of groundwater phosphorus to surface water eutrophication. *In*: *Proceedings of the 19th Annual Groundwater Seminar, International Association of Hydrogeologists (Irish Group)*, Portlaoise, 20–21 April 1999.

LEWIS, M. A., LILLY, A. & BELL, J. 2000. Groundwater vulnerability mapping in Scotland: modifications to classifications used in England and Wales. *This volume*.

LOGAN, P. 1980. *The holy wells of Ireland*. Colin Smythe, Buckinghamshire.

LOVELOCK, P. E. R. 1977. Aquifer properties of Permo-Triassic sandstones in the United Kingdom. *Bulletin of the Geological Survey of Great Britain* No 56.

MALCOM, R. & SOULSBY, C. 2000. Modelling the potential impact of climate change on a shallow coastal aquifer in northern Scotland. *This volume*.

MATHER, J. D. 1997. The history of research into radioactive waste disposal in the UK and the selection of a site for detailed investigation. *Environmental Policy & Practice*, **6**, 167–177.

MCMILLAN, A. A., HEATHCOTE, J. A., KLINCK, B. A., SHEPLEY, M. G., JACKSON, C. P. & DEGNAN, P. J. 2000. Hydrogeological characterisation of the onshore Quaternary sediments at Sellafield using the concept of domains. *Quarterly Journal of Engineering Geology and Hydrogeology*, **33**(4).

MISSTEAR, B. D. R., DALY, E. P. D., DALY, D. & LLOYD, J. W. 1980. *The groundwater resources of the Castlecomer Plateau*. Geological Survey of Ireland Report **RS 80/3**.

MISSTEAR, B. D. R. & DALY, D. 2000. Groundwater protection in a Celtic region: the Irish example. *This volume*.

NATIONAL RIVERS AUTHORITY. 1994. *Abandoned mines and the water environment*. National Rivers Authority, Bristol.

ROBINS, N. S. 1990. *Hydrogeology of Scotland*. HMSO, London.

——1999. Groundwater occurrence in the Lower Palaeozoic and Precambrian rocks of the UK: implications for source protection. *Water and Environment Management*, **13**, 447–453.

——in press. Groundwater quality in Scotland: major ion chemistry of the key groundwater bodies. *Hydrology and Earth System Sciences* (in press).

——, SHAND, P. & MERRIN, P. D. 2000. Shallow groundwater in drift and Lower Palaeozoic bedrock: the Afon Teifi valley in west Wales. *This volume*.

SOULSBY, C., MALCOLM, R. & MALCOLM, I. 2000. Groundwater in headwaters: hydrological and ecological significance. *This volume*.

WRIGHT, G. 1999. How many wells are there in Ireland? *The GSI Groundwater Newsletter*, **35**.

——2000. QSC graphs – an aid to classification of data-poor aquifers in Ireland. *This volume*.

——, ALDWELL, C. R., DALY, D. & DALY, E. 1982. *Groundwater resources of the Republic of Ireland*. Vol. 6 in the European Community Atlas of Groundwater Resources, SDG, Hanover.

——, MISSTEAR, B. D. R., GALLAGHER, V., O'SUILLEABHAIN, D. & O'CONNOR, P. 1999. Avoca mines: uncontrolled acid mine drainage in Ireland. *In*: *Proceedings of the International Congress of the International Mine Water Association 'Mine, Water and Environment for the 21st Century: Mine/Quarry Waste Disposal and Closure'*, Sevilla, Spain, 13–17 September 1999.

YOUNGER, P. L. & LAPIERRE, A. B. 2000. 'Uisge Mèinne': mine water hydrogeology in the Celtic lands, from Kernow (Cornwall, UK) to Ceap Breattain (Cape Breton, Canada). *This volume*.

Groundwater in headwaters: hydrological and ecological significance

C. SOULSBY, R. MALCOLM & I. MALCOLM

Department of Geography, University of Aberdeen, Aberdeen AB24 3UF, UK (e-mail: c.soulsby@abdn.ac.uk)

Abstract: The hydrological and ecological significance of groundwater has generally been under-estimated in headwater catchment studies within the Celtic regions. The paper presents data from headwater catchments in both upland and lowland settings in northern Scotland to address this gap in our understanding. Research in the 10 km^2 Allt a' Mharcaidh catchment in the western Cairngorms has demonstrated that a range of groundwater sources in various drift deposits can account for *c.* 50% of annual runoff, even in a high altitude headwater stream. Despite the traditional assumption that upland catchments have limited aquifer storage, oxygen isotope studies of groundwater imply mean water residence times of up to five years which indicate a range of groundwater sources in montane environments. Moreover, hydrogeochemical reactions in the saturated zone appear to regulate stream water chemistry at moderate and low flows. In such montane environments, groundwater discharges at springs create unique wetland habitats which often form the source of headwater streams and affect riparian areas. In lowland catchments the hydrological significance of groundwater is equally important. In addition, recent studies in a salmon spawning stream in the Newmills Burn, Aberdeenshire has shown that aquifer–stream interactions in hyporheic zones are crucial in maintaining habitat conditions conducive to the survival of salmonid eggs, and the subsequent population of salmon streams. It is concluded that interdisciplinary studies incorporating hydrogeological investigations are fundamental to a proper understanding of the hydrology and functional ecology of catchment systems in the Celtic regions.

It is being increasingly recognized that groundwater plays an important role in the hydrology, hydrogeochemistry and ecology of headwater streams (Winter 1995). In the Celtic regions this is as true in upland headwaters, where impermeable geology and thin peaty soils have been traditionally posited as reasons to discount groudwater influence (Wade *et al.* 1999), as it is in lowland areas, where baseflow dominance by groundwater has always been recognised (Robins 1990). Recent work, most notably in Wales (Hill & Neal 1997; Neal *et al.* 1997) and Scotland (Cook *et al.* 1991; Edmunds & Savage 1991; Jenkins *et al.* 1994; Soulsby *et al.* 1998*a*), has demonstrated how groundwater inputs to streams can significantly influence the quantity and quality of baseflows and storm runoff. Moreover, it is clear that groundwater–surface water interactions are important in maintaining diverse habitats in freshwater ecosystems, both in terms of physical and chemical characteristics, which in turn underpin the biodiversity of headwater stream systems (Grieve *et al.* 1995; Brunke & Gonser 1997; Soulsby *et al.* 1998*b*; Soulsby *et al.* 1999). Particular interest has focused on the hyporheic zone within stream channels, sometimes called the transitional zone, or ecotone, where groundwater and surface water mix (Hynes 1983; Jones & Holmes 1996).

This paper highlights some of these increasingly important issues in the Celtic regions by presenting a synthesis of recent research efforts in two headwater catchments in northern Scotland. These are the Allt a' Mharcaidh in the western Cairngorms and the Newmills burn in the lower Don catchment, Aberdeenshire. The aims of the paper are: (a) to assess the extent to which complex groundwater flow systems have an important influence on the hydrology and hydrochemistry of headwater streams in mountainous terrain underlain by crystaline (granitic) solid geology; (b) to examine the hydroecological importance of groundwater–surface water interactions by an investigation of headwater streams that form important salmon spawning habitats; and (c) to identify pressing research needs that must be addressed if understanding in the Celtic regions is not to lag behind other parts of the world where, to date, research efforts have been much more extensive. It is argued that

From: ROBINS, N. S. & MISSTEAR, B. D. R. (eds) *Groundwater in the Celtic Regions: Studies in Hard Rock and Quaternary Hydrogeology*. Geological Society, London, Special Publications, **182**, 19–34. 1-86239-077-0/00/

such work needs to be carried out within an interdisciplinary framework if an holistic understanding of catchment systems is to be realized.

Study areas

Allt a' Mharcaidh

The Allt a' Mharcaidh is one of the most intensively studied catchments in Scotland, lying within the Cairngorms, the most extensive mountain plateau in the UK and a conservation site of international importance (Gordon *et al.* 1998). Over the past 15 years a range of hydrological and ecological studies has examined the functioning of the catchment, particularly in terms of its response to acidification and climate change (Harriman & Ferrier 1990). The catchment lies on the western edge of the Cairngorm mountains and drains into the River Feshie, which in turn drains into the Spey. It covers some 10 km^2 and has an altitudinal range of 330–1111 m (Fig. 1). The catchment is underlain by granite of Lower Red Sandstone age, which is associated with the late stages of the Caledonian Orogeny. The granite is massive, though weathered fractures are evident where it is exposed. In most places the granite is covered by various locally-derived drift deposits resulting from glacial and periglacial processes. The mineralogy of both granite and drift is dominated by plagioclase, K-feldspar, biotite, chlorite, kaolinite and quartz. At higher altitudes (>*c.* 800 m) frost shattered debris provides the parent material for freely draining alpine soils (Fig. 1b). This large upland plateau provides an extensive

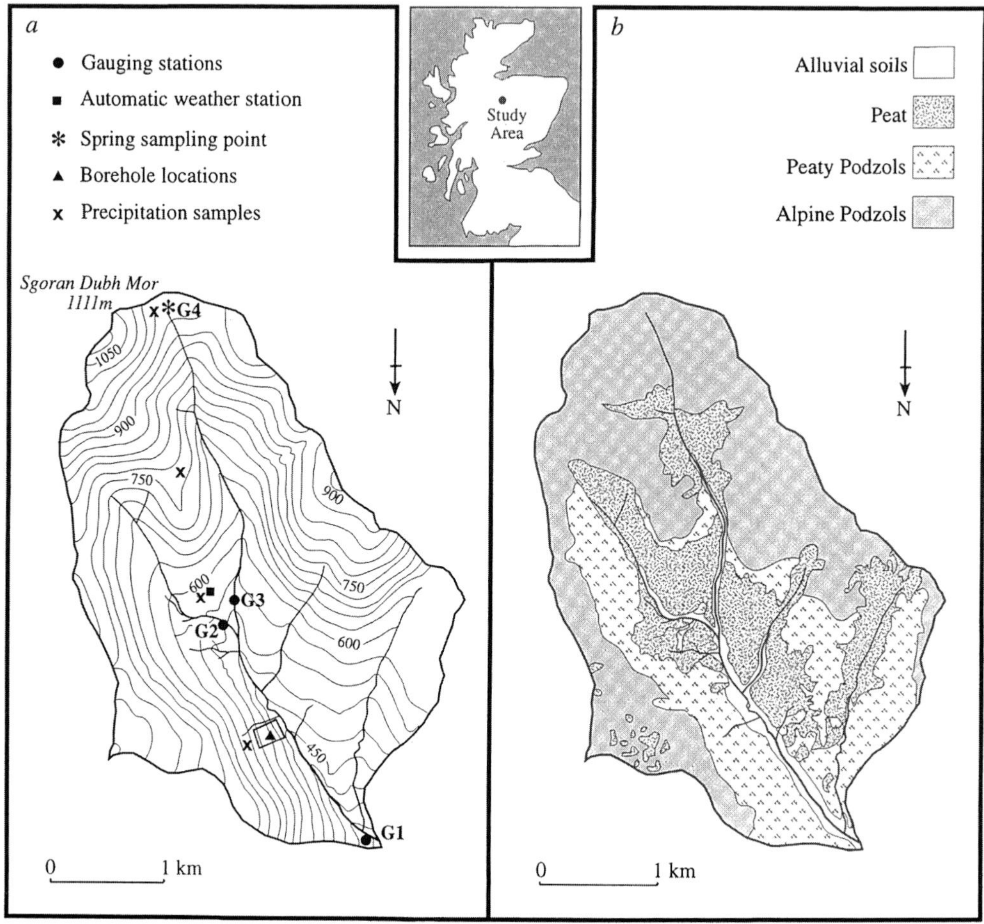

Fig. 1. The Allt a' Mharcaidh experimental catchment showing location, (**a**) topography and instrumentation, and (**b**) soil coverage.

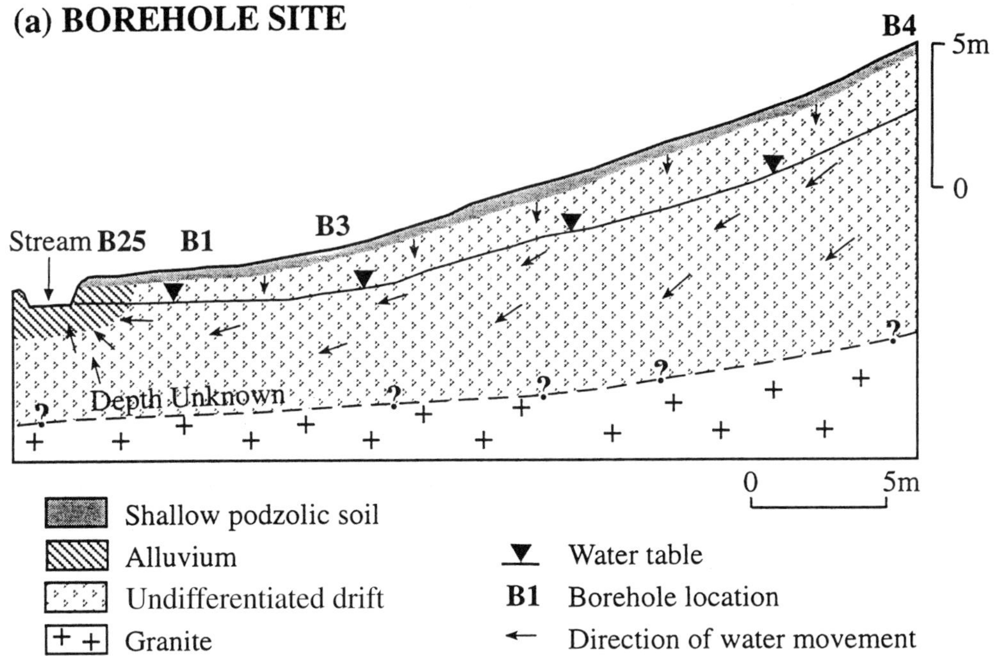

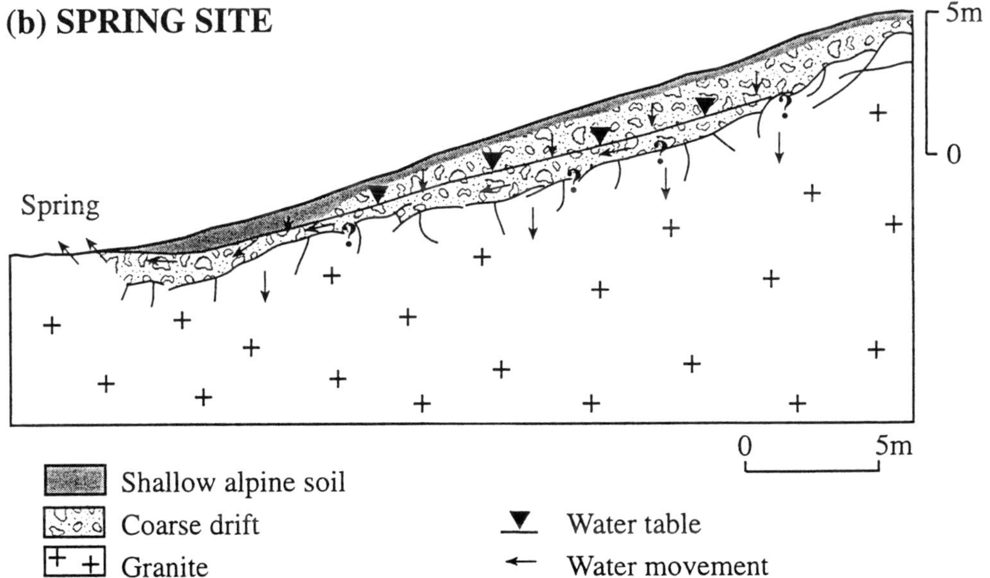

Fig. 2. Conceptual model of two major groundwater discharge zones in the Allt a' Mharcaidh in (**a**) riparian areas and (**b**) at montane spring sites.

recharge area with much recharge appearing to contribute to large groundwater flow systems at the drift/bedrock interface, which subsequently discharge either directly into the riparian area or stream bed (Fig. 2a) or indirectly into the peat soils which dominate the lower slopes of the catchment. However, more localized groundwater flow systems discharge at a number of spring sites which often form the source of first order streams throughout the Cairngorm region (Fig. 2b). The steeper transit slopes that link the plateau with the valley floor are dominated by humus iron podzols, which deflect subsurface storm flow laterally in the organic soil horizon or transmit groundwater at the soil–drift–bedrock interface

The climate of the Allt a' Mharcaidh is subarctic and is characterized by cold winters (Mean January temperature is 1.2°C) and mild summers (Mean July temperature is 10.3°C). An average precipitation of 1200 mm falls each year with much of this falling as snow in the winter months (Soulsby *et al.* 1997*a*), though inter-annual variation in snow pack accumulation is great (Dunn & Langan 1998). Annual runoff is around 850 mm, with the annual baseflow index ranging between *c.* 0.4 and 0.55.

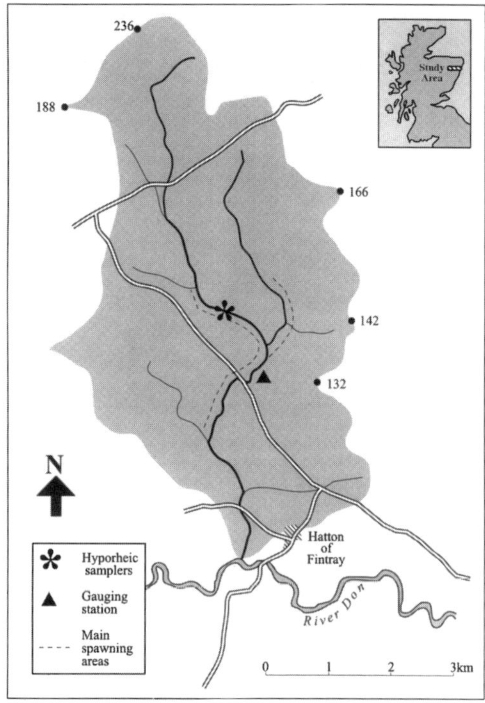

Fig. 3. Newmills burn experimental catchment showing location of groundwater-surface water interaction study.

Newmills burn

The Newmills burn, although at 12 km^2 a similar size to Allt a' Mharcaidh, is a lowland headwater stream that drains an agricultural catchment which has an altitudinal range of 45–230 m (Fig. 3). Such headwater streams are very typical of the fertile eastern lowlands of Scotland and such streams often form tributaries of the major east coast salmon rivers. The geology of the catchment is dominated by psammite and pelite. Extensive and complex drift deposits cover the catchment which range from fluvioglacial sands and gravels to lodgement tills. These give rise to poorly drained gleyed soils and imperfectly drained brown soils. Consequently, much of the catchment is affected by agricultural underdrainage giving it a rather 'flashy' hydrological response to large precipitation events, though for most of the year low to moderate flows dominate. Like Allt a' Mharcaidh, the range of drift deposits probably results in a range of groundwater flow paths with differing velocities and residence times, some of which are intercepted by artificial drains.

Mean annual precipitation in the Newmills catchment is around 800 mm, with annual evaporation estimated at 400 mm. Mean January and

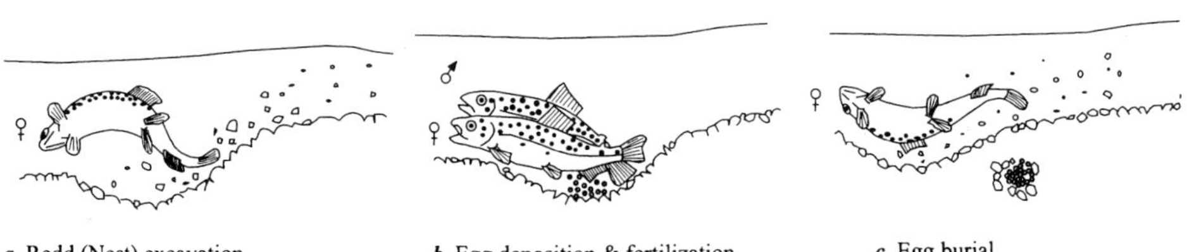

Fig. 4. Simplified diagram of the creation of redds by salmonids in spawning gravels.

July temperatures are 5 and 16°C, respectively, while snowfall is uncommon. The stream has been canalized as a result of agricultural drainage schemes, but it retains a gravel bed structure dominated by simple pool-riffle sequences (Moir 1999). It is extensively used by migratory salmonids as a spawning stream where female fish excavate redds in the stream bed in the early winter (November–December) and bury eggs that hatch the following spring (Fig. 4). Genetic analysis has revealed that the redds are mainly constructed by Sea Trout (*Salmo trutta*), though Atlantic salmon (*Salmo salar*) also occur (Youngson & MacLaren 1998)

Methodology

Allt a' Mharcaidh

The groundwater studies in the Allt a' Mharcaidh have been described in detail elsewhere (Soulsby *et al.* 1998*a*, 1999, 2000) therefore only brief details follow. The catchment was established in 1985 as part of surface water acidification studies (Harriman & Ferrier 1990). A stream gauging station with continuous pH, conductivity and temperature monitoring was installed at the catchment outfall along with an Automatic Weather Station (AWS) at 560 m (Fig. 1). Within the acidification studies, emphasis was placed on understanding hydrological pathways dominating the stormflow response of the stream where acid ($pH < 5.0$) water was observed (Jenkins *et al.* 1993; Wheater *et al.* 1993). Groundwater was assumed to maintain relatively well buffered baseflows which had a pH of around 6.5–7.0. Thus, runoff from peaty soils and snowmelt episodes was extensively studied using hydrometric techniques, but the potential importance of groundwater was not fully recognized until tracer studies (using oxygen isotopes, deuterium and other natural tracers) demonstrated important groundwater inputs into the stream during storm events (Ogunkoya & Jenkins 1993; Jenkins *et al.* 1994). As part of these tracer studies, ten shallow (up to 4 m deep) boreholes were drilled in two 200 m long transects perpendicular to the valley bottom (Figs 1a, 2a), some 2 km upstream of the gauging station (Soulsby *et al.* 1998*a*). In addition, groundwater discharging from a spring at the source of the stream was monitored. For periods between 1990 and the present, weekly water levels in some of the boreholes were recorded and discharge from the spring in the upper catchment monitored. In addition to hydrometric observations, regular (bi-weekly) streamwater samples were collected along with weekly samples of precipitation, spring water and groundwater. These samples were all returned to the laboratory and analysed for pH, alkalinity and a range of major and minor determinands (using standard methods). In addition, the ratio of oxygen isotopes ($^{18}O/^{16}O$) were measured and expressed in delta units, $\delta^{18}O$ (‰) defined in relation to SMOW (standard mean ocean water). The analytical precision of $\delta^{18}O$ determinations was approximately ±0.1‰ (see Soulsby *et al.* (2000) for further details on analytical procedures).

Newmills burn

Interest in salmonid spawning behaviour in the Newmills burn was the main impetus for its establishment as an experimental site (Youngson & MacLaren 1998). However, stream flows and stream chemistry have been monitored since the end of 1997 (Soulsby *et al.* 2000*a*). In the 1998/99 spawning season, groundwater–surface water interactions were monitored around salmonid redds. Nests of piezometers and hyporheic water samplers (Fig. 5) were installed to examine the nature of aquifer stream exchanges

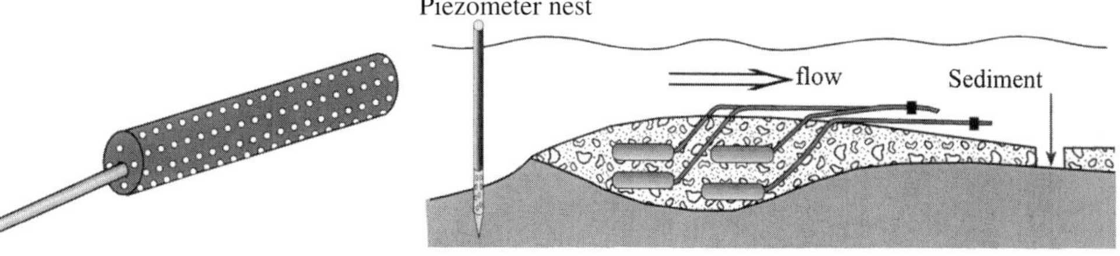

Fig. 5. Sampling hyporheic waters in salmon spawning gravels: (**a**) hyporheic samplers and location in simulated redd in (**b**) horizontal cross-section.

and assess how these affected the hydrochemistry of water flowing through spawning gravels. Samples were collected from the stream and the upper (0.1 m depth) and lower (0.3 m depth) hyporheic zone at weekly intervals between January and March and monthly thereafter. The samplers were buried in simulated redds, the physical structure (size, depth and sedimentary characteristics) of which was determined by direct excavation of over 50 redds between 1997 and 1999. In addition, fine sediment infiltration rates into the stream bed were monitored using small baskets inserted into the stream bed which were monitored weekly as described by Soulsby *et al.* (2000*a*).

Upland catchments: groundwater and catchment hydrology

Hydrometric and Hydrogeochemical observations

Hydrometric observations at the boreholes show that shallow groundwater is found within 2 m of the soil surface in the lower valley slopes (Soulsby *et al.* 1998*a*). This groundwater lies within undifferentiated drift deposits of variable thickness. The water table is responsive to precipitation and snowmelt inputs, though the nature of the response is highly localized (Fig. 6). In the ten boreholes monitored, water table fluctuations exceeded 1 m in some boreholes whilst others varied by less the 0.3 m. This probably indicates the combined effect of groundwater flow paths, that are highly preferential, and marked spatial variation of aquifer properties (storativity, permeability etc.) in various drift deposits. Nevertheless, there was clear evidence that rapid influx of recharge water could occur during specific hydrological events, suggesting active groundwater flow systems recharged by downslope movement of meteoric water or vertical movement of recent precipitation. The boreholes close to the stream bank (B25) showed relatively low fluctuations, probably indicating a boundary outflow control on groundwater fluxes into the stream through the bed and banks (cf. Fig. 2a).

The chemistry of groundwater sampled from individual boreholes was also spatially variable, though all sampled groundwaters were dilute with low alkalinity (Table 1). Atmospherically derived solutes dominated the ionic content of precipitation and groundwaters. However, compared to precipitation, groundwater was enriched in Si and base cations as a result of weathering reactions, which also generated alkalinity and raised the pH. Nevertheless, the concentrations remained low, implying that relatively shallow water movement through decalcified soils and drifts characterized the groundwater flow system (cf. Edmunds & Savage 1991). Temporal variations in groundwater chemistry also occurred, though these were much less marked than for precipitation (see ranges in Table 1). In general, concentrations were higher during the summer and lowest during late winter/early spring. Groundwater discharging from the spring in the upper catchment showed similar, though less marked, increases in weathering-related ions (Table 1).

Geochemical mass balance calculations using the NETPATH model (see Plummer *et al.* 1991) implied that plagioclase dissolution was the dominant weathering reaction in the various drift deposits, though K–feldspar and biotite weathering were also predicted with kaolinite precipitation occurring (Soulsby *et al.* 1998*a*). Weathering rates during the summer were estimated as being approximately 30% higher than during the winter months (Soulsby *et al.* 1999)

The chemical composition of the Allt a' Mharcaidh stream varies with discharge (Jenkins *et al.* 1993; Soulsby *et al.* 1997*b*). At high flows, runoff from overland flow in the peat soils in the lower catchment or snowmelt dominates, producing acidic (pH <5.5) stream water with diluted base cation concentrations. At low flows, where groundwater inputs predominate, the pH is higher (*c.* 7.0) with higher concentrations of Ca, Si and other weathering derived species. The hydrological controls on stream water chemistry are reflected in the relationship between stream flow and Acid Neutralizing Capacity (ANC), which is shown in Fig. 7. Such hydrogeochemical signatures in stream water may be viewed, to a first approximation, as representing a mix of soil water and groundwater sources (Wade *et al.* 1999). Flow variant species (such as ANC, Ca and Si) were used by Soulsby *et al.* (1998*a*) as tracers in a geochemically-based hydrograph separation using End Member Mixing Analysis (EMMA). This used the chemistry of groundwater and soil water end members in a two component mixing model (Neal *et al.* 1997). Although different tracers and different assumptions led to slightly different results, a consistently high proportion of annual runoff (40–60%) could be attributed to groundwater sources (Fig. 8). Thus, far from simply providing baseflows, active groundwater within the catchment clearly exerts a strong influence on both the hydrology and hydrochemistry of the Allt a' Mharcaidh, despite its montane character.

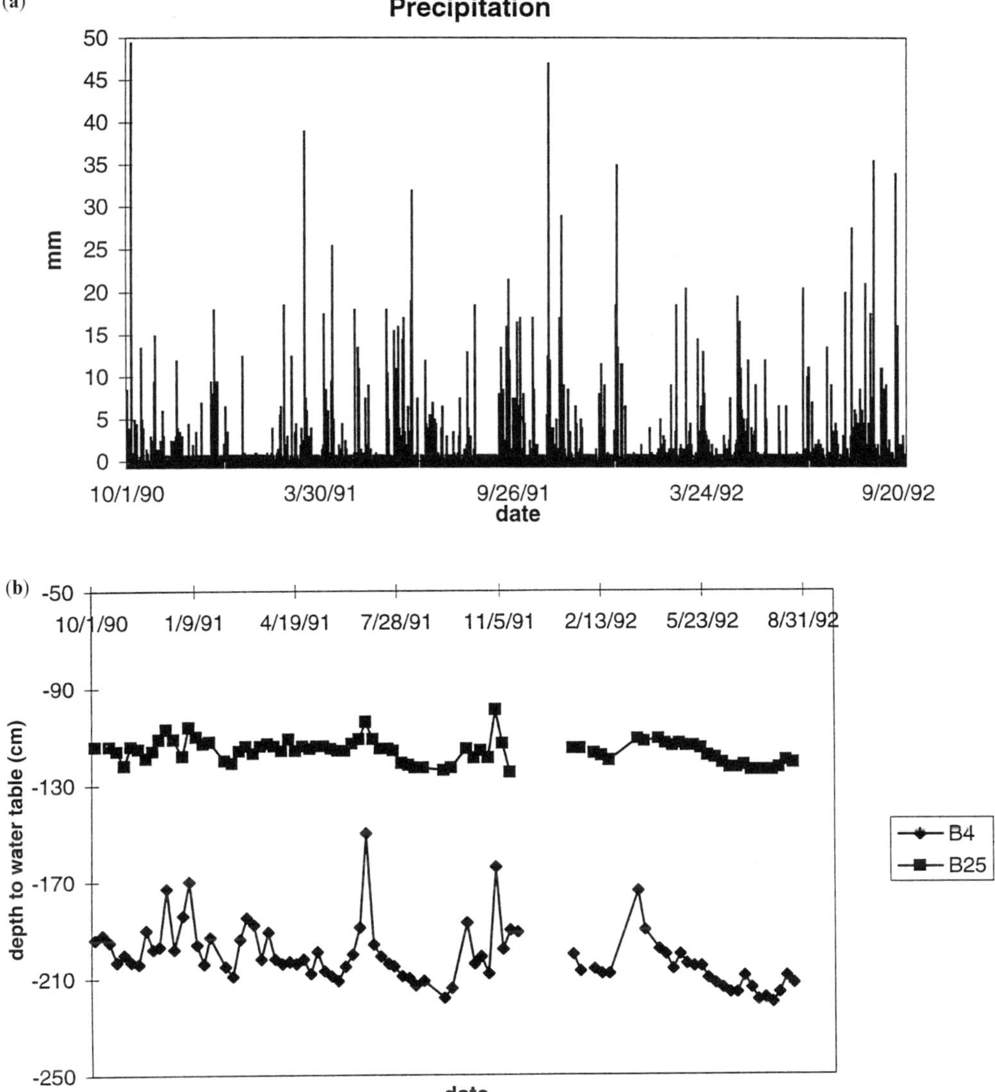

Fig. 6. Hydrometric measurements of depth to water table in response to (**a**) precipitation events in the Allt a' Mharcaidh. (**b**) Traces show B4 borehole on hillslope 20 m from stream and B25 borehole 5 m from stream channel (cf. Fig. 2a).

Isotope studies

Stable oxygen isotope variations (Table 2) have also been examined to understand the linkages between catchment precipitation, soil waters, groundwaters and stream flows (Kendall *et al.* 1995). Whilst δ^{18}O levels in precipitation are seasonally variable, with δ^{18}O depleted precipitation characterizing the winter and δ^{18}O enriched precipitation characterizing the summer, stream water variations exhibit a markedly damped response (Fig. 9a). Nevertheless, despite strong mixing within the catchment, seasonal variations in δ^{18}O in stream waters were observed, with the depleted snowmelt signal in the spring being evident in stream waters, as were the occasional signals of summer events where precipitation was enriched in δ^{18}O (Fig. 9b). Variations in δ^{18}O in groundwater in both the boreholes and spring waters

Table 1. *Chemical composition (mean and range) of precipitation, groundwater, spring water and stream water in the Allt a' Mharcaidh catchment*

	Rainfall	Borehole B4	Borehole B25	Spring	Stream
pH	4.8	7.22	7.02	5.79	6.47
	4.0–5.4	6.27–7.72	6.73–7.4	5.44–6.06	5.69–6.99
ANC	−1	104	36	18	59
	−3–0	32–178	1.290	20–59	24–117
Na	52	171	141	78	126
	5–230	87–313	116–182	53–121	98–154
K	7	16	10	4	6
	1–55	7–36	7–17	3–9	4–15
Mg	16	45	27	14	23
	2–68	24–66	15–41	10–31	12–35
Ca	4	59	36	22	46
	21–107	18–375	12–181	6–53	30–88
Cl	64	129	151	68	98
	6–258	59–338	94–237	34–83	76–150
SO_4	53	52	51	40	44
	13–187	32–89	17–81	20–63	29–68
NO_3	27	1	<0.1	2	<1
	4–102	<0.1–32	<0.1–1	<0.1–4	<1–3
SiO_2-Si	<0.02	5.92	5.92	3.35	5.32
	<0.01–0.07	3.72–7.51	3.72–7.51	2.18–3.73	2.92–7.21

All units in $\mu eq\,l^{-1}$except pH and SiO_2-Si ($mg\,l^{-1}$)

are even more damped (Table 2), though slight seasonal patterns could be observed. The mean isotopic concentration of both stream water and groundwater was much closer to that of winter precipitation than summer precipitation. This implied that winter precipitation and the subsequent snowmelt period is the main groundwater recharge in the catchment.

These seasonal variations in $\delta^{18}O$ isotope levels were used by Soulsby *et al.* (2000) to model mean

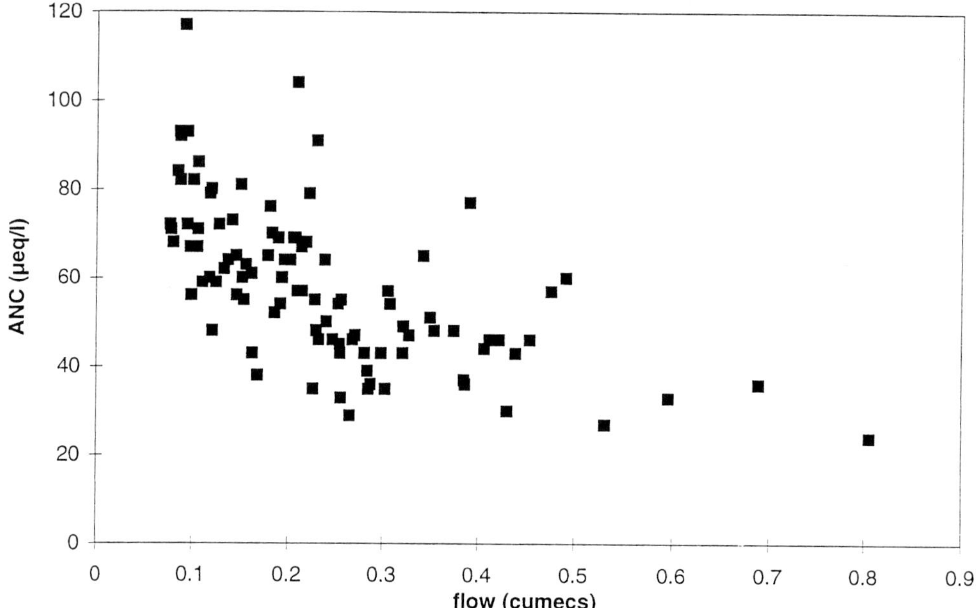

Fig. 7. ANC variations with flow in Allt a' Mharcaidh.

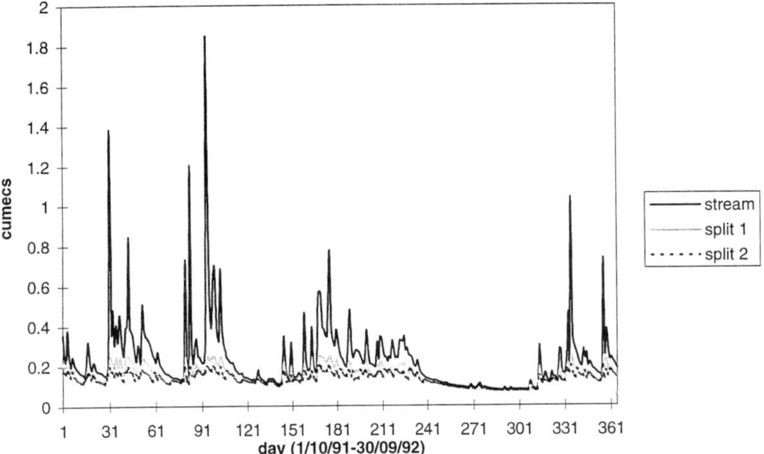

Fig. 8. Modelled range of likely groundwater contributions to stream flow estimated by a 2-component mixing model for chemical hydrograph separation.

water residence times using the exponential, well-mixed model described by Maloszewski *et al.* (1983). The amplitude of seasonal peaks and troughs in $\delta^{18}O$ levels in various catchment waters were modelled using fitted sine wave functions (Fig. 10). These were combined with a well-mixed model (cf. Stewart & McDonnell 1991) to estimate mean residence times of <6 months for soil waters contributing to storm runoff, *c.* 2 years for shallow spring water and *c.* 5 years for deeper groundwaters that sustains baseflows (Table 2). Although the model fits were statistically significant for precipitation, soil water and spring water, the simple sine wave models obviously provide no more than a first approximation of water residence times, though they are consistent with other tracer studies in similar headwater catchments in terms of order of magnitude estimates (Maloszewski *et al.* 1983; DeWalle *et al.* 1997; Vitvar & Balderer 1997). The models also provide no real insight into the probability distribution of residence times that will be produced by the spectrum of flow paths in both the catchment soils and drift deposits. Intuitively, these can be expected to range between <1 day to >100 years on the basis of topography and likely variability in the permeability of various drift deposits in the catchment (Robins 1990).

Despite these limitations, the groundwater studies in the Allt a' Mharcaidh clearly show the hydrological and hydrochemical importance of groundwater in the Cairngorm mountains. The past emphasis on acidification and near-surface

Table 2. *Isotope composition ($\delta^{18}O$) of precipitation, soil water, borehole water, spring water and stream water in the Allt a Mharcaidh catchment. Mean and range values (‰) and mean residence times estimated by Soulsby et al. (2000)*

	Mean	Minimum	Maximum	Amplitude	Residence time (years)
Inputs					
Winter precipitation	−8.84	−20.93	−4.55	1.47*	–
Summer precipitation	−6.82	−16.04	−2.47	–	–
Stream water					
G1	−9.56	−10.44	−8.45	0.2	–
Groundwater					
G1 Baseflow	−9.44	−9.69	−9.16	0.02	5
Spring	−9.19	−9.74	−8.86	0.1	2
Borehole	−9.03	−9.74	−8.45	0.15	
Soil water					
G1 catchment	−8.28	−9.91	−5.87	0.96	>0.6

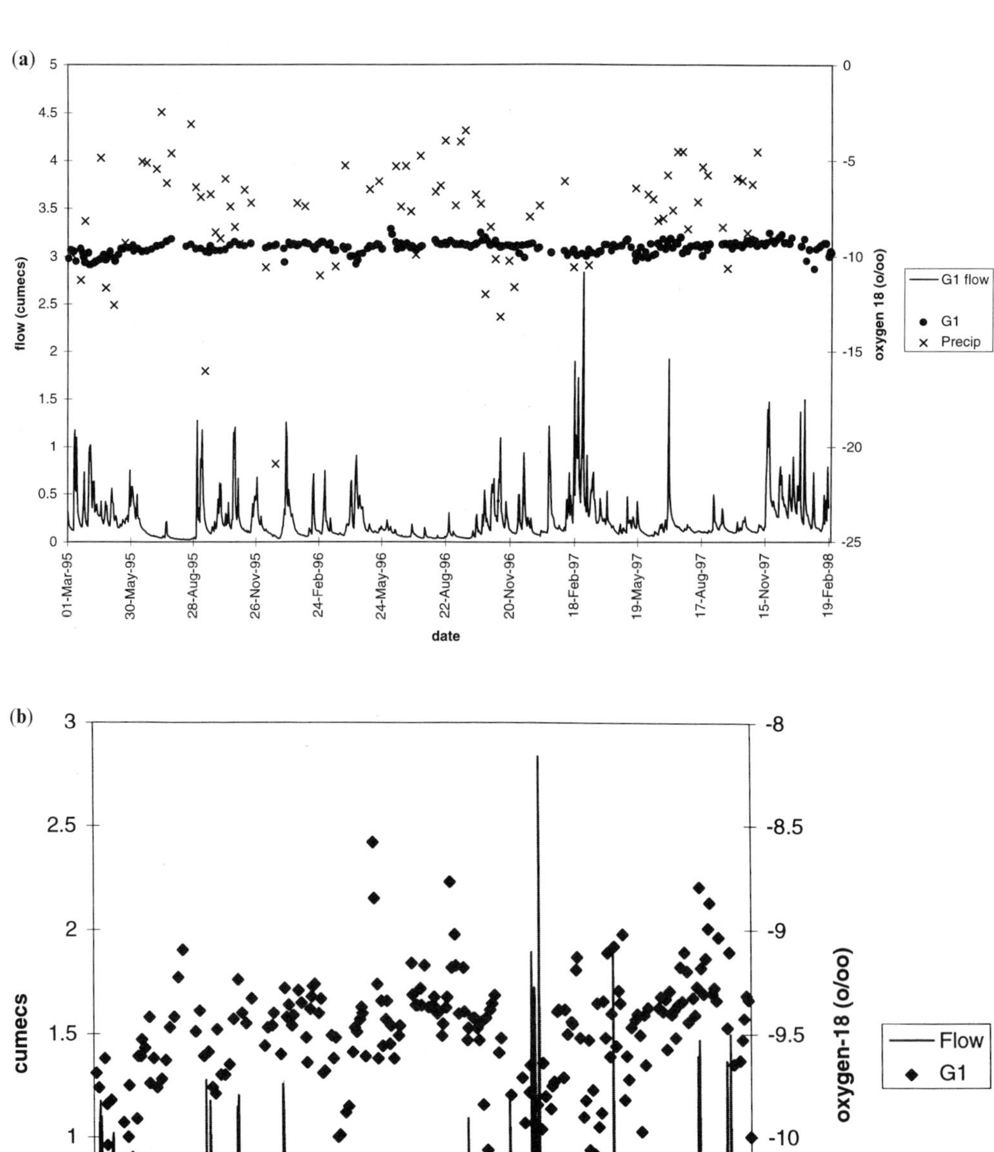

Fig. 9. Annual variations in ^{18}O levels in (**a**) precipitation and (**b**) streamfow in the Allt a' Mharcaidh.

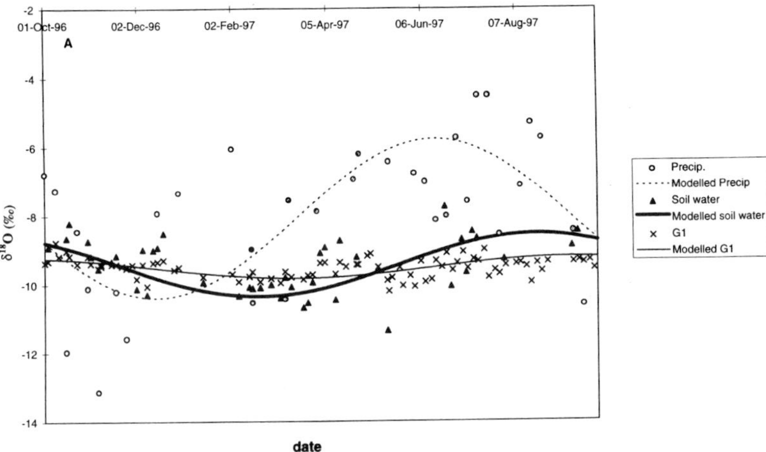

Fig. 10. Modelled sine wave functions fitted to seasonal variations in ^{18}O in precipitation, streamwater and groundwater in the Allt a' Mharcaidh.

hydrological pathways that produce acid storm runoff has assumed a far too passive role for groundwater which both hydrogeochemical and isotope studies imply is erroneous. Clearly the discharge of diverse groundwater flow paths into the stream bed and riparian area is important as implied by other studies in Scotland and elsewhere (Hooper *et al.* 1998; McDonnell *et al.* 1998). Moreover, the discharge of groundwater at high altitude spring sites is a particular feature of the Cairngorms and produces wetland habitats with a high conservation value, particularly for their bryophyte flora. These wetlands are important in functional units within montane ecosystems in the Cairngorms as well as forming the sources of many first order streams (Rodwell 1991; Soulsby *et al.* 1999).

Lowland catchments: groundwater and freshwater ecology

The hyporheic environment

In contrast to the Allt a' Mharcaidh catchment, the lowland Newmills burn appears to conform to more classical models of groundwater behaviour. The hydrological regime of the Newmills burn reflects two dominant sets of hydrological processes. For most of the year, the flow is maintained by groundwater discharges which sustain baseflows. Under such conditions, piezometers show an upwards hydraulic gradient in the stream bed environment reflecting slow groundwater discharge into the stream (Fig. 11a). This baseflow dominance is interspersed by very flashy responses to hydrological events, where overland flow from saturated riparian soils and preferential flow through networks of agricultural drains rapidly route storm flows into the stream network. The higher stream stage during storm events can result in a reversal of the hydraulic gradient in the hyporheic zone causing streamwater to flow into river gravels. During high flow events the stream has extremely high suspended sediment concentrations (>1000 mg l^{-1}) as a result of soil erosion within the catchment. These fines readily infiltrate into the stream bed, and generally comprise 25% (by mass) of the stream sediments (Moir 1999). Fine sediments close to the surface of the stream bed are readily mobilized during storm events and then re-deposited as flows subside; this process is likely to alter the permeability of the stream bed as flows change (Younger *et al.* 1993).

Consequently, the hyporheic environment of the Newmills burn clearly shows a stratification of groundwater and stream water, separated by a mixing zone in the upper few decimetres of the stream gravels (Table 3). In general, stream waters are relatively well-oxygenated (mean concentration of 11.3 mg l^{-1}) compared with lower hyporheic zone waters (mean 7.4 mg l^{-1}). In contrast, hyporheic waters are enriched in alkalinity and calcium, with higher conductivities than the stream. The upper hyporheic zone (at 0.1 m depth) appears to represent part of an area of marked chemical gradients between stream water and groundwater, where complex mixing occurs.

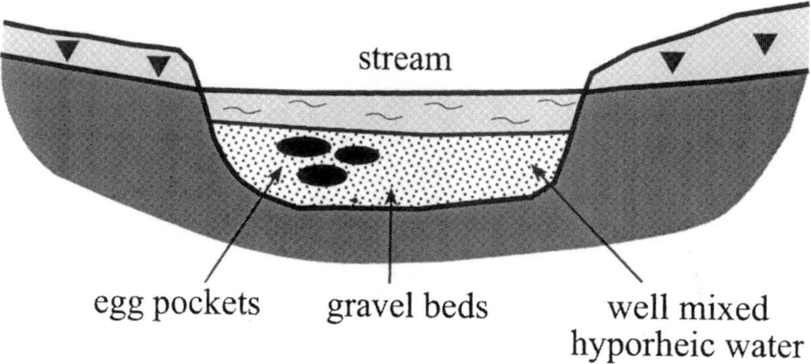

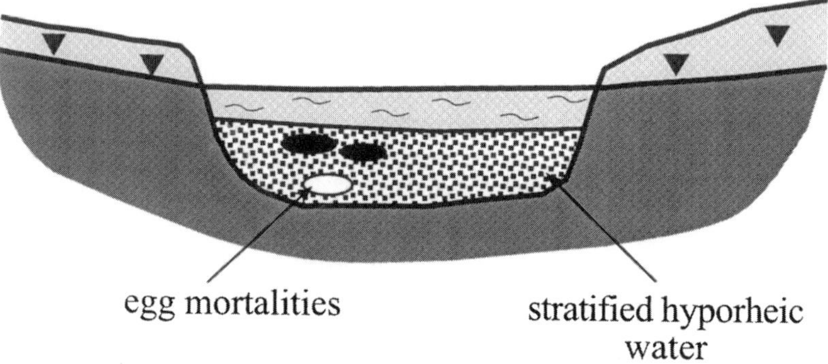

Fig. 11. Groundwater–surface water interactions in salmon spawning gravels after spawning and prior to hatching.

Table 3. *Chemical composition (mean and range in mg l^{-1}) of stream water, upper and lower hyporheic water in the Newmills burn*

	Stream	Upper hyporheic water	Lower hyporheic water
pH	7.25	7.19	7.16
	7.01–7.53	7.0–7.38	6.98–7.34
Dissolved O_2	11.26	8.65	7.35
	9.30–12.90	7.05–11.95	4.50–11.30
Alkalinity	34.92	37.38	42.6
	23.0–44.0	30.5–47.0	33–59.5
Na	20.73	20.05	18.78
	15.3–40.0	15.25–36.80	15.05–28.3
Ca	15.97	16.92	18.36
	14.5–19.1	14.75–21.4	15.0–24.1

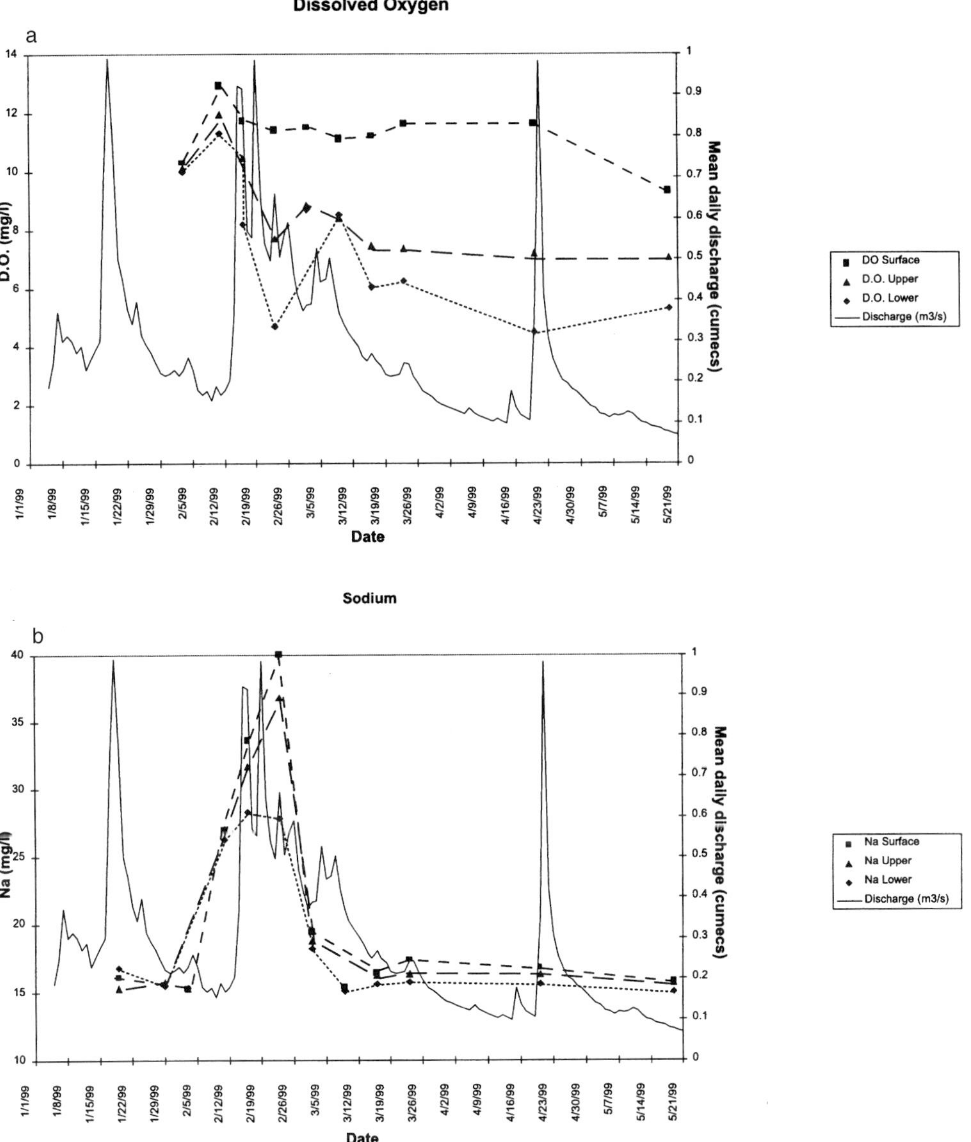

Fig. 12. The chemistry of streamwater and hyporheic water in salmon spawning gravels (**a**) dissolved oxygen and (**b**) sodium.

The hydrochemical responses shown in Fig. 12 indicate how the groundwater–surface water interactions in the hyporheic zone of the Newmills burn are extremely dynamic and exert a strong control on the chemical composition of water within salmon redds. Between the cessation of spawning activity (late December) and the end of fry emergence (late spring), flows in the Newmills burn varied. Rainfall in mid-January produced a marked flood peak, after which flows declined during a cold period which eventually saw around 30 mm of precipitation fall as snow which remained lying on the catchment for around a week. In mid-February, a major snowmelt event caused flows to rise again, after which they fell for much of March and April, before a brief marked spate occurred at the end of April.

The creation of the redd removed much of the fine sediment out of the gravel bed. Consequently, mixing of stream water and hyporheic water is thorough, and the chemical composition of the stream water and upper and lower hyporheic water is similar. However, over time the chemistry of hyporheic waters diverges from that of the stream with the effect being most marked in the lower (0.3 m deep) hyporheic zone (Fig. 12). This is clearly reflected in the responses of dissolved oxygen where concentrations in stream water remain relatively high, but fall off in both the lower and upper hyporheic zones (Fig. 12a). This probably reflects the combined effects of fine sediment influx, reducing the permeability of the upper stream bed and the increasing influence of more reduced groundwater entering the hyporheic zone (Soulsby *et al.* 2000*b*). However, it may also reflect the in-washing of organic sediments into the stream bed, which subsequently decompose reducing dissolved oxygen concentrations, particularly as fine organic sediments normally infiltrate deeper into spawning gravels than coarser sand deposits (Soulsby *et al.* 2000*a*).

The sodium response shows that interactions in the hyporheic zones are complex (Fig. 12b). At the start of the monitoring period Na concentrations in the stream and hyporheic waters were similar. However, the snowmelt event in February simulated a tracer experiment, as road salts were washed into the stream from road surfaces as the melt proceeded. The Na response clearly shows the movement of this 'tracer' in the stream, but also within the hyporheic zone, implying a reversal of hydraulic gradients when the stream stage is elevated (by approximately 1 m) and the finer sediments on the stream bed are mobilized. The more dampened response in the deeper hyporheic sampler, the deeper penetration of the mixing zone into the stream gravels which is short lived as stream flows subside; fine sediment infiltration occurs and streamward hydraulic gradients once again prevail, with Na concentrations returning to pre-event levels.

Ecological implications

Clearly the survival of salmonid eggs and the recruitment of juveniles to the stream rely on the maintenance of suitable conditions within stream gravels between spawning (November–December) and fry emergence (April–May). The discharge of groundwater with lower oxygen concentrations into the spawning gravels may thus have a limiting effect on egg survival (Sowden & Bower 1985; Wood & Armitage 1997). The dissolved oxygen content of the lower hyporheic samplers falls to around 5 $mg\,l^{-1}$, a level where an impact on mortalities might be expected (Rubin & Glimsater 1996). Within the Newmills burn, egg survival rates range between 0 and 100%; there is circumstantial evidence that this may be related to variations in groundwater inputs to the stream. Youngson (pers. comm.) has excavated over 50 redds in the burn and noted that egg mortalities tended to increase with depth in the redd, where groundwater influences would be expected to be greatest. However, overall, median survival rates in redds are approximately 80%, thus it may be the ability of stream waters to mix with groundwater in the hyporheic zone that maintains adequate oxygen levels to ensure egg survival, in the majority of cases, in this particular stream.

Investigation of spawning habitats in the Newmills burn is just one example of a situation where groundwater has a key ecological significance. Although the work presented here represents only a preliminary assessment of the role of groundwater–surface water interactions in a headwater stream, it is clear that further investigation is needed to provide more substantial understanding. For example, groundwater–stream water exchanges strongly influence the thermal regime of streams, and subsequently strongly influence primary and secondary production (Brunke & Gonser 1997). This broader role of groundwater in freshwater ecosystems is only just beginning to be realized in the Celtic regions, and it is likely to provide a fruitful area of research over the next decade.

Conclusion

It is becoming clear that active groundwater flow systems in headwater catchments are strongly regulated by topography, soil characteristics and management, the nature of drift deposits, and the presence of fracture zones in the upper weathering zones of rocks with low primary porosity. Thus, it is likely that a range of groundwater sources, with different velocities and residence times, contribute to stream flows. In many instances, preferential flow is likely to be extremely important, and thus it is the subsequent effects on stream hydrology, hydrochemistry and ecology that will prove complex.

It is difficult to envisage how such complexity can be accurately incorporated into physically-based models of catchment hydrology (Bevan 2000). Nevertheless, it is clear from hydrogeochemical work that distinct groundwater sources can be identified for many upland catchments and provide an important input into streams.

These need to be realistically incorporated into semi-distributed hydrological models (e.g. Dunn *et al.* 1998), particularly if the transfer of pollutants through catchment systems is to be modelled with any degree of accuracy. Clearly, given the hydrological and hydrogeochemical significance of this groundwater end member, this represents a major challenge to contemporary catchment studies.

Our understanding of the ecological importance of groundwater–surface water interactions lags behind that of other areas, most notably North America, where much more research has been done in this field (Fraser & Williams 1998). Nevertheless, an increasing number of studies, including those presented here, are demonstrating how an understanding of groundwater systems can elucidate some of the key processes which maintain the biodiversity of freshwater ecosystems. It is exciting to see a new field of hydrogeological investigation developing, though it is one where traditional expertise will need to contribute to interdisciplinary research with other specialists in order to develop the holistic understanding of environmental systems that is needed to underpin sustainable management (Naiman *et al.* 1995).

For those involved in such endeavours in the Celtic regions, it is encouraging to know that we are part of a long tradition. In the Gaelic legends of the Fenian cycle, dating back to AD 300–400, eating the Salmon of Knowledge from the river Boyne was said to give possession of 'knowledge of everything in the world, past, present and future' (Heaney 1994). These fish lived in pools near the source of the Boyne, where groundwater upwelling occurred: clearly, interest in groundwaters and headwaters in the Celtic regions is nothing new!

The Allt a' Mharcaidh studies form part of the NICHE initiative. The work reported here has been variously funded by the Leverhulme Trust, NERC, the Royal Society, the DETR and the Scottish Office. The invaluable help of Jo and Molly Porter in collecting much of the data and Bob Ferrier in allowing access to the data is gratefully acknowledged. The Newmills burn investigation has been funded by Fisheries Research Services and the University of Aberdeen through the Aberdeen Research Consortium. Alan Youngson started the work and Pat Donald kindly allows access to the site.

References

BEVAN, K. in press. Calibration, validation and equifinality in hydrological modelling. *In*: ANDERSON, M. G. & BATES, P. (eds) *Uncertainty in geomorphological modelling*. Wiley, Chichester.

BRUNKE, M. & GONSER, T. 1997. The ecological significance of exchange processes between rivers and groundwater. *Freshwater Biology*, **37**, 1–33.

COOK, J. M., EDMUNDS, W. M. & ROBINS, N. S. 1991. Groundwater contribution to an acid upland lake (Loch Fleet, Scotland) and the possibilities for amelioration. *Journal of Hydrology*, **125**, 111–128.

DEWALLE, D. R., EDWARDS, P. J., SWISTOCK, B. R., ARAVENA, R. & DRIMMIE, R. J. 1997. Seasonal hydrology of three Appalachian forest catchments. *Hydrological Processes*, **11**, 1895–1906.

DUNN, S. M. & LANGAN, S. J. 1998. Climate, snow and the water resources of a major Scottish river. *In*: LEMMELA, N. & HELENIUS, N. (eds) *Proceedings of the Second International Conference on Climate and Water*. Espoo, Finland, 178–187.

DUNN, S. M., MCALISTER, E. & FERRIER, R. C. 1998. Development of a distributed catchment scale model for the river Ythan, NE Scotland. *Hydrological Processes*, **11**, 1895–1906.

EDMUNDS, W. M. & SAVAGE, D. 1992. Geochemical characteristics of groundwater in granites and related crystalline rocks. *In*: WILKINSON, W. B. & DOWNING, R. A. (eds) *Applied Groundwater Hydrology*. Oxford University Press, Oxford. 266–282.

FRASER, B. G. & WILLIAMS, D. D. 1998. Seasonal boundary dynamics of a groundwater/surface water ecotone. *Ecology*, **79**, 2019–2031.

GORDON, J. E., THOMPSON, D. B. A., HAYNES, V., BRAZIER, V. & MACDONALD, R. 1998. Environmental sensitivity and conservation management in the Cairngorm mountains, Scotland. *Ambio*, **27**, 335–344.

GRIEVE, I. C., GILVEAR, D. G. & BRYANT, R. G. 1995. Hydrochemical and water source variations across a floodplain mire, Insch marshes, Scotland. *Hydrological Processes*, **9**, 99–110.

HARRIMAN, R. & FERRIER, R. C. 1990. Scottish catchment studies. *In*: MASON, B. J. (ed.) *The Surface Water Acidification Programme*. Cambridge University Press, Cambridge, 8–17.

HEANEY, M. 1994. *Over nine waves: a book of Irish legends*. Faber, London.

HILL, T. & NEAL, C. 1997. pH, alkalinity and conductivity in runoff and groundwater for the Upper River Severn catchment. *Hydrology and Earth Systems Science*, **1**, 697–715.

HOOPER, R. P., AULENBACH, B. T., BURNS, D. A., MCDONNELL, J. J., FREER, J., KENDALL, C. & BEVAN, K. 1998. Riparian control of stream-water chemistry: implications for hydrochemical basin models. *IAHS Publications*, **248**, 451–458.

HYNES, H. B. N. 1983. Groundwater and stream ecology. *Hydrobiologia*, **100**, 93–99.

JENKINS, A., FERRIER, R. C. & HARRIMAN, R. 1993. Meltwater chemistry and its impact on stream water quality. *Hydrological Processes*, **8**, 335–349.

JENKINS, A., FERRIER, R. C., HARRIMAN, R. & OGUNKOYA, Y. O. 1994. A case study in catchment hydrochemistry: conflicting interpretations from hydrological and chemical observations. *Hydrological Processes*, **8**, 335–349.

JONES, J. B. & HOLMES, R. M. 1996. Subsurface-surface interactions in stream ecosystems. *Trends in Ecology and Evolution*, **11**, 239–242.

KENDALL, C., SKLASH, M. G. & BULLEN, T. D. 1995. Isotope tracers of water and solute sources in catchments. *In*: TRUDGILL, S. T. (ed.) *Solute Modelling in Catchment Systems*. Wiley, Chichester, 261–303.

MALOSZEWSKI, P., RAUERT, W., STICHLER, W. & HERRMAN, A. 1983. Application for flow models in alpine catchment area using tritium and deuterium data. *Journal of Hydrology*, **66**, 319–330.

MCDONNELL, J. J., MCGLYNN, B. L., KENDALL, K., SHANLEY, J. & KENDALL, C. 1998. The role of near-stream riparian zones in the hydrology of steep upland catchments. *International Association of Hydrological Sciences Publications*, **248**, 173–180.

MOIR, H. J. 1999. *Characterization of salmon spawning gravels in a Scottish upland stream*. PhD thesis, University of Aberdeen.

NAIMAN, R., MAGNUSSON, J. J., MCKNIGHT, D. & STANFORD, J. A. 1995. *The Freshwater Imperative*. Island Press, New York.

NEAL, C., HILL, T., HILL, S. & REYNOLDS, B. 1997. Acid neutralization capacity measurements in surface and groundwaters in the upper River Severn, Plynlimon: from hydrograph splitting to water flow pathways. *Hydrology and Earth Systems Science*, **3**, 687–696.

PLUMMER, L. N., PRESTEMON, E. C. & PARKHURST, D. L. 1991. An interactive code (NETPATH) for modelling net geochemical reactions along a flow path. *USGS Water Resources Inventory*, 91–4078. USGS.

OGUNKOYA, O. O. & JENKINS, A. 1993. Analysis of storm hydrograph and flow separations using a three-component hydrograph separation model. *Journal of Hydrology*, **142**, 71–88.

ROBINS, N. S. 1990. *Hydrogeology of Scotland*. HMSO, London.

RODWELL, D. 1991. *British Plant Communities: Mires and Heaths*. Cambridge University Press, Cambridge.

RUBIN, J. F. & GLIMSATER, C. 1996. Egg-to-fry survival of the sea trout in some streams of Gotland. *Journal of Fish Biology*, **53**, 84–92.

SOULSBY, C., MALCOLM, I. & YOUNGSON, A. In press *b*. Hydrochemistry of the hyporheic zone in salmon spawning gravels: a preliminary assessment in a small regulated stream. *Regulated Rivers* (in press).

——, MOIR, H., CHEN, M. & GIBBINS, C. 1998*b*. Assessing the impact of groundwater development on Atlantic salmon spawning habitat: integrating hydrogeological and hydroecological models. *In*: KIRKBY, C. & WHEATER, H. (eds) *Hydrology in a Changing Environment*. Wiley and Sons, Chichester.

——, YOUNGSON, A. F., MOIR, H. J. & MALCOLM, I. A. In press *a*. Fine sediment influence on salmonid spawning habitat in a lowland agricultural stream. *Science of the Total Environment*.

——, HELLIWELL, R. C., FERRIER, R. C., JENKINS, A. & HARRIMAN, R. 1997*a*. Seasonal snowpack influence on the hydrology of a sub-arctic catchment in Scotland. *Journal of Hydrology*, **192,** 17–32.

——, MALCOLM, R., HELLIWELL, R. C., FERRIER, R. C. & JENKINS, A. 1999. Influence of springs on the hydrogeochemistry of mountain streams in the Cairngorms, Scotland. *Hydrology and Earth Systems Science*, **3**, 409–420.

——, ——, ——, —— & —— 2000. Isotope hydrology of the Allt a' Mharcaidh catchment, Cairngorms, Scotland: implications for hydrological flow paths and water residence times. *Hydrological Processes*, **14**, 747–762.

——, TURNBULL, D. A., HIRST, D., LANGAN, S. J. & OWEN, R. 1997*b*. Reversibility of stream acidification in the Cairngorm region of Scotland. *Journal of Hydrology*, **195**, 291–311.

——, CHEN, M., FERRIER, R. C., HELLIWELL, R. C., JENKINS, A. & HARRIMAN, R. 1998*a*. Hydrogeochemistry of shallow groundwater in an upland Scottish catchment. *Hydrological Processes*, **12**, 1111–1118.

SOWDEN, T. K. & BOWER, G. 1985. Prediction of rainbow trout embryo survival in relation to groundwater seepage and particle size of spawning substrates. *Transactions of the American Fisheries Society*, **114**, 804–812.

STEWART, M. K. & MCDONNELL, J. J. 1991. Modelling base flow soil water residence times from deuterium concentrations. *Water Resources Research*, **27**, 2681–2693.

VITVAR, T. & BALDERER, W. 1997. Estimation of mean residence times and runoff generation by stable isotope measurements in a small pre-alpine catchment. *Applied Geochemistry*, **12**, 787–796.

WADE, A. J., NEAL, C., SOULSBY, C., SMART, R. P., LANGAN, S. J. & CRESSER, M. S. 1999. Modelling streamwater quality under varying hydrological conditions at different spatial scales. *Journal of Hydrology*, **217**, 266–283.

WHEATER, H. S., TUCK, S., FERRIER, R. C., JENKINS, A., KLEISSEN, F. M., WLKER, T. A. B. & BECK, M. B. 1993. Hydrological flow paths at the Allt a' Mharcaidh catchment: an analysis of plot and catchment scale observations. *Hydrological Processes*, **7**, 359–371.

WINTER, T. 1995. Recent advances in understanding the interaction of groundwater and surface water. *Review of Geophysics*, **Supplement**, 985–984.

WOOD, P. J. & ARMITAGE, P. D. 1997. Biological effects of fine sediment in the lotic environment. *Environmental Management*, **21**, 203–217.

YOUNGER, P. L., MACKAY, R. J. & CONNORTON, B. J. 1993. Streambed sediment as a barrier to groundwater pollution: insioghts from fieldwork in the river Thames basin. *Journal of of the Institution of Water and Environmental Management*, **7**, 577–585.

YOUNGSON, A. F. & MCCLAREN, I. 1998. *Relocation of naturally spawned salmonid ova as a countermeasure to patchiness in adult spawning distribution*. Scottish Fisheries Research Report **No 61**.

'*Uisge Mèinne*': mine water hydrogeology in the Celtic lands, from *Kernow* (Cornwall, UK) to *Ceap Breattain* (Cape Breton, Canada)

PAUL L. YOUNGER[1] & ANTHONY B. LaPIERRE[2]

[1] *Reader in Water Resources, Water Resource Systems Research Laboratory, Department of Civil Engineering, University of Newcastle, Newcastle upon Tyne NE1 7RU, UK (e-mail: p.l.younger@ncl.ac.uk)*
[2] *Department of Civil Engineering, Dalhousie University (Daltech), Halifax, Nova Scotia, Canada*

Abstract: Mining has historically made a major contribution to the economies of all the contemporary Celtic lands. At the start of the third Millennium, the bulk of mining activity has ceased, and problems associated with hydrogeological changes in abandoned mines are now common to all these lands. In global terms an unusually high proportion of mining in the Celtic lands has been by underground methods, a fact which is reflected in the range of hydrogeological problems encountered in these countries. Recent experiences in these countries offer insights which should be useful elsewhere when currently-active deep mines are eventually abandoned. Particular lessons are drawn from case studies and analyses of previously unpublished data from *Alba* (Scotland), *Ceap Breattain* (Cape Breton, Canada), *Cymru* (Wales) and *Kernow* (Cornwall). These lessons are:

(1) The importance of recognizing the predominance of mined features in the post-closure hydrogeology of abandoned mines.
(2) Dynamic temporal changes in hydrogeological behaviour arise from collapse of mined voids, caused by fluvial erosion by rapidly-flowing mine waters and/or by pneumatic fracturing by mine gases compressed in pockets during mine water rebound; these changes can have significant implications for human safety and environmental protection.
(3) Net-acidic mine waters are generally restricted to situations in which high-sulphur strata are present in (i) recently-flooded deep-mine workings (ii) shallow partially flooded mine workings and (iii) perched groundwater systems in spoil heaps and opencast backfill.
(4) Net-alkaline mine waters are associated with (i) low-sulphur strata in any hydrogeological setting and (ii) high-sulphur strata at depth in long-flooded workings. In practice, this means that the net-alkaline mine waters are far more abundant than the net-acidic.
(5) The presence of limestone in a mined sequence is not on its own a guarantee that mine waters will be net-alkaline; the patterns of groundwater flow (which determine the transport of limestone dissolution products through the mined system) must also be favourable.
(6) Mine water often becomes hydrochemically stratified during rebound. However, when discharge from a mined system commences, this stratification can break down, resulting in discharges considerably poorer in quality than would have been inferred by sampling the uppermost waters alone.

Mining and subsequent metal-working has been a major economic and cultural activity amongst the Celtic peoples from the earliest times (Shepherd 1993). The ancient Celts were remarkably fond of gold and silver jewellery, and indeed the two principal phases of Celtic cultural dispersion (the Hallstatt and La Tène eras) are mainly recognized on the basis of their characteristic artistic metalwork (Ross 1986). In this paper we consider the present-day hydrogeological legacy of the millennia-old mining industry of the Celtic lands, which both fed the Celtic love of precious metals, and supported the economy of the British Empire in its heyday. Although the Celts ranged widely over continental Europe in their early history (see, for

From: Robins, N. S. & Misstear, B. D. R. (eds) *Groundwater in the Celtic Regions: Studies in Hard Rock and Quaternary Hydrogeology*. Geological Society, London, Special Publications, **182**, 35–52. 1-86239-077-0/00/ $15.00

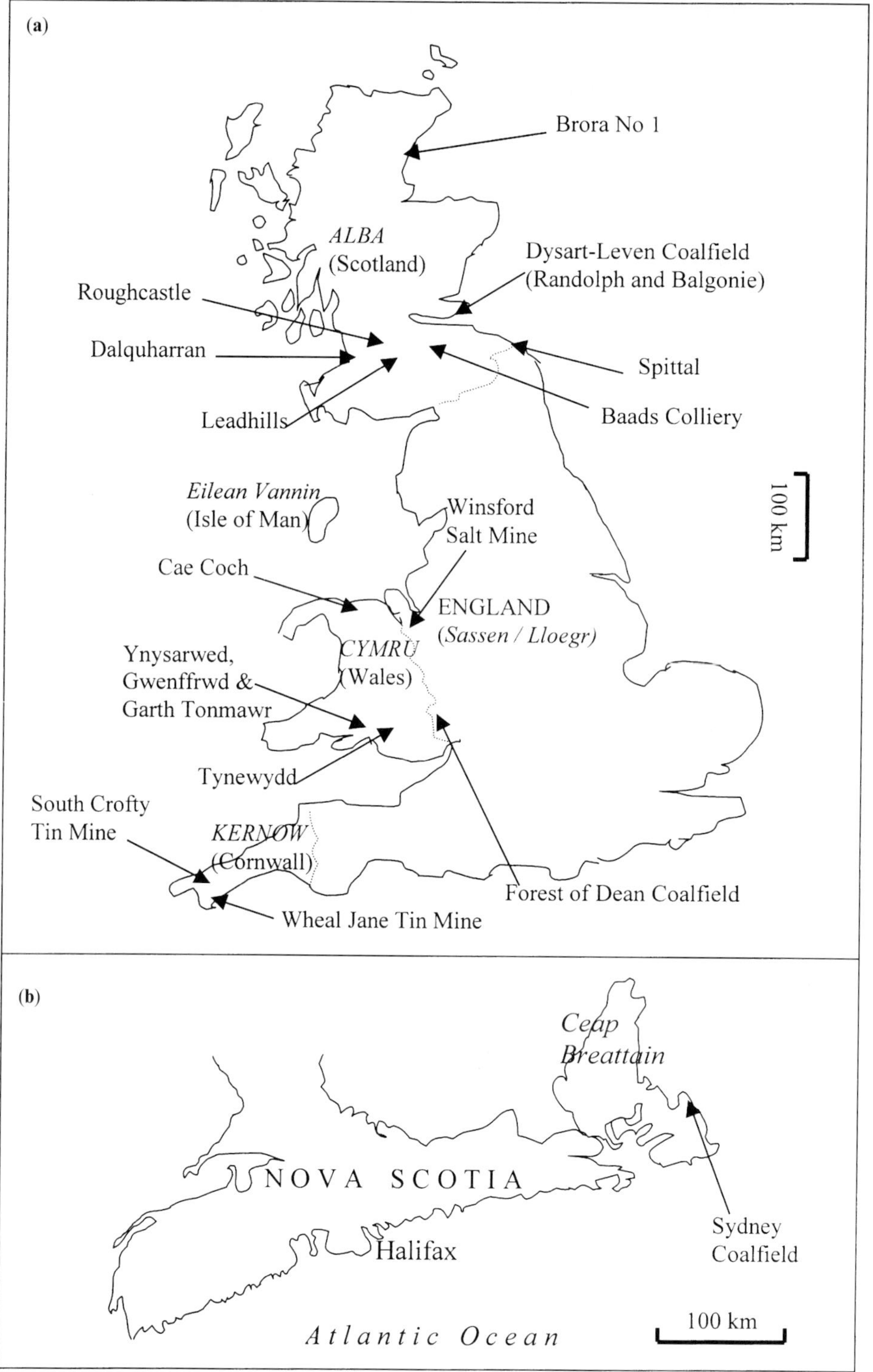

Fig. 1. Location maps for sites mentioned in the text. (**a**) A map showing the Celtic lands of Britain, namely *Alba* (Scotland), *Cymru* (Wales), *Kernow* (Cornwall), and *Eilean Vannin* (Isle of Man). The Celtic language names are also given for England in Scots Gaelic (*Sassen*) and Welsh (*Lloegr*) respectively. The specific sites marked on the map are those discussed in detail in the text. (**b**) Map of Nova Scotia, Canada, showing the location of the *Gaidhealtachd* of *Ceap Breattain* (Cape Breton Island), within which lies the Sydney Coalfield study area.

instance, Herm 1976; Hubert 1993), the focus of this volume is on the remnant nations of Celtic ancestry along the northwestern sea-board of Europe. These nations are sometimes collectively labelled (somewhat ambiguously) the 'Celtic Fringe'. In fact the Celtic languages and their cultural inheritance are not restricted to this 'fringe', for the emigration of entire communities to the New World in the 19th Century resulted in the establishment of enduring Welsh- and Gaelic-speaking communities in Patagonia and Nova Scotia, respectively. For the purposes of the present paper, the focus will be on the mining hydrogeology of the following Celtic lands (Fig. 1):

(a) *Kernow* (Cornwall)
(b) *Cymru* (Wales)
(c) *Alba* (Scotland)
(d) *Ceap Breattain* (Cape Breton Island, the *Gaidhealtachd* of Nova Scotia, Canada; Fig. 1b)

For want of data and space we do not consider Ireland, Brittany, the Isle of Man or Galicia (north-west Spain) in any detail, though all of these have had important mining industries with considerable environmental impacts (e.g. Gray 1996; O'Brien 1996; Dhonau & Wright 1998; Monterroso & Macias 1998). As the third Millennium AD begins, mining has all but ceased in the Celtic lands (with the exception of Ireland, where a renaissance of base-metal sulphide mining is underway; Dodds *et al.* 1994; Dhonau & Wright 1998). It has often been found elsewhere that the hydrogeological changes which accompany mine abandonment have considerable environmental significance (e.g. Younger 1998*a*). Experiences of such problems are now common to all the Celtic territories (e.g. Henton 1974, 1979, 1981; Robins 1990; Cain *et al.* 1994; Hamilton *et al.* 1994; Reddish *et al.* 1994; Gray 1996; Robins & Younger 1996; Bowen *et al.* 1998; Younger 1999*a*). It is the purpose of this paper to draw some general hydrogeological lessons of potentially global relevance from the Celtic experience. As an unusually high proportion of mining in the Celtic lands has been by underground methods (i.e. by 'deep mines', as opposed to 'surface (or opencast) mines'), the experiences distilled below are especially relevant in countries with deep-mining industries.

Data and methods

The data used in this study were obtained from two sources:

(a) the archives of public bodies (the Environment Agency (EA) for sites in Cornwall and Wales, the Scottish Environment Protection Agency for sites in Scotland);
(b) by direct measurement by staff of the Departments of Civil Engineering at the University of Newcastle (UK) and Dalhousie University (Canada).

All analyses were made using the standard methods of the American Public Health Association, following long-established QA/QC procedures, which have been publicly accredited in accordance with the latest norms (most recently, the UKAAS (UK Analytical Accreditation Scheme)). pH, conductivity, Eh and temperature were determined in the field using daily-calibrated electronic meters. The Newcastle team determined alkalinity in the field using a HACH digital titrator. Dissolved metals were determined by atomic absorption spectrophotometry (AAS) and inductively coupled plasma (ICP) with optical and mass spectrometry. Anions were determined by ion chromatography (Dionex 500 machine). The consistency of results between the different laboratories providing data to this study has been established during previous studies (see Younger 1998*a*, 1999*a*; La Pierre 1999).

Selected Celtic experiences with the physical hydrogeology of abandoned deep mines

General observations on the hydrogeology of deep-mined strata

The hydrogeology of abandoned deep mines has recently been reviewed in great detail in a research report prepared for the Environment Agency (England and Wales) by Younger & Adams (1999). Supplementary information on the behaviour of ancient adit systems in Cornwall (as well as northern England) has been presented by Younger (1998*b*). The general comments given here represent a précis and partial update of these earlier sources, with particular emphasis on those aspects of deep mine hydrogeology illustrated by recent experiences in the Celtic lands.

There are three principal factors to consider in relation to the hydrogeological behaviour of abandoned deep mines:

(a) the natural hydrogeology of the country rock within which the mine has been excavated;

(b) the nature and degree of alteration of the hydrogeological behaviour of the country rock by subsidence and fracturing induced by mining;
(c) the hydrogeological behaviour of the mined voids themselves.

With regard to the natural hydrogeology of the country rock, it is worth noting that the majority of productive mines have been developed in strata which are not amongst the most permeable, therefore have not been subjected to the same degree of hydrogeological investigation as the major public supply aquifers. The avoidance of major aquifers by miners is not surprising, for (as Kesserû 1995, has pointed out) miners have just as much interest as water resource managers in minimizing water inflows into active mines (the former for reasons of safety and minimizing the costs of dewatering, the latter for reasons of conservation of water resources). The recently- and currently-worked base metal sulphide mines of Ireland represent something of an exception to this general rule, having been excavated in Carboniferous Limestone which is widely used for private and public water supplies in central and western areas of the country (Dodds *et al.* 1994; Dhonau & Wright 1998). Yet even this exception 'proves the rule' as the operators of Lisheen mine in County Tipperary discovered to their cost in early 1999 when underground work was halted (it is still hoped temporarily) by unexpectedly prolific groundwater inflows which overwhelmed the installed dewatering capacity. Hence, whether the mine is in poorly documented low permeability strata or within a highly permeable aquifer, the lesson remains that the country rock within which mines are developed is rarely sufficiently well-characterized that it is amenable to accurate, deterministic predictions of future behaviour.

The processes by which subsidence and associated fracturing induced by mining alter the hydrogeological behaviour of the surrounding country rock have been the subject of substantial research efforts in the UK, principally in the context of planning and safe working of mines beneath the sea bed. (Although most UK undersea mining was undertaken in north-east England, workings also extended offshore in the Celtic lands, notably in Fife (Scotland; Younger *et al.* 1995), at Point of Ayr Colliery, North Wales (Younger 1996) and in north Cornwall (most notably at Levant Mine; Dines 1956)). Engineers working on behalf of the former UK state mining corporation British Coal developed a certain 'orthodoxy' in relation to the effects of longwall mining on the permeability of the overlying strata (Orchard 1975; Singh & Atkins 1983; Aston & Whittaker 1985). Figure 2 represents a simplified synthesis of the orthodox conceptual model for permeability development above longwall panels (for further discussion see Younger & Adams 1999). It should be noted that this orthodox conceptual model remains somewhat controversial in mining geology circles (Dumpleton, British Geological Survey, pers. comm., 1998; K. Whitworth, International Mining Consultants Ltd, pers. comm. 1–3–1999), although no alternative model has yet been proposed in the open literature. Nevertheless, in the majority of circumstances it can safely be assumed that 'unsupported' mining techniques (i.e. those which allow progressive collapse of the roof in worked out areas, which includes longwall coal workings) result in extensive fracturing of the roof strata (Fig. 2), usually inducing an increase of two to three orders of magnitude in the permeability of the roof strata (Singh & Atkins 1983). Exceptions to this increase in permeability can be expected where:

(a) the roof strata include swelling clays (which expand to fill new pore space upon extension);
(b) the workings are vertical or very steeply inclined (as in many Cornish, Welsh and Scottish metal vein mines);
(c) the country rock is extremely competent (e.g. the Cornish granites) and thus resists collapse indefinitely.

Where underground workings are excavated using a 'supported' method of mining (such as room-and-pillar) there may be little or no fracturing (and hence little or no increase in permeability) in the roof strata. Perhaps the most extreme illustration of this principle comes from north-west England (near the border with Wales) where room-and-pillar workings in Winsford Salt Mine (Fig. 1a) record *zero* groundwater ingress (G. Hall, Salt Union Ltd, pers. comm., 14–10–1998), despite the workings being overlain by saturated Quaternary sands. While the mechanical properties of halite-rich strata will clearly be different from those of coal-bearing strata, the principle that careful, supported methods of mining can greatly reduce the likelihood of water ingress is in principle applicable in a wide range of rock types.

Relatively few direct, formal studies have been made of the hydrogeological behaviour of open, flooded mine voids, and fewer still in the Celtic mining districts (Younger & Adams 1999). One of the earliest well-documented studies was that of Aldous & Smart (1988),

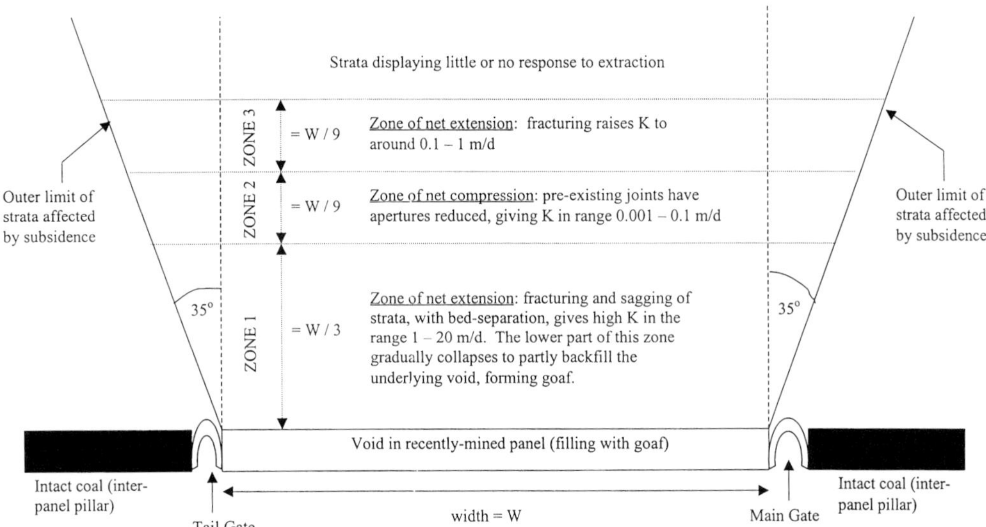

Fig. 2. Schematic diagram showing the 'orthodox' conceptual model for the development of zones of altered permeability above a recently-extracted longwall coal panel (after Singh & Atkins 1983; Younger & Adams 1999).

who applied speleological tracer techniques to abandoned workings in the Forest of Dean Coalfield (Aldous *et al.* 1986), near the Welsh border in south-west England. These studies revealed flow velocities as high as $16\,km\,d^{-1}$ above the water table (essentially by open channel flow), while roadways below the water table (which are consequently subject to lower hydraulic gradients) display velocities approaching $0.5\,km\,d^{-1}$. Given that the roadways in question typically have diameters of several metres, such velocities imply that flow below the water table is still turbulent (cf. Ford & Williams 1989, p. 145). Similar studies of roadways below a very low-gradient water table in the USA revealed velocities of $3\text{–}20\,m\,d^{-1}$ (Aljoe & Hawkins 1994). In the latter case, the velocities imply laminar flow, although imposition of steep gradients (e.g. during a pumping or injection test) could easily induce turbulent conditions. One important corollary of this is that turbulent flow might predominate in roadways and similar features during groundwater rebound, when hydraulic gradients are likely to be steep, with laminar flow becoming more common when rebound is complete and shallow gradients are established.

The importance of mined features in post-closure hydrogeology

Recent experiences in the Celtic mining districts of Britain have illustrated the predominance of mined features in the hydrogeological behaviour of abandoned mines. It is convenient to consider these experiences under two categories:

(1) hydrogeological processes during mine water 'rebound' (i.e. during the flooding of the voids after the cessation of dewatering);
(2) hydrogeological behaviour after rebound is complete.

Hydrogeological experiences during mine water rebound

During mine water rebound, the differences in water level between one set of workings and another may be very large. For instance, in the Dysart-Leven Coalfield of East Fife (Fig. 1a), Younger *et al.* (1995) reported water table elevations differing by as much as 300 m between inland and coastal 'ponds' in the old workings. Where flow paths become established between adjoining ponds (as often occurs when the water table rises to drown some previously dry old roadway), flow in response to such extreme head gradients is inevitably turbulent (Sherwood & Younger 1997), and may be so powerful that it causes rapid erosion of the mined voids. Consider for instance the mine water rebound curve as recorded in the Randolph Shaft of the Dysart-Leven Coalfield (Fig. 3). The short-lived peak on the rebound curve (Fig. 3b) around January 1986 corresponded to a sudden increase of water arriving at the dewatering pumps of the adjoining Frances Colliery

(Fig. 3a) which lies to the south-east of Randolph. At the same time, the suspended solids content of the Frances waters increased markedly. It is considered that this small peak on the rebound curve corresponds to a 'backing-up' of water behind some obstruction between Randolph and Frances Collieries. The drop in head after the peak corresponds to the opening up of this obstruction by erosion, which caused the simultaneous increases in flow and suspended solids at Frances. Similarly the second peak on the rebound curve (towards the end of 1986; Fig. 3b) coincided with a peak in suspended solids encountered in the pumped waters of Michael Colliery (to the north-east of Randolph; Fig. 3a), suggesting further erosion during flow between the collieries.

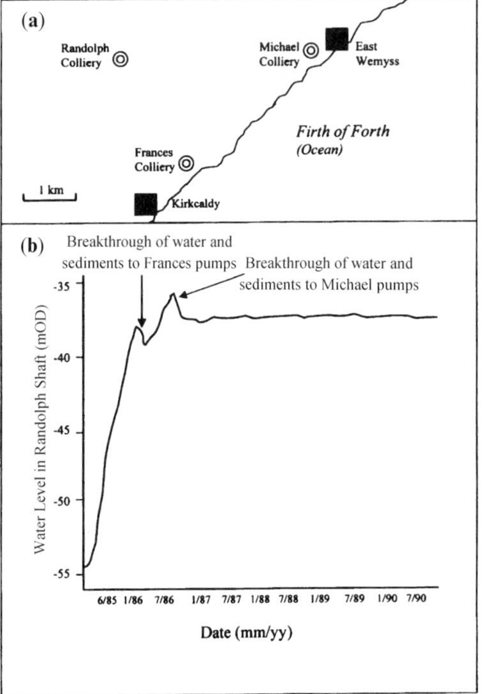

Fig. 3. (**a**) Location map showing the relative positions of the Randolph, Frances and Michael shafts in the Dysart-Leven Coalfield of Fife, (Scotland). (For overall location of Dysart-Leven Coalfield, see Figure 1a). Circular symbols are major mine shafts; black squares are urban settlements. (**b**) Observed rebound (water table recovery) curve for the Randolph Shaft in 1985, following removal of dewatering pumps. The stable level attained eventually corresponds to the level of a roadway connection to Michael Colliery through which the entire recharge to the Randolph workings is transmitted eastwards.

On 13th January 1992, during the final stages of the post-abandonment flooding of the Wheal Jane tin mine in Cornwall (Fig. 1a), physical erosion by rapidly flowing mine water gave rise to a particularly spectacular surface water pollution plume. Although earlier accounts of this event have invoked failure of a man-made plug in the portal of an old mine entrance called the Nangiles Adit (e.g. Hamilton *et al.* 1994; NRA 1994; Banks *et al.* 1997; Bowen *et al.*, 1998), recent underground exploration in this adit revealed no trace of any plug in the portal, but strong evidence that the mine water had been impounded behind a pile of roof-fall debris (Gatley *et al.* 1998*a*). It appears that rapid erosion of flow paths through the debris pile occurred (essentially by piping) once sufficient head of mine water had accumulated behind the pile. The resultant outrush of some 50 Ml of acidic, metalliferous waters in less than 24 hours led to the infamous propagation of a dramatic red plume in the Fal Estuary, drawing media attention to the pollution hazard posed by the mine (Banks *et al.* 1997), and thus providing the basis for sustained public pressure for remedial action, which has already cost the UK government more than £18M over the 7 years to 1999.

At the nearby South Crofty tin mine (also in Cornwall) dewatering ceased in March 1998 (Adams & Younger 1999). Although the ore body at South Crofty contains little pyrite, and hence does not have the same problematic water quality as Wheal Jane, public concern is naturally heightened following the Wheal Jane experience. Extensive underground exploration was therefore undertaken at South Crofty (Gatley *et al.* 1998*b*) to establish the most likely flowpaths of mine water in the shallow subsurface after the completion of flooding. It was found that the deep mine water will migrate via the North Roskear Shaft into the Dolcoath Deep Adit and thence to the Red River. It was also established that this migration will occur via an old roadway which had been partly back-filled with waste rock during adit maintenance operations in the 1950s. To avoid the risk of mine water building up (as it did at Wheal Jane) and bursting out in a spectacular manner, the backfill material was mined out of the roadway before South Crofty was finally abandoned.

Monitoring of the rise of water level in the deep workings of South Crofty has continued in tandem with the development of a physically-based model of the flooding process (Adams & Younger 1999). Although erosion of flow paths can cause temporary peaks and troughs in mine water recovery curves (as seen at Randolph Colliery; Fig. 3b), the more usual expectation is a monotonic rise. This was the recovery pattern expected at South Crofty, where the veins are sub-vertical and have been widely exploited over considerable vertical intervals. However, in early August 1998 an unexpected perturbation in the mine water recovery curve occurred, with measured water levels in the main shaft (which are made manually, at approximately monthly intervals) dropping to about 20 m below the level recorded in June 1998. This drop in water levels followed an unusual subsurface event in late July 1998 (M. Owen, pers. comm., January 1999). The South Crofty office received a number of phone calls from local residents enquiring whether the mine had reopened, for they had heard and felt what they assumed was a large subsurface explosion (indistinguishable from the noise and vibrations associated with large longhole stoping detonations used while the mine was active). The South Crofty mining engineers had also heard and felt the blast, but knew that it could not be a man-made explosion. Comparison of the mine water levels with stope plans suggested the following explanation (Fig. 4). A large open stope at around the 340 fathom level in the mine had been left such that it terminated upwards in a large, closed cupola (Fig. 4a). The rise of the water table was expected to trap gas in this cupola (Fig. 4b). It is hypothesized that the blast sensed on the surface occurred when the gas pressure exceeded the strength of the overlying rock, which was spontaneously pneumatically fractured (Fig. 4c). Once fractured, the cupola allowed mine water to drain into previously isolated workings, accounting for the drop in water level in the shaft. Although it will never be possible to raise this hypothesis beyond this anecdotal level, it remains a credible process explanation to add to those outlined above in relation to Randolph Colliery (Fig. 3).

In this context, it is worth noting also the well-documented occurrence of pockets of compressed air in flooded mines in the Celtic coalfields. The most notable case occurred at Tynewydd Colliery (Fig. 1a) in the Rhondda Valley, South Wales in 1877 (Llewellyn 1992), when a development heading intersected old flooded workings, causing an inrush of water. Although four miners were drowned immediately, ten others survived below the water table for a week, in air pockets which were trapped by the rising water table in isolated up-dip workings. One of the ten trapped miners was sadly killed at the moment of rescue, when the release of compressed air forced him into the hole dug from above by the rescue brigade, fracturing his skull. The remaining nine men survived, although all suffered from the bends following the rapid decompression (Llewellyn 1992).

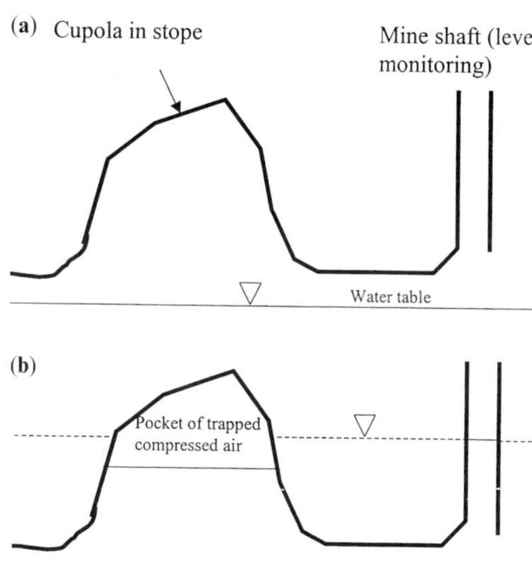

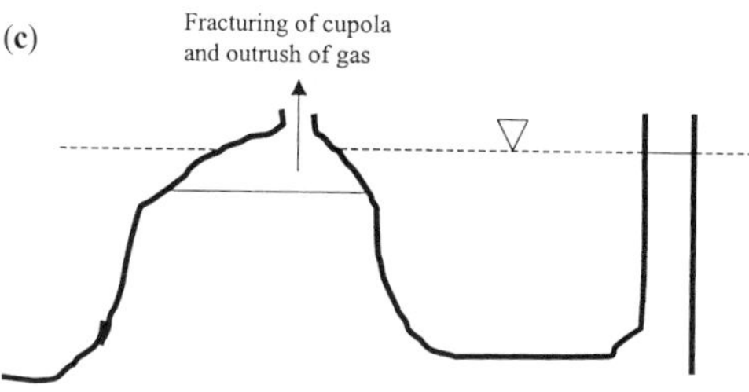

Fig. 4. Possible process of pneumatic fracturing of a cupola in a large open stope, as is postulated to have occurred during mine water rebound in South Crofty Tin Mine, Cornwall, in late July 1998 (translated and modified from Younger, 1999*b*).

Similar pockets of compressed methane are a hazard to drillers investigating the hydrogeology of abandoned coalfields (D. Gowans, Drilcorp Ltd, pers. comm., 1998). It is therefore suggested that possible cupolas in flooded workings should be identified by careful inspection of mine plans as part of the risk assessment for drilling contracts in flooded mine-workings in coalfield areas.

Post-rebound hydrogeological experiences

After mine water recovery is complete, flow would normally be expected to be predominantly laminar, reflecting the low hydraulic gradients which typically develop in very permeable ground. Previously unpublished tracer test results from flooded coal mine workings at

Ynysarwed, SouthWales (Fig. 1a) provide corroboration of this expectation. An acidic mine water discharge began to flow from the Ynysarwed Drift (National Grid Reference SN 807017) in March 1993. This drift was used to access a large area of workings in the No 2 Rhondda Seam until 1938. Nearby, at Blaenant–Cefn Coed Colliery (SN 785033), deeper workings accessed by shafts and an inclined drift remained operational until May 1990, when dewatering was finally discontinued. After rebound was complete, the large flow (up to $3.5\,\mathrm{Ml\,d^{-1}}$) of acidic, ferruginous mine water from the Ynysarwed Drift caused severe pollution of 12 km of the Neath Canal (Younger 1994, 1997; Ranson & Edwards 1997). Subsequent investigations of remedial options included evaluation of the possibility of inducing *in situ* sulphate reduction within the mine workings by injecting suitable reactants. To assess the hydraulic feasibility of this proposal, the Environment Agency undertook two tracer tests at the site in July and November 1995 (Edwards 1996), the second of which is presented here. (Very little tracer was recovered in the first test, for unknown reasons). Figure 5 shows the layout of the site and the breakthrough curve obtained. The test was undertaken by slug injection (at 15:15 hrs on 16th November 1995) of 1 kg of rhodamine WT dye into a borehole (SN 805020) accessing flooded workings alongside the Ynysarwed Drift, some 400 m in from the portal (Fig. 5a). Although available mine plans are not clear, these workings are presumed to have been created by room-and-pillar methods, with later secondary extraction of the pillars resulting in wide areas filled with goaf (collapsed roof strata). A Rock and Tyler QMP auto-sampler was installed at the Drift portal to collect a sample every 8 hours. Tracer recovery was not very high (a total of 5% of the tracer mass over the full period of monitoring); nevertheless, the breakthrough curve (Fig. 5b) repays consideration. The relatively short distance from the borehole to the drift portal (400 m) is reflected in the arrival of the leading edge of the tracer slug after only 7 days. However, the commencement of the main rising limb of the breakthrough curve does not become established until day 20, with the peak concentration being attained after 40 days. These first arrival and peak concentrations correspond to linear velocities on the order of 57 and $10\,\mathrm{m\,d^{-1}}$, respectively. These velocities are of the same order of magnitude as those found by Aljoe & Hawkins (1994) in Pennsylvania, and given the dimensions of the roadways at Ynysarwed they imply predominantly laminar flow conditions (cf. Ford & Williams 1989, p. 145). The marked attenuation of the Ynysarwed breakthrough curve (Fig. 5b; with background fluorescence not being attained until March 1996, about 130 days after the tracer injection) can be interpreted in terms of substantial diffusional exchange fluxes between mobile and immobile volumes of water (cf. Younger & Elliot 1995). The operation of such a major dispersive process in these mine workings is consistent with the high dispersivity values (longitudinal dispersivity (i.e. in the direction of advection), $\alpha_L = 60\,\mathrm{m}$; transverse dispersivity (i.e. normal to the direction of advection), $\alpha_T = 20\,\mathrm{m}$) which were invoked in a numerical model for the entire Blaenant–Ynysarwed system, which was used to approximate the flushing of iron (Younger *et al.* 1996; Younger 1997).

Even where a mine has been abandoned for many years, unexpected perturbations in the hydraulics of the flooded mine voids can prompt a renewed phase of turbulent flow, with problems of erosion similar to those which are more often associated with the initial rebound period. For instance, erosion of mine voids by rapidly flowing water has led to urban flooding problems in the small mining village of Spittal (Fig. 1a) on the Anglo-Scottish border. Around midnight on the 24th of June 1998, water suddenly began to surcharge the main storm sewer in Spittal High Street, and quickly flooded 19 homes with ochreous, sediment-laden water, which continued to flow as a torrent for 17 hours. Subsequent investigations (involving hydrochemical analysis, CCTV inspection of the sewer and searching of local archives; Younger 1999*a*) led to the conclusion that a substantial head of mine water had built up behind a roof fall in old coal workings which had been accessed by a now-buried stone-arched adit (NU005514; observed on the CCTV) which was abandoned in 1820. When the head exceeded some critical limit, collapse and/or rapid erosion of the roof fall debris gave rise to the turbulent discharge and the consequent flooding.

At the village of Leadhills (Fig. 1a), in Scotland's Southern Uplands, post-abandonment changes in mine stability led to a serious threat of flooding (Schmolke 1998). Prior to August 1991, an old drainage adit known as the Gripps Level (approximate grid reference NS887141) drained some $23\,\mathrm{Ml\,d^{-1}}$ of water from abandoned lead mines. A roof-fall in the Gripps Level at that time halted the discharge, and led to the impoundment of a vast volume of mine water in the old workings. A head of 25 m built up behind the roof fall, and eventually

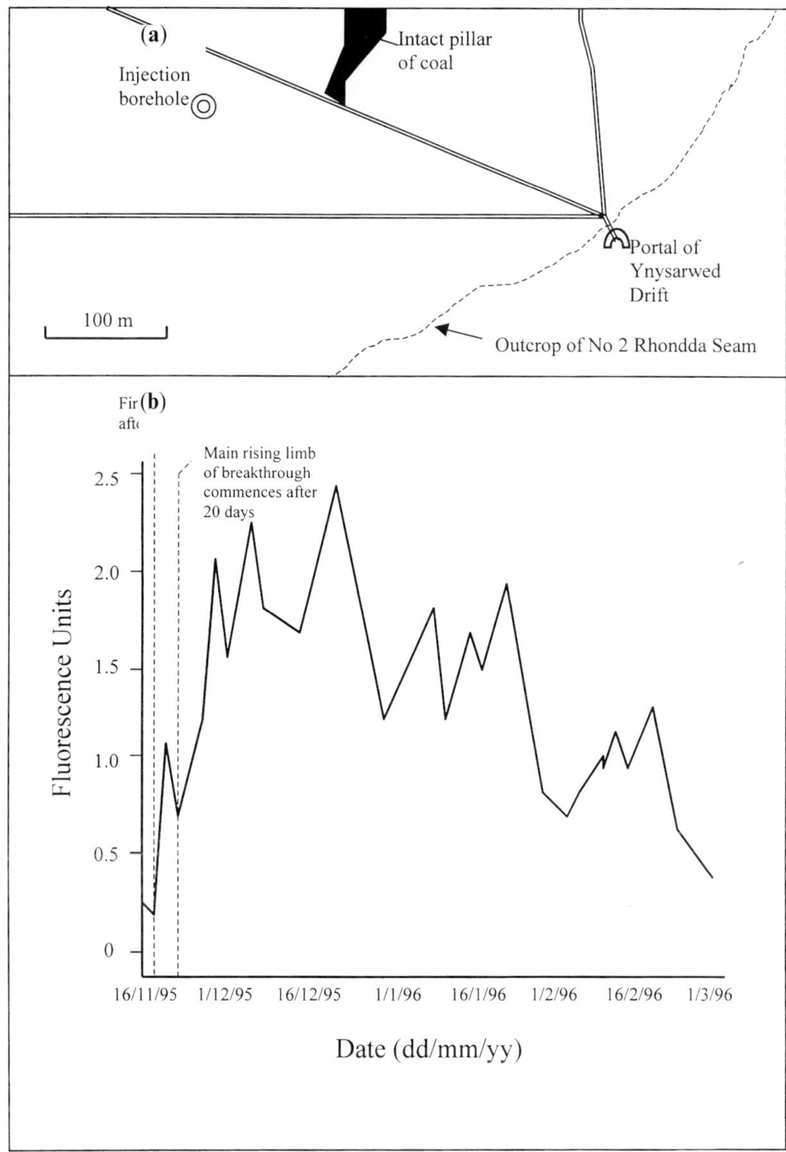

Fig. 5. (**a**) Plan showing the layout of the injection borehole (into which rhodamine WT was introduced), surrounding workings and the major roadways (double lines) of the Ynysarwed Drift mine, radiating away from the portal. Note that all of the subsurface to the left of the No 2 Rhondda Seam outcrop (marked) is recorded as worked, apart from the pillar indicated. (**b**) Breakthrough curve of rhodamine WT dye at the Ynysarwed Drift portal following injection into the borehole on 16th November 1995.

water began to emerge from old air shafts along the Level. More alarmingly, a large tension crack opened up in the hillside, and this also began to yield water. The risk of a catastrophic failure of these newly-flooded workings remains, posing a substantial risk to life and property downstream of the mine site (Schmolke 1998).

An event with echoes of both Spittal and Leadhills (albeit with different consequences) occurred during the winter of 1998/99 in South Wales, at the site of the Gwenffrwd mine water discharge (SS802972; Fig. 1a). The Gwenffrwd discharge had flowed unimpeded for 80 years from an abandoned coal drift mine near the village of Tonmawr, contributing substantially to the pollution of the River Pelenna. In the spring of 1998 the discharge was fitted with a very successful passive treatment system (Ranson *et al.* 1998; Younger 1998*c*). Together with the remediation of two neighbouring discharges, this passive system brought a substantial improvement in the quality of the River Pelenna (Ranson *et al.* 1998). The winter of 1998/99 was the wettest ever recorded in South Wales. After one particularly heavy downpour, residents in Tonmawr village became alarmed when it was noticed that the Gwenffrwd discharge had ceased to flow. A potentially dangerous build-up of mine water in old workings above the village (as at Leadhills) was feared. However, subsequent investigations revealed that the discharge had simply relocated, and was now coming out of a topographically lower mine adit (Whitworth B; SS799974), a few hundred yards further up the valley. Although this meant that the danger of flooding in Tonmawr village could thankfully be discounted, the re-positioning of the discharge meant that the recently-commissioned passive treatment system was almost completely bypassed, so that the pollution of the River Pelenna increased again. At the time of writing, the possible diversion of the mine water back around the hillside to the treatment system is under investigation. In the meantime, studies of mine plans supported by limited excavations around the Gwenffrwd portal have provided a hydrogeological explanation for this occurrence (C. Ranson, Neath Port Talbot County Borough Council, pers. comm., 1999). The bulk of the water which formerly flowed from the Gwenffrwd mine was actually sourced in the lower workings of Whitworth B. Impoundment of the mine water behind roof-fall debris had raised the head in the workings so that the water decanted from the higher man-way drift of Gwenffrwd. The exceptional flows of the wettest winter on record had finally eroded an efficient flow pathway through the roof-fall debris in the Whitworth B workings, allowing the mine water to drain down to the lower portal.

Some Celtic experiences in the hydrogeochemistry of abandoned deep mines

Mine water chemistry

Some of the earliest detailed studies of mine water quality were made in the Celtic countries. For instance Brown (1977) documented tin mine drainage impacts on invertebrate communities in Cornwall. Scullion & Edwards (1980) investigated the effects of coal mine effluents on the fish fauna of the Taff Bargoed river in South Wales. In Scotland, Henton (1979, 1981) documented the pollution of the River Ore following the flooding of the underlying coalfield (a case subsequently updated by Robins 1990, and Younger 1999*a*). Hydrogeochemical studies have become increasingly common in the Celtic coalfields during the last decade in response to widespread mine closures (e.g. Robins 1990; Younger *et al.* 1995; Robins & Younger 1996; Chen *et al.* 1997; Younger 1997, 1999*a*; Sadler & Rees 1998; LaPierre 1999; Wood *et al.* 1999). These studies have significantly advanced the generic conceptual understanding of the hydrogeochemical changes which occur during and after the flooding of deep mine workings. Since comprehensive syntheses of these findings have recently been published (Younger 1998*a*, 1999*a*, 2000; Younger & Adams 1999), there is no need to rehearse them here. Rather, this section aims to highlight some basic geological controls on water quality which should be of generic value in the management of mine waters.

It has frequently been pointed out that the commonly-used term 'acid mine drainage' is often a misnomer, since relatively few polluting discharges from mines have a consistently low pH (e.g. Henton 1981; Robins 1990; Younger 1995). In reality, mine water quality spans a continuum of compositions, most conveniently defined on the basis of the acid-base balance (Hedin *et al.* 1994; Younger 1995). Total acidity in mine waters is a reflection of the availability not just of protons (represented by pH) but also of 'acid' metals (Fe, Mn, Al, Zn etc) which readily form hydroxide minerals at surface temperatures and pressures, and can thus consume alkali during acidity titrations. Total alkalinity in mine waters is primarily a reflection of the bicarbonate concentration. Hence acidity and alkalinity are not mutually exclusive categories in mine waters, and it is therefore convenient to define two principal classes of mine water:

- net-acidic mine waters (acidity > alkalinity)
- net-alkaline mine waters (alkalinity > acidity)

(For further discussion of this concept, the interested reader is referred to Hedin *et al.* 1994; Younger 1995, 1998*a*; Banks *et al.* 1997).

Geological controls on mine water chemistry

Younger (1999*a*) has shown that in Scotland, net-alkaline mine waters are fifty times more abundant (by volume) than net-acidic mine waters.

However, where net-acidic discharges *do* occur, they are usually far more environmentally damaging than net-alkaline discharges (Jarvis & Younger 1997). Hence it is of considerable value to environmental managers to be able to predict which future discharges are likely to be net-acidic and which should be net-alkaline. Similarly, it is of value to know whether a currently net-acidic discharge will always remain so, or whether it will follow the pattern shown by many other such waters and become net-alkaline over time (Younger 1997, 2000).

Existing data from the Celtic mining regions are instructive in this regard. Table 1 lists a number of net-acidic mine waters in the Celtic lands and documents their geological field relations in broad terms. Table 2 repeats the exercise for net-alkaline mine waters.

It is apparent from Table 1 that net-acidic mine waters are generally restricted to situations in which high-sulphur strata (>2.5 weight percent S) are present in:

(i) recently-flooded deep-mine workings;
(ii) shallow, partially flooded mine workings; and
(iii) perched groundwater systems in spoil heaps and opencast backfill.

This is logical, as all of these settings are susceptible to ready ingress of meteoric water and atmospheric oxygen, which are the two

Table 1. *Net-acidic mine waters in UK Celtic lands and their hydrogeological settings*

Site name	Celtic land*	Grid ref	pH	Total Fe (mg/l)	Total acidity (meq/l)	Hydrogeological setting	Source of further info
Brora No 1	S	NC 898042	3.6	8	1.82	Overflowing shaft accessing shallow flooded working in a high sulphur coal seam	Younger (1999*a*)
Baads East	S	NT 005612	2.8	550	48.4	Colliery spoil heap hosting perched water table system	Younger (1999*a*)
Randolph Colliery	S	NT 303957	3.7	43	31.32	Colliery spoil heap hosting perched water table system	Younger (1999*a*)
Balgonie Colliery	S	NT 306990	4.3	62	6.4	Colliery spoil heap hosting perched water table system	Younger (1999*a*)
Dalquharran Mine	S	NS 266017	5.5	150	5.5	Drift accessing extensive shallow and deep workings in a high sulphur coal seam	Robins (1990); Robb (1994); Marsden *et al.* (1997)
Roughcastle	S	NS 847796	5.7	89	3.56	Backfilled coal opencast in communication with extensive underground fireclay workings	Bullen Consultants (1999)
Ynysarwed	W	SN 809018	5.8	200	7.2	Drift accessing extensive shallow and deep workings in a moderate sulphur coal seam	Younger (1994, 1997); Ranson & Edwards (1997)
Garth Tonmawr Drift Mine	W	SS 799973	5.2	35	1.26	Drift accessing extensive shallow workings in a high-S coal seam	Younger (1997)
Cae Coch	W	SH 775653	2.5	1630	62	Adit accessing shallow workings in a pyritic copper ore body	McGinness and Johnson (1993)
Wheal Jane	C	SW 773425	3.7	250	12	Adit accessing shallow workings in a pyritic tin/zinc ore body	NRA (1994) Bowen *et al.* (1998)

*C, Cornwall; S, Scotland; W, Wales.

most important triggers of acid generation by pyrite weathering.

By contrast, Table 2 reveals that net-alkaline mine waters are associated with:

(i) low-sulphur strata (<1.5 weight percent S) in any hydrogeological setting; and
(ii) high-sulphur strata, where these lie far below the water table in workings which have been flooded for a long time (usually a few decades).

The fact that the bulk of the total volume of deep mine workings falls into one of these latter two categories explains why net-alkaline mine waters are volumetrically far more abundant than net-acidic mine waters.

It is also clear that many deep mine waters are initially net-acidic after completion of rebound, reverting to net-alkaline status some time later, following a period of flushing of the flooded mine voids. Younger (2000) has used data from all the UK Celtic lands and England to argue that the transition to net-alkaline status usually takes place within a period of time around four times as long as the time it took for the deep mine workings to flood up to surface level following the cessation of dewatering.

What difference does the availability of limestone make?

Although many of the Scottish mine waters drain strata which belong to a stratigraphic division known as the 'Limestone Coal Group' (Namurian), limestones are very rarely found in

Table 2. *Net-alkaline mine waters in UK Celtic lands and their hydrogeological settings*

Site name	Celtic land*	Grid ref	pH	Total Fe (mg/l)	Alkalinity (meq/l)	Hydrogeological setting	Source of further info
Lathallan Mill	S	NO 465063	6.1	10.8	3.64	Overflowing shaft accessing deep, long-flooded coal workings.	Younger (1999*a*)
Star Road (Markinch)	S	NO 296025	6.5	4.0	3.46	Overflowing shaft accessing deep, long-flooded coal workings.	Younger (1999*a*)
Kames Colliery	S	NS 685262	5.8	14	4.64	Overflowing shaft accessing deep, long-flooded coal workings.	Younger (1999*a*)
Pool Farm	S	NS 987542	6.3	8	1.96	Adit accessing shallow workings in a low-sulphur coal seam	Younger (1999*a*)
Cuthill	S	NS 990628	5.5	37	4.94	Adit accessing shallow workings in a low-sulphur coal seam	Younger (1999*a*)
Gwynfi	W	SS 892973	5.5	8	1.1	Adit accessing shallow workings in a low-sulphur coal seam	EA archives
Morlais	W	SN 572023	6.7	63	4.5	Overflowing shaft accessing deep, long-flooded workings in a moderate- to high-S coal seam	EA archives
Bryn	W	SS 817922	6.9	0.4	0.7	Adit accessing shallow workings in a low-sulphur coal seam	EA archives
Goytre	W	SS 787897	6.8	2.5	1.72	Adit accessing shallow workings in a low-sulphur coal seam	EA archives
Dolcoath Deep Adit	C	SW 648418	7.1	0.5	0.6	Adit accessing deep and shallow workings in a low-sulphur tin/copper ore body	EA archives
Geevor Deep Adit	C	SW 372349	6.4	0.17	0.3	Adit accessing deep recently-flooded workings in a low-sulphur tin ore body	EA archives

* C, Cornwall; S, Scotland; W, Wales.

Table 3. *Net-acidic mine waters in the Sydney Coalfield of Cape Breton Island, Nova Scotia, Canada, and their hydrogeological settings*

Site name	Latitude and longitude	pH	Total Fe (mg/l)	Total acidity (meq/l)	Hydrogeological setting
No 1A Colliery	46°11′52″N 60°03′10″W	5.6	33	0.9	Pipe connecting to adit accessing shallow workings in Phalen Seam (total S = 3.2 wt%)
No 11 Colliery	46°11′03″N 59°57′57″W	3.8	0.6	0.48	Inclined drift portal to shallow workings in the Emery Seam (total S = 2.7 wt%)
No 24a Mine Water	46°10′30″N 59°57′05″W	3.1	7.6	4.74	Spoil heap of waste rock from workings in the Emery Seam
Morrison's Pond	46°19′31″N 60°19′02″W	2.9	49	5.1	Land drain in area of shallow workings in the Hub Seam (total S = 2.3 wt%)
Shaft 1B at 198m depth	46°13′06″N 59°58′44″W	6.4	150	8.5	Deep mine shaft discrete sample at depth. Water in shaft still rising. Accesses deep workings in the Phalen Seam (total S = 3.2 wt%)

close proximity to coal seams in Scotland. (The stratigraphic term arises from the fact that the coal bearing sequence is bounded above and below by prominent limestones, whereas none occur within the sequence itself; Francis 1991.) By contrast in the Sydney Coalfield of Nova Scotia, calcretes and other forms of limestone are relatively common in the seam roof setting. While this might be considered sufficient to guarantee that mine water discharges in Cape Breton will be net-alkaline, a significant number of discharges in Cape Breton are in fact net-acidic (Table 3) (even though these are volumetrically less prolific than the net-alkaline discharges; Table 4). Current data do not make it clear if the net-acidic discharges in Cape Breton are from mines which only began to overflow at the surface recently. Nevertheless, it is clear

Table 4. *Net-alkaline mine waters in the Sydney Coalfield of Cape Breton Island, Nova Scotia, Canada, and their hydrogeological settings*

Site name	Latitude and longitude	pH	Total Fe (mg/l)	Alkalinity (meq/l)	Hydrogeological setting
No. 4	46°10′34″N 59°56′28″W	6.6	5.7	0.56	Low-flow seepage from workings in the Phalen and Emery Seams (3.2 and 2.7 wt% S respectively), abandoned in 1961
No. 8	46°12′54″N 59°59′39″W	7.6	0.1	3.6	Old workings (abandoned 1914) in the Harbour Seam (2.5 wt% S) discharging (probably via an old adit) to a land drain
No. 25	46°11′52″N 60°03′10″W	6.7	21	1.7	Overflowing borehole near old shaft which accesses flooded deep workings (abandoned 1959) in the high sulphur (3.9 wt% S) Gardiner Seam.
Old Harbour Discharge	46°11′39″N 59°57′08″W	7.5	1.5	4.9	Old workings (abandoned 1914) in the Harbour Seam (2.5 wt% S) discharging (probably via an old adit) to a land drain
Shaft 1B at 122 m depth	46°13′06″N 59°58′44″W	7.4	1.5	3.92	Deep mine shaft – discrete sample at depth. Water in shaft still rising. Accesses deep workings in the Phalen Seam (total S = 3.2 wt%)
Shaft 1B at 168 m depth	46°13′06″N 59°58′44″W	6.8	19	1.92	Deep mine shaft – discrete sample at depth. Water in shaft still rising. Accesses deep workings in the Phalen Seam (total S = 3.2 wt%)

that the availability of limestone in the sequence is not on its own a sufficient guarantee of net-alkalinity.

Figure 6 compares the major ion chemistry of uncontrolled discharges from flooded deep coal mines in Scotland with waters from flooded deep coal mines in the Sydney Coalfield, Cape Breton. In broad terms the Scottish and Cape Breton mine waters are similar, in that they are predominantly of Ca–SO_4 facies. However, in the Cape Breton mine waters the sulphate ion generally accounts for less than 80% (as equivalents) of the total anions, whereas in Scotland it usually accounts for between 80 and 100% of the total anions. In terms of absolute concentrations, this difference is also evident, with Scottish mine waters having a mean sulphate concentration of 470 mg/l, whereas the Cape Breton mine waters have a mean sulphate of only 208 mg/l. In neither case is the sulphate sufficiently high

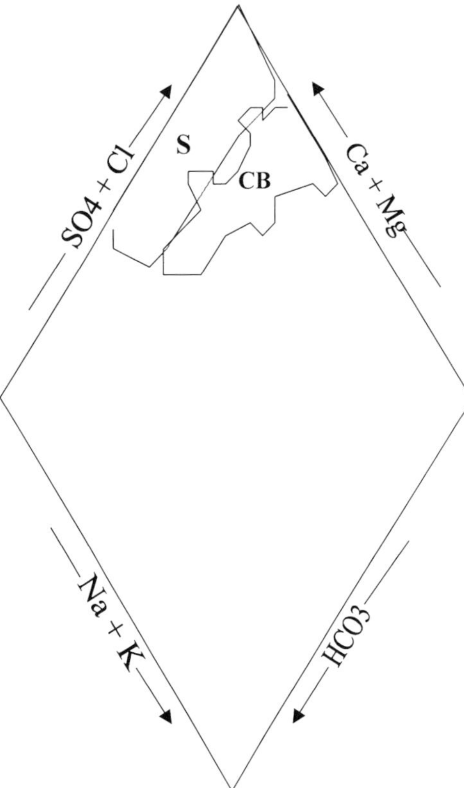

Fig. 6. Upper diamond of a Piper diagram, showing the plotting fields of uncontrolled minewater discharges in Scotland (S) and Cape Breton (CB). The Scottish field is adapted from Younger (1999*a*), the Cape Breton field from LaPierre (1999).

that a gypsum solubility control on concentrations might be invoked (a control which would clearly have a logical coupling to the availability of calcite). An alternative explanation for this contrast could be that less pyrite was oxidized per unit volume of water in the Cape Breton mines than in the Scottish mines. This would be consistent with the elevated pH and alkalinity of recharge waters in Cape Breton where these pass through limestone. Interpretation of alkalinity values must be approached with caution, however, as the mean alkalinity of the Scottish mine waters (at 186 mg/l as $CaCO_3$; Younger 1999*a*) is considerably greater than that in the Cape Breton mine waters (50 mg/l as $CaCO_3$; the balance of the anions in the Cape Breton mine waters is chloride, averaging 51.6 mg/l and reflecting the marine influence on mine water quality), despite the fact that the higher sulphate concentrations in the Scottish waters indicate more extensive pyrite oxidation than in Cape Breton. The coincidence of high alkalinities and high sulphates can be explained in the manner expounded by Wood *et al.* (1999), in relation to mine water discharges in Scotland: '... if waters were initially highly acidic, reaction with carbonates ... will result in a correspondingly high level of dissolved HCO_3 after neutralization'. By contrast, weakly acidic mine waters, with correspondingly lower sulphate concentrations (as in Cape Breton), provoke less alkalinity generation in the process of attaining neutral pH.

This brief comparison serves to demonstrate that the presence of limestone in a worked sequence is no guarantee of net-alkaline drainage. It thus follows that hydrogeological factors, such as groundwater flow patterns and the order of encounter of strata and groundwater (cf. Freeze & Cherry 1979), are at least as important.

Vertical hydrochemical profiles in shafts during rebound

The obvious way to reduce uncertainty about the risk of pollution after flooding of deep mine workings is to undertake water sampling during rebound using open shafts and boreholes. A rigorous approach to such an exercise would demand that samples be obtained at discrete vertical intervals, to delineate any variations in water quality over depth. Results from two such shaft sampling exercises (one in Cornwall, the other in Cape Breton) are compared on Fig. 7. It is immediately apparent in both cases that the better quality water is found at the top of the water column. This phenomenon has been widely termed 'stratification' in mining circles (see for instance Ladwig *et al.* 1984), even though step changes in the different water quality parameters do not always coincide (Fig. 7).

The stratification of water quality during rebound implies that mechanical mixing of the water column in the mine is minimal. This in turn implies, for instance, that there are few lateral inflows and outflows at depth. As long as flow remains sluggish and laminar, there is little to disturb the stratification. However, when discharge from a mined system commences, turbulent flow in the vicinity of shafts and open roadways can cause substantial mixing of the mine water body, leading to a breakdown in the stratification. Hence the quality of a mine water discharge from a formerly stratified system is more likely to resemble a mixture of all depth intervals rather than the better quality water previously found at the top of the water column. For instance, when the Wheal Jane mine first overflowed, the water was considerably poorer in quality than the uppermost waters marked on Fig. 7a; it in fact resembled the median of the values shown.

This has two practical implications:

(1) Mine water discharges may well be considerably poorer in quality than would have been inferred by sampling the uppermost waters alone. Hence depth-sampling is highly advisable in studies of mine water rebound.
(2) Where stratification is identified, it would be imprudent to assume that future discharges at the surface will resemble the uppermost waters in the rebounding water column. Rather, surface discharges are likely to have a quality approaching that of the mean concentrations found over the full depth of the mine.

Conclusions

Recent experiences in the Celtic lands of Britain (Cornwall, Wales and Scotland) and in Cape Breton (Nova Scotia) yield the following lessons:

(1) The importance of recognizing the predominance of mined features in the post-closure hydrogeology of abandoned mines.

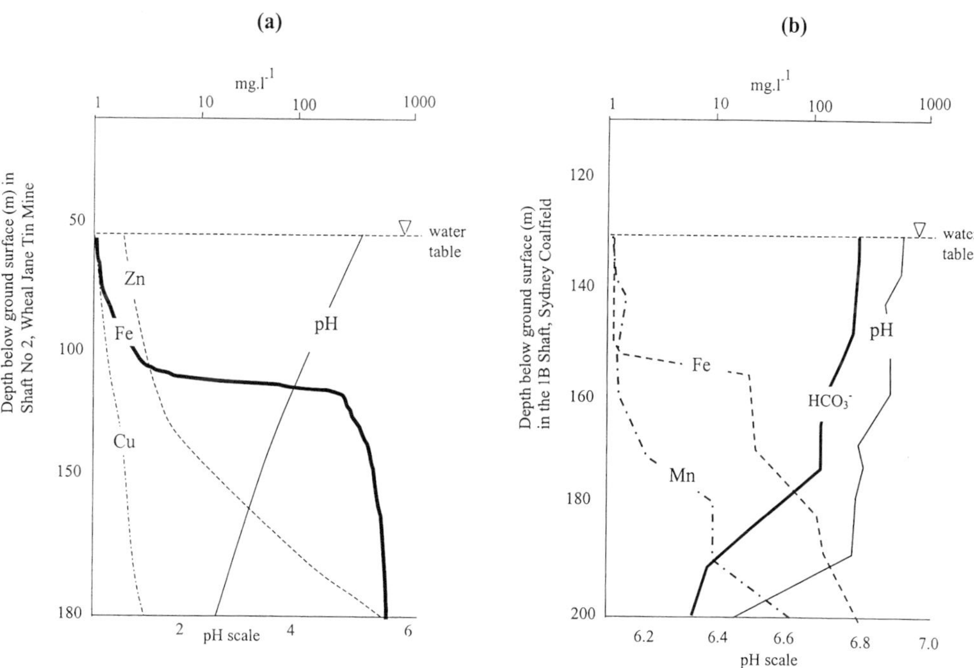

Fig. 7. Variation of selected hydrochemical parameters over depth during mine water rebound in (**a**) Wheal Jane tin mine, Cornwall (November 1991), and (**b**) The 1B Shaft of the Sydney Coalfield, Cape Breton, Canada (August 1997).

(2) Dynamic temporal changes in hydrogeological behaviour arise from collapse of mined voids, due to fluvial erosion by rapidly-flowing mine waters and/or by pneumatic fracturing by mine gases compressed in pockets during mine water rebound (the latter having significant implications for human safety and environmental protection).

(3) Net-acidic mine waters are generally restricted to situations in which high-sulphur strata are present in (i) recently-flooded deep-mine workings, (ii) shallow flooded mine workings and (iii) perched groundwater systems in spoil heaps and opencast backfill.

(4) Net-alkaline mine waters are associated with (i) low-sulphur strata in any hydrogeological setting and (ii) high-sulphur strata at depth in long-flooded workings. In practice, this means that they are far more abundant than net-acidic mine waters.

(5) The presence of limestone in a mined sequence is not on its own a guarantee that mine waters will be net-alkaline; the patterns of groundwater flow (which determine the transport of limestone dissolution products through the mined system) must also be favourable.

(6) Significant vertical variations in mine water quality over depth have been found by means of depth-sampling in abandoned mine shafts. Where water quality is strongly stratified, it is inferred that few mechanical mixing processes are operative. However, when discharge from a mined system commences, turbulent flow in the vicinity of shafts and open roadways can cause substantial mixing, leading to a breakdown in the stratification of the water body, and a mine water discharge quality which is more akin to a mixture of all depth intervals in the formerly stratified water column rather than the 'skimming off' of the (usually better quality) top water.

We would like to acknowledge the following individuals who provided data and discussions to the work presented here: Russell Adams, Adam Jarvis, Charlotte Nuttall and Julia Sherwood at the University of Newcastle; Jim Wright and Paul Edwards of the EA; Cath Ranson of Neath Port Talbot County Borough Council; Mark Owen, formerly of South Crofty plc and now of the Camborne School of Mines; Dr Tom Martel, Dr David Hansen, Dr Martin Gibling and Heather Cross of Dalhousie University. The Cape Breton work was funded by the Carboniferous Hydrogeological Research Project of Dalhousie University, and the UK work was funded from a variety of Environment Agency sources and Neath Port Talbot County Borough Council. I finally salute warmly Colm O'Boyle, Professor of Celtic at the University of Aberdeen, for indulging my wish to use the *gadhlig* correctly in this paper.

References

ADAMS, R. & YOUNGER, P. L. 1999. Modelling the timing of rebound at South Crofty. Report to the Environment Agency, South West Area. Department of Civil Engineering, University of Newcastle and NUWATER Consulting Services Ltd, Newcastle Upon Tyne, May 1999. 26pp.

ALDOUS, P. J. & SMART, P. L. 1988. Tracing groundwater movement in abandoned coal mined aquifers using fluorescent dyes. *Ground Water*, **26**, 172–178.

——, —— & BLACK, J. A. 1986. Groundwater management problems in abandoned coal-mined aquifers: a case study of the Forest of Dean, England. *Quarterly Journal of Engineering Geology*, **19**, 375–388.

ALJOE, W. W. & HAWKINS, J. W. 1994. Application of aquifer testing in surface and underground coal mines. *In*: *Proceedings of the 5th International Mine Water Congress*, Nottingham, UK, 17–24 September 1994, **1**, 3–21.

ASTON, T. R. C. & WHITTAKER, B. N. 1985. Undersea longwall mining subsidence with special reference to geological and water occurrence criteria in the north-east of England. *Mining Science and Technology*, **2**, 105–130.

BANKS, D., YOUNGER, P. L., ARNESEN, R.-T., IVERSEN, E. R. & BANKS, S. D. 1997. Mine-water chemistry: the good, the bad and the ugly. *Environmental Geology*, **32**(3), 157–174.

BOWEN, G. G., DUSSEK, C. & HAMILTON, R. M. 1998. Pollution resulting from abandonment and subsequent flooding of Wheal Jane Mine in Cornwall, UK. *In*: MATHER, J., BANKS, D., DUMPLETON, S. & FERMOR, M. (eds) *Groundwater contaminants and their migration*. Geological Society, London, Special Publications, **128**, 93–99.

BROWN, B. E. 1977. Effects of mine drainage on the River Hayle, Cornwall. (A) Factors affecting concentrations of copper, zinc and iron in water, sediments and dominant invertebrate fauna. *Hydrobiologia*, **52**, 221–233.

BULLEN CONSULTANTS 1999. *Roughcastle Workings Study report*. Draft report to the Scottish Environment Protection Agency, Edinburgh, Scotland.

CAIN, P., FORRESTER, D. J. & COOPER, R. 1994. Water inflows at Phalen Colliery in the Sydney Coalfield, and their relation to interaction of workings. *In*: *Proceedings of the 5th International Minewater Congress*, Nottingham, UK, 17–24 September 1994, **1**, 83–95.

CHEN, M., SOULSBY, C. & YOUNGER, P. L. 1997. Predicting water quality impacts from future minewater outflows in an urbanized Scottish catchment. *In*: CHILTON, P. J., *et al.* (eds) *Groundwater in the urban environment, Volume 1: Problems, processes and management*. (Proceedings of the XXVII Congress of the International Association of Hydrogeologists, Nottingham, UK, 21–27 September 1997). A. A. Balkema Publishers, Rotterdam, 383–388.

DHONAU, N. B. & WRIGHT, G. R. 1998. Zinc-lead mines in Ireland: Approaches to groundwater control and protection. *In*: *Proceedings of the Symposium on Mine Water and Environmental Impacts*, Johannesburg, South Africa, 7–13th September 1998, **2**, 279–290.

DINES, H. G. 1956. *The metalliferous mining region of south-west England. Volume 1*. Memoirs of the Geological Survey of Great Britain.

DODDS, J. E., DALY, E. P. & CULLEN, K. 1994. Relationship between mining and ground water within the dolomites of Ireland. *In*: *Proceedings of the 5th International Minewater Congress*, Nottingham, UK, 17–24 September 1994, **1**, 185–195.

EDWARDS, P. J. 1996. *Investigations into the residence times of groundwaters in the Lower Ynysarwed mineworkings*. Unpublished note to file (ref 2/EQ/MW/Y16, 22–8–96), Environment Agency, Swansea.

FORD, D. C. & WILLIAMS, P. W. 1989. *Karst Geomorphology and Hydrology*. Chapman & Hall, London.

FRANCIS, E. H. 1991. Carboniferous. *In*: CRAIG, G. Y., (ed.), *Geology of Scotland, (Third Edition)*. Geological Society, London. 347–392.

FREEZE, R. A., & CHERRY, J. A. 1979. *Groundwater*. Prentice Hall, New Jersey.

GATLEY, S. T. & OWEN, M. & ROGERS, C. 1998*a*, *Nangiles Adit Inspection*. Unpublished Report by South Crofty plc to the Environment Agency, Exeter.

——, ——, —— & BOON, R. 1998*b*. *Dolcoath Deep Adit. Report on its condition and the South Crofty Mine decant location*. Unpublished report of South Crofty plc, Pool, Cornwall.

GRAY, N. F. 1996. Field assessment of acid mine drainage contamination of surface and ground water. *Environmental Geology*, **27**, 358–361.

HAMILTON, R. M., BOWEN, G. G., POSTLETHWAITE, N. A. & DUSSEK, D. J. 1994. The abandonment of Wheal Jane, a tin mine in SW England. *In*: *Proceedings of the 5th International Minewater Congress*, Nottingham, UK, 17–24 September 1994, **2**, 543–551.

HEDIN, R. S., NAIRN, R. W. & KLEINMANN, R. L. P. 1994. Passive Treatment of Coal Mine Drainage, *US Bureau of Mines Information Circular 9389*. US Department of the Interior, Bureau of Mines, Pittsburgh, PA.

HENTON, M. P. 1974. Hydrogeological problems associated with Waste Disposal into Abandoned Coal Workings. *Water Services*, **78**, 349–352.

——1979. Abandoned Coalfields: Problem of Pollution. *The Surveyor*, **153**, 9–11.

——1981. The Problem of Water Table Rebound after Mining Activity and its Effects on Ground and Surface Water Quality. *In*: VAN DUIJVENBOODEN, W., GLASBERGEN, P. & VAN LELYVELD, H. (eds) *Quality of Groundwater*. Elsevier, The Netherlands, 111–116.

HERM, G. 1976. *The Celts. The people who came out of the darkness*. (First published in German by Econ Verlag 1975). English Translation. Weidenfeld and Nicolson Ltd, London.

HUBERT, H. 1993. *The history of the Celtic people*. (A one volume facsimile edition of 'The Rise of the Celts' and 'The Greatness and Decline of the Celts', originally published as separate volumes in 1934). Bracken Books, London.

JARVIS, A. P. & YOUNGER, P. L. 1997. Dominating chemical factors in mine water induced impoverishment of the invertebrate fauna of two streams in the Durham Coalfield, UK. *Chemistry and Ecology*, **13**, 249–270.

KESSERÛ, Z. 1995. New approaches and results on the assessment of risks due to undermining for mines safety and for protection of water resources. *In*: *Proceedings of the International Conference on 'Water Resources at Risk'*, held by the American Institute of Hydrology and the International Mine Water Association, Denver, Colorado, May 14th–18th 1995, IMWA 53–IMWA 72.

LADWIG, K. J., ERICKSON, P. M., KLEINMANN, R. L. P. & POSLUSZNY, E. T. 1984. *Stratification in water quality in inundated anthracite mines, eastern Pennsylvania*. US Bureau of Mines Report of Investigations **RI-8837**.

LAPIERRE, A. B. 1999. *Characterization of discharges from abandoned mines of the Sydney Coalfield, Nova Scotia*. Unpublished MSc Thesis, Dalhousie University.

LLEWELLYN, K. 1992. *Disaster at Tynewydd. An account of a Rhondda mine disaster in 1877*. (Second Edition). Church in Wales Publications, Penarth.

MARSDEN, M., HOLLOWAY, D. & WILBRAHAM, D. 1997. The Position in Scotland. *In*: BIRD, L., (ed.), *Proceedings of the UK Environment Agency Conference on 'Abandoned Mines: Problems and Solutions'*, Tapton Hall, University of Sheffield, 20th–21st March 1997, 76–84.

MCGINNESS, S. & JOHNSON, D. B. 1993. Seasonal variations in the microbiology and chemistry of an acid mine drainage stream. *Science of the Total Environment*, **132**, 27–41.

MONTERROSO, C. & MACIAS, F. 1998. Drainage waters affected by pyrite oxidation in a coal mine in Galicia (N. W. Spain): composition and mineral stability. *Science of the Total Environment*, **216**, 121–132.

NRA. 1994. *Abandoned mines and the water environment*. Report of the National Rivers Authority of England and Wales. Water Quality Series No 14. HMSO, London.

O'BRIEN, W. 1996. *Bronze Age copper mining in Britain and Ireland*. Shire Archaeology No 71. Shire Publications, Buckinghamshire, UK.

ORCHARD, R. J. 1975. Working under bodies of water. *The Mining Engineer*, **170**, 261–270.

RANSON, C. M. & EDWARDS, P. J. 1997. The Ynysarwed experience: active intervention, passive treatment and wider aspects. *In*: YOUNGER, P. L. (ed.) *Minewater treatment using wetlands*. Proceedings of a National Conference held 5th September 1997, the University of Newcastle, UK, 151–164.

RANSON, C. M., REYNOLDS, N. & SMITH, A. C. 1998. Minewater treatment in Neath and Port Talbot. *In*: FOX, H. R., MOORE, H. M. & MCINTOSH, A. D (eds) *Land Reclamation: Achieving Sustainable Benefits*. Balkema, Rotterdam, 497–507.

REDDISH, D. J., YAO, X. L., BENBIA, A., CAIN, P. & FORRESTER, D. J. 1994. Modelling of caving over the Lingan and Phalen mines in the Sydney Coalfield Cape Breton. *In*: *Proceedings of the 5th International Minewater Congress*, Nottingham, UK, 17–24 September 1994, **1**, 105–124.

ROBB, G. A. 1994. Environmental consequences of coal mine closures. *Geographical Journal*, **160**, 33–40.

ROBINS, N. S. 1990. *Hydrogeology of Scotland*. HMSO for British Geological Survey, London. 90pp.

—— & YOUNGER, P. L. 1996. Coal abandonment – mine water in surface and near-surface environment: some historical evidence from the United Kingdom. *In*: *Proceedings of the Conference 'Minerals, Metals and the Environment II'*, Prague, Czechoslovakia, 3–6 September 1996.

ROSS, A. 1986, *The Pagan Celts*. B. T. Batsford Ltd, London.

SADLER, P. & REES, A. 1998. Use of probabilistic methods for sizing a wetland to treat mine water discharging from the former Polkemmet Colliery, West Lothian, Scotland. *In*: *Proceedings of the International Mine Water Association Symposium on 'Mine Water and Environmental Impacts'*, Johannesburg, South Africa, 7th–13th September 1998.

SCHMOLKE, C. M. R. 1998. Leadhills, Scotland–Flood and pollution threat from disused lead workings. *In*: FOX, H. R., MOORE, H. M. & MCINTOSH, A. D. (eds) *Land Reclamation: Achieving Sustainable Benefits*. Balkema, Rotterdam.

SCULLION, J. & EDWARDS, R. W. 1980. The effect of pollutants from the coal industry on the fish fauna of a small river in the South Wales Coalfield. *Environmental Pollution (Series A)*, **21**, 141–153.

SHEPHERD, R. 1993. *Ancient Mining*. Institution of Mining and Metallurgy/Elsevier Applied Science, London.

SHERWOOD, J. M. & YOUNGER, P. L. 1997. Modelling groundwater rebound after coalfield closure. *In*: CHILTON, P. J. *et al.* (eds) *Groundwater in the urban environment, Volume 1: Problems, processes and management*. A. A. Balkema Publishers, Rotterdam.

SINGH, R. N. & ATKINS, A. S. 1983. Design considerations for mine workings under accumulations of water. *International Journal of Mine Water*, **4**, 35–36.

WOOD, S. C., YOUNGER, P. L. & ROBINS, N. S. 1999. Long-term changes in the quality of polluted minewater discharges from abandoned underground coal workings in Scotland. *Quarterly Journal of Engineering Geology*, **32**, 69–79.

YOUNGER, P. L. 1994. Minewater Pollution: The Revenge of Old King Coal. *Geoscientist*, **4**(5), 6–8.

——1995. Hydrogeochemistry of minewaters flowing from abandoned coal workings in the Durham coalfield. *Quarterly Journal of Engineering Geology*, **28**(4), S101–S113.

——1996. *Environmental Consequences of a Cessation of Pumping at Point of Ayr Colliery, North Wales*. Environment Agency Report, Cardiff. NUWATER Consulting Services Ltd, University of Newcastle.

——1997. The Longevity of Minewater Pollution: A Basis for Decision-Making. *Science of the Total Environment*, **194/195**, 457–466.

——1998*a*, Coalfield abandonment: geochemical processes and hydrochemical products. *In*: NICHOLSON, K. (ed.) *Energy and the Environment. Geochemistry of Fossil, Nuclear and Renewable Resources*. Society for Environmental Geochemistry and Health. McGregor Science, Aberdeenshire, 1–29.

——1998*b*, Adit hydrology in the long-term: observations from the Pb-Zn mines of northern England. *In*: *Proceedings of the International Mine Water Association Symposium on 'Mine Water and Environmental Impacts'*, Johannesburg, South Africa, 7th–13th September 1998.

——1998*c*. Design, construction and initial operation of full-scale compost-based passive systems for treatment of coal mine drainage and spoil leachate in the UK. *In*: *Proceedings of the International Mine Water Association Symposium on 'Mine Water and Environmental Impacts'*, Johannesburg, South Africa, 7th–13th September 1998.

——1999*a*. Mine water pollution in Scotland: nature, extent and preventative strategies. *Science of the Total Environment* (in press).

——1999*b*. Pronóstico del ascenso del nivel freático en minas subterráneas y sus consecuencias medioambientales. *Boletín Geológico Minero*, **110**, 407–422.

——2000. Predicting temporal changes in total iron concentrations in groundwaters flowing from abandoned deep mines: a first approximation. *Journal of Contaminant Hydrology*, **44**, 47–69.

—— & ADAMS, R. 1999. *Predicting mine water rebound*. Environment Agency R&D Technical Report **W179**.

—— & ELLIOT, T. E. 1995. Chalk Fracture System Characteristics: Implications for Groundwater Flow and Solute Transport. *Quarterly Journal of Engineering Geology*, **28**, S39–S50.

——, BARBOUR, M. H. & SHERWOOD, J. M. 1995. Predicting the consequences of ceasing pumping from the Frances and Michael Collieries, Fife. *In*: BLACK, A. R. & JOHNSON, R. C. (eds) *Proceedings of the Fifth National Hydrology Symposium*, British Hydrological Society, Edinburgh, 4–7th September 1995.

——, PENNELL, R. & COWPERTWAIT, P. S. P. 1996. *Evaluation of Temporal Changes in the Quality of the Lower Ynysarwed Minewater Discharge, Neath Valley, South Wales*. Report to the National Rivers Authority, Swansea, South Wales.

Groundwater protection in a Celtic region: the Irish example

BRUCE D. MISSTEAR[1] & DONAL DALY[2]

[1] *Department of Civil, Structural & Environmental Engineering, University of Dublin, Trinity College, Dublin 2, Ireland (e-mail: bmisster@tcd.ie)*
[2] *Groundwater Section, Geological Survey of Ireland, Beggars Bush, Haddington Road, Dublin 4, Ireland*

Abstract: One of the key environmental objectives of the proposed EU Water Framework Directive is that Member States must prevent the deterioration of groundwater quality. A national groundwater protection scheme for Ireland has been published recently. This scheme shows certain broad similarities to the groundwater protection policy for England and Wales, incorporating the concepts of groundwater vulnerability, source protection zones and responses to potentially polluting activities. However, the Irish scheme is different in several important respects, reflecting the different hydrogeological conditions and pollution concerns in Ireland. Some of these hydrogeological conditions and pollution concerns are common to the other Celtic regions. A major feature of the Irish scheme is the importance given to subsoil permeability in defining groundwater vulnerability. At present, the subsoil permeability is classified in qualitative terms as high, moderate or low. For the protection scheme to be defensible, it is essential to adopt a systematic and consistent approach for assigning subsoil units to these permeability categories. In mapping groundwater vulnerability, it is also useful to take account of secondary indicators such as groundwater recharge potential, natural and artificial drainage density and vegetation characteristics. Another important issue in Ireland is the protection of groundwater in karst areas, since these areas are especially vulnerable to contamination.

The protection of groundwater against contamination is an important environmental objective within the European Union. The 1980 Groundwater Directive (80/68/EEC) introduced controls on discharges of specific contaminants to groundwater. The current EU proposal for a Water Framework Directive on water is much more wide-ranging, and proposes that 'Member States shall draw up and make operational within a comprehensive River Basin Management Plan the programmes of measures envisaged as necessary, in order to ... prevent deterioration of groundwater quality' (European Commission 1996). A policy for the protection of groundwater in England and Wales was published by the National Rivers Authority in 1992, and updated by the Environment Agency in 1998. This policy has been used as a basis for the preparation of groundwater protection schemes in Northern Ireland (Department of the Environment for Northern Ireland 1994, 1999) and Scotland (Scottish Environment Protection Agency 1997).

A groundwater protection scheme for the Republic of Ireland has been published recently (Department of the Environment and Local Government *et al.* 1999*a*). This scheme shows certain similarities to the groundwater protection policy for England and Wales, but it is different in several important respects, reflecting the different hydrogeological conditions and pollution concerns in Ireland. Some of these differences are relevant to groundwater protection in the other Celtic regions of Britain and Ireland.

The objectives of this paper are to:

(a) set out the current situation with regard to groundwater protection in the Republic of Ireland;
(b) compare the Irish scheme with the groundwater policies in England and Wales and with the other regions of the United Kingdom, explaining the underlying reasons for the main differences;
(c) discuss the limitations of the Irish scheme and the ways in which these limitations are being addressed.

The main aquifers in the Republic of Ireland are found in Quaternary deposits and in Palaeozoic bedrock formations, including Carboniferous limestones and sandstones, Old Red

From: ROBINS, N. S. & MISSTEAR, B. D. R. (eds) *Groundwater in the Celtic Regions: Studies in Hard Rock and Quaternary Hydrogeology*. Geological Society, London, Special Publications, **182**, 53–65. 1-86239-077-0/00/ $15.00

Sandstone, and Ordovician volcanics. Except for very limited exposures, the Permo-Triassic sandstones and Cretaceous Chalk that comprise the major aquifers in England (Downing 1993), are absent in the Republic of Ireland (hereinafter, for simplicity, the Republic of Ireland will be referred to as Ireland, whereas Northern Ireland will be given its full title). The main aquifers in Ireland are usually classed as those that can provide well yields in excess of $400\,m^3/d$ – relatively modest yields in comparison to those from the Sherwood Sandstone, Jurassic Limestone or Chalk of England. Nevertheless, groundwater is a very important source of water supply in Ireland, providing about 25% of drinking water supplies nationally and close to 100% in some rural regions (Daly 1993; Environmental Protection Agency 1999). Interestingly, the total number of water supply wells in Ireland is estimated to be about 200 000 (Wright 1999), about double the number quoted for England and Wales (Environment Agency 1998) where the population is more than ten times greater. These statistics highlight the fact that groundwater development in Ireland comprises a very large number of individual wells and small group or public water supply schemes; the larger urban areas rely mainly on surface water supplies. In this respect, it is similar to the other Celtic regions of Wales, Scotland and Northern Ireland.

The groundwater contaminants of most concern in Ireland are those associated with farmyards (manure, silage, etc.) and septic tank systems (Daly 1993). Although these contamination sources tend to be localized ('point sources'), they are many in number and hence the problems can be frequent and widespread. A recent national water quality survey reported that coliform contamination of groundwater was relatively widespread (Environmental Protection Agency 1999). Until recently, there was less concern in Ireland about the 'diffuse sources' of agricultural pollution that have been such a major issue in England over the past 20 years (Department of the Environment 1988). However, there is growing evidence in Ireland for increases in nitrate levels in groundwater resulting from the application of organic and inorganic fertilizers (Page & Keyes 1999). Trace organic contaminants such as chlorinated solvents, fuel oils and pesticides have received much less attention in Ireland than in England and Wales (Environment Agency 1997). Nevertheless, a survey in the early 1990s found trace organic contaminants such as trichloroethene, tetrachloroethene and refined mineral oils in groundwaters at several sites (K. T. Cullen & Co. 1994). With the current growth of computer, pharmaceutical and other high technology industries in Ireland, trace organic and other industrial contaminants may become more of an issue in the future.

Groundwater protection in Ireland

The Irish groundwater protection scheme was developed over a period of 20 years from the late 1970s, initially by the Geological Survey of Ireland (GSI), and more recently by the GSI in conjunction with the Department of the Environment and Local Government (DoELG) and the Environmental Protection Agency (Creighton *et al.* 1979; Daly 1995; DoELG *et al.* 1999*a*). The scheme is based partly on the experience of groundwater protection in England and Wales (e.g. Selby & Skinner 1979; National Rivers Authority (NRA) 1992; Environment Agency 1998), partly on available guidance from the USA (e.g. US Environmental Protection Agency 1994) and elsewhere (e.g. Office fédéral de la protection de l'environment 1982), all adapted for the Irish situation.

As in the NRA/Environment Agency policy, the concepts of risk assessment and risk management underlie the Irish scheme (Misstear *et al.* 1998). In the risk assessment element, the source of contamination, the groundwater vulnerability and the value of the groundwater (be it an aquifer, well or spring) are evaluated in terms of a source–pathway–receptor model. Once the nature and extent of the risk have been assessed, a response to that particular risk can be prepared which might involve a code of practice for landspreading of organic wastes or an engineered barrier for a landfill site. This is the risk management element. The application of these concepts of risk assessment and risk management to groundwater protection is shown schematically in Fig. 1.

The two main components of the Irish scheme are land surface zoning (risk assessment) and response matrices for potentially polluting activities (risk management). Land surface zoning comprises:

(a) mapping the groundwater vulnerability;
(b) defining protection areas around major wells and springs;
(c) classifying the value of the groundwater resource.

The main features of land surface zoning for groundwater protection in Ireland are compared with those in Britain and Northern Ireland in Table 1, and are discussed below.

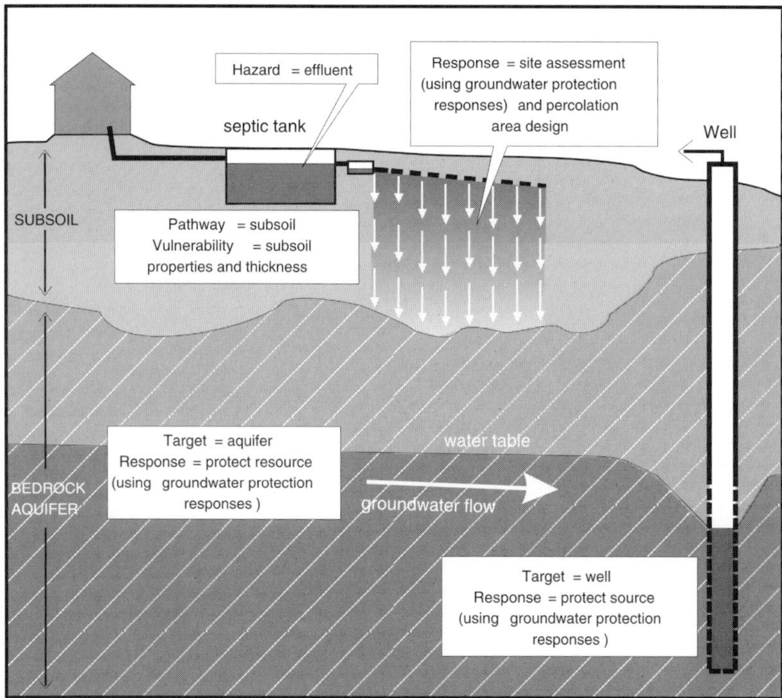

Fig. 1. Schematic diagram showing how the elements of risk are applied to groundwater protection (after DoELG *et al.* 1999*a*).

Groundwater vulnerability

The background to defining groundwater vulnerability in Ireland is given in Daly & Warren (1998). Vulnerability is 'a term used to represent the intrinsic geological and hydrogeological characteristics that determine the ease with which groundwater may be contaminated by human activities'. Vulnerability mapping is 'the technique of assessing the geological and hydrogeological factors and displaying these on a map in a manner that is understandable and useful'.

The vulnerability of an aquifer to contamination may depend on many factors including the leaching characteristics of the topsoil, the permeability and thickness of the subsoil, the presence of an unsaturated zone, the type of aquifer and the amount and form of recharge. In Ireland, groundwater vulnerability is determined mainly according to the thickness and permeability of the subsoil that underlies the topsoil. The nature of the recharge is also important, especially in karstic areas where 'point recharge' may occur through swallow holes or sinking streams. The unsaturated zone of bedrock aquifers is not relied upon for attenuating contaminants, since virtually all bedrock formations in Ireland possess secondary rather than primary porosity. However, the vulnerability classification in sand and gravel aquifers does take account of an unsaturated zone.

Subsoil is equivalent to the term 'drift' widely used in Britain. The main subsoils found in Ireland are glacial tills, alluvial, glaciofluvial and lacustrine deposits, and peat. A simplified classification of groundwater vulnerability based on subsoil characteristics is shown in Table 2. A more detailed classification is included in Daly & Warren (1998). A limitation of the current vulnerability ratings is that they include general, non-quantitative statements about subsoil permeability. This issue is discussed in more detail later in the paper.

The main differences between the approaches adopted for groundwater vulnerability mapping in Ireland on the one hand, and Britain and Northern Ireland on the other, are:

(a) the topsoil is not taken into account in the Irish scheme;
(b) the subsoil (drift) is the main feature of the vulnerability maps in Ireland, whereas it is generally only a secondary feature on vulnerability maps in Britain and Northern Ireland.

Table 1. *Comparison of the main features of land surface zoning for groundwater protection in the Republic of Ireland, England and Wales, Scotland and Northern Ireland*

	Republic of Ireland*	England and Wales†	Scotland‡	Northern Ireland§
Groundwater vulnerability classification		Topsoil (high, intermediate and low leaching potential)	Topsoil (high, intermediate and low leaching potential)	Topsoil (high, intermediate and low leaching potential)
	Subsoil characteristics (thickness and permeability)	Drift (usually shown as undifferentiated stipple on maps)	Drift (usually shown as undifferentiated stipple on maps)	Drift (usually shown as undifferentiated stipple on maps)
	Depth of unsaturated zone (sand and gravel aquifers only)	Depth of unsaturated zone (not shown on maps)	Depth of unsaturated zone (not shown on maps)	Depth of unsaturated zone (not shown on maps)
	Nature of recharge (point or diffuse)			
Vulnerability maps	1:50 000 scale	1:100 000 scale	1:100 000 scale	1:250 000 scale (regional map)
Source protection zones	Inner Source Protection Area (SPA) (100 day travel time)	Zone I (Inner) (50 day travel time, minimum 50 m radius)	Zone I (Inner) (50 day travel time, minimum 50 m radius)	Zone I (Inner) (50 day travel time, minimum 50 m radius)
		Zone II (Outer) (400 day travel time)	Zone II (Outer) (400 day travel time)	Zone II (Outer) (400 day travel time)
	Outer SPA (zone of contribution)	Zone III (Source Catchment) (zone of contribution)	Zone III (Source Catchment) (zone of contribution)	Zone III (Source Catchment) (zone of contribution)
Aquifer classification	Regionally important	Major	Highly permeable	Highly permeable
	Locally important	Minor	Moderately permeable	Moderately permeable
	Poor	Non-aquifers	Weakly permeable	Weakly permeable

Notes:
* Department of the Environment and Local Government *et al.* (1999*a*).
† National Rivers Authority (1992) and Environment Agency (1998).
‡ Scottish Environment Protection Agency (1997).
§ Department of the Environment for Northern Ireland (1994 and 1999).

Table 2. *Groundwater vulnerability classification based on subsoil characteristics*

Vulnerability category	Subsoil characteristics
Extreme	Areas of outcropping bedrock
	Areas with thin (less than 3 metres) subsoil
High	Areas with 3–10 metres of moderately permeable subsoil
	Areas with 3–5 metres of low permeability subsoil
Moderate	Areas with over 10 metres of moderately permeable subsoil
	Areas with 5–10 metres of low permeability subsoil
Low	Areas with thick (over 10 metres) low permeability subsoil

The leaching potential of the topsoil is one of the most important parameters in classifying groundwater vulnerability in Britain and Northern Ireland. This is largely a response to the widespread concerns in the 1970s and 1980s about diffuse pollution from agricultural fertilizers, especially nitrate (Robins *et al.* 1994). The classification of soil leaching potential does take into account other types of pollutants, including pesticides and liquids, but cannot be used directly for non-aqueous phase pollutants such as chlorinated solvents and fuel hydrocarbons (Environment Agency 1998; Palmer & Lewis 1998). In Ireland, the main problems traditionally have been from point sources such as septic tank systems, where the contaminants may enter the ground below the soil zone. Hence the topsoil is not included in the vulnerability classification. However, as noted earlier, there are growing concerns in Ireland about diffuse contamination sources such as fertilizers, and these types of hazards are considered within the risk response element of the groundwater protection scheme described later in the paper.

On groundwater vulnerability maps in Britain and Northern Ireland the subsoil (drift) is treated in a number of different ways. Drift deposits are generally shown as a black stipple ornament, without details of their composition and thickness (NRA 1995*a*; Palmer & Lewis 1998). Where the drift deposits are known locally to be sufficiently thick and of low enough permeability to protect the underlying aquifers, then those aquifers are not shown on the maps. Alternatively, permeable drift may be classed as a Minor Aquifer or as part of an underlying Major Aquifer where there is hydraulic continuity.

Vulnerability maps in Ireland are being produced on a county by county basis at a scale of 1:50 000 scale (or 1:63 360 where the base map is 1 inch to 1 mile). Larger scale (1:10 000 or 1:10 560 scale) vulnerability maps are produced around source protection areas. The maps are prepared by the GSI in association with the relevant local authority. In England and Wales, a national series of 53 vulnerability maps has been prepared for the Environment Agency by the Soil Survey and Land Research Centre and the British Geological Survey. These maps are published in both paper and CD-ROM formats, at a scale of 1:100 000 (Environment Agency 1998). A similar scale has been adopted by the Scottish Environment Protection Agency (SEPA) for its series of vulnerability maps. These will cover the main Scottish aquifers only (SEPA 1997). Three regional groundwater vulnerability maps are also available: England and Wales (1:1 000 000 scale), Scotland (1:625 000) and Northern Ireland (1:250 000).

Source protection areas

The concept of source protection areas used in Ireland is similar to that applied in Britain and elsewhere (US Environmental Protection Agency 1994; NRA 1995*b*). In the Irish groundwater protection scheme two source protection areas (SPAs) are delineated around each well: an Inner Protection Area, defined by a 100-day travel time within the aquifer, and an Outer Protection Area, encompassing the whole source catchment area. In an earlier version of the scheme a third SPA was delineated as the site around the well or spring (Daly 1995), but this has been discarded as it was deemed unnecessary and difficult to define precisely.

The Inner Protection Area aims to protect the well against the effects of human activities that might have a rapid effect on the source, especially microbiological pollution. This is similar to the Inner Zone 1 in Britain (SEPA 1997; Environment Agency 1998). A 100-day travel time is allowed for bacteria die-off, rather than the 50 days incorporated in groundwater protection schemes in Britain, and in several other countries (Van Waegeningh 1985). This conservative, 100-day figure, is considered to be appropriate for Ireland in view of the heterogeneity of the aquifers, and the possibility that some bacteria, viruses and other microorganisms may survive for longer than 50 days in groundwater.

The Outer Protection Area covers the total source catchment area, or zone of contribution of the well or spring. It is the area needed to support the long-term abstraction, and hence is strongly dependent on the amount of recharge. This is equivalent to the Source Catchment Zone III in Britain. There is no equivalent to the intermediate 400-day Outer Zone II in Ireland.

The techniques used to define SPAs in Ireland are generally the same as those used by the Environment Agency in England and Wales (NRA 1995*b*; Burgess & Fletcher 1998). The four types of methods currently used are:

(a) calculated fixed radii;
(b) analytical methods;
(c) hydrogeological mapping;
(d) numerical models such as FLOWPATH, MODFLOW and other commercially available models.

Arbitrary radii (normally of 1000 m) have been used for protecting groundwater sources in Ireland in the past (Daly 1995), but they are no longer recommended where sufficient data are available to apply one of the methods listed above. Arbitrary radii were also formerly applied to public water supply sources in parts of Britain (e.g. Selby & Skinner 1979), but now are generally only retained for small sources where a fixed-radius of 50 m is currently in use (Burgess & Fletcher 1998).

In Ireland, owing to the heterogeneity of the bedrock aquifers, the relatively few high-yielding wells and the often sparse database, there is less reliance on modelling techniques to delineate SPAs than in England and Wales: numerical models have been applied to about 15 well sources to date, whereas 2-dimensional models such as FLOWPATH have been used for over 80% of the 1000 or more source protection zones defined in England and Wales (Burgess & Fletcher 1998). Correspondingly greater emphasis is given to simple analytical methods and hydrogeological mapping in Ireland. Where models are used, the Outer Protection Area includes factors of safety to allow for the heterogeneity of the aquifer, seasonal/annual variations in recharge and possible future increases in well abstraction rates. The latter is especially important because there is no groundwater abstraction licensing system in Ireland, and hence no reliable way of predicting long-term groundwater withdrawals. In England and Wales, a somewhat more sophisticated approach has been adopted for dealing with uncertainty, involving the concepts of the Zone of Uncertainty, Zone of Confidence and Best Estimate Zone for each Source Protection Zone (Burgess & Fletcher 1998).

Aquifer classification

The classification of aquifers in Ireland is shown in Table 3. The table summarizes the criteria used for the classification, and gives examples of aquifers within each of the main categories. The three main aquifer categories are: Regionally Important Aquifers, Locally Important Aquifers and Poor Aquifers.

In view of the large number of small groundwater abstractions in Ireland, many of which are located in low permeability bedrock formations, it was important to avoid the use of terms such as non-aquifer or aquiclude in the classification. Hence the term Poor Aquifer has been adopted to cover, for example, Lower Palaeozoic metamorphic formations that might only be capable

Table 3. *Aquifer classifications in the Republic of Ireland*

	Regionally Important (R)	Locally Important (L)	Poor (P) Aquifers
Sub-categories	• Karstified aquifers where conduit flow is dominant (Rk)	• Sand/gravel (Lg)	• Generally unproductive except for local zones (Pl)
	• Fractured bedrock aquifers (Rf)	• Generally moderately productive (Lm)	• Generally unproductive (Pu)
	• Extensive sand/gravel aquifers (Rg)	• Moderately productive only in local zones (Ll)	
Criteria for main aquifer categories	Areal extent >25 km^2 Well yields >400 m^3/d Specific capacities >40 m^3/d/m Occurrence of large springs	Areal extent <25 km^2 Well yields 100–400 m^3/d (or higher yields where aquifer extent is <25 km^2)	Well yields generally <100 m^3/d (and <40 m^3/d in Pu sub-category)
*Examples**	Quaternary sand and gravel (where extensive)	Quaternary sand and gravel (local deposits)	
	Wexford Limestone Formation	Westphalian sandstones	
	Ballyadams/Culahill Limestone	Calp limstones (Midlands/east)	Ballysteen Limestone (Midlands)
	Waulsortian Limestone (South)	Waulsortian Limestone (North)	
	Kiltorcan Sandstone	Old Red Sandstone (except Kiltorcan Sandstone)	
	Ordovician volcanics (Waterford)		Lower Palaeozoic sediments
			Precambrian and Lower Palaeozoic granites and metamorphics

Notes
* Adapted from Wright (this volume).

of providing well yields of a few m^3/d. The groundwater protection schemes of Scotland and Northern Ireland have followed a similar reasoning, with the adoption of the term Weakly Permeable Aquifer for the least productive formations (Table 1).

The term Regionally Important rather than Major Aquifer is used in Ireland because many of the better Irish aquifers are capable of sustaining well yields of only 500 to $1000\,m^3/d$. These well yields would not be regarded as very large for the Major Aquifers in England such as the Sherwood Sandstone and Chalk.

The groundwater protection scheme

Groundwater protection zones

The aquifer classification is used to prepare maps of Resource Protection Areas (RPAs). These RPAs, together with the SPAs delineated for individual sources, are combined with the vulnerability maps to produce Groundwater Protection Zones (GPZ therefore has a different meaning in Ireland to Britain, where it is often used as a synonym for source protection zone). The sequence of data analysis and map preparation is shown schematically in Fig. 2 and the integration of an SPA map and a vulnerability map is illustrated in Fig. 3.

Groundwater protection zones can also be displayed in the form of a matrix. As on the maps, each protection zone is assigned a code that indicates both the value of the groundwater (aquifer category, major well source or spring) and the vulnerability of the groundwater to contamination. Thus zone SI/M represents an Inner Source Protection Area with moderate groundwater vulnerability.

Groundwater protection responses

The matrix format is also used to illustrate the groundwater protection response for any potentially polluting activity within a groundwater protection zone. The response matrix is the risk management element of the Irish groundwater protection scheme. It has some similarity with the acceptability matrix approach included in the groundwater protection policies in Britain (SEPA 1997; Environment Agency 1998). The major difference is that the Irish matrix incorporates groundwater vulnerability as well as source protection areas and aquifer class. In Britain, groundwater vulnerability is considered within the policy statements.

By consulting a response matrix, a planner or developer can determine:

(a) whether or not such a development is likely to be acceptable on that site;
(b) the further investigations that may be necessary to reach a final decision; and
(c) the planning or licensing conditions that may be necessary for that development.

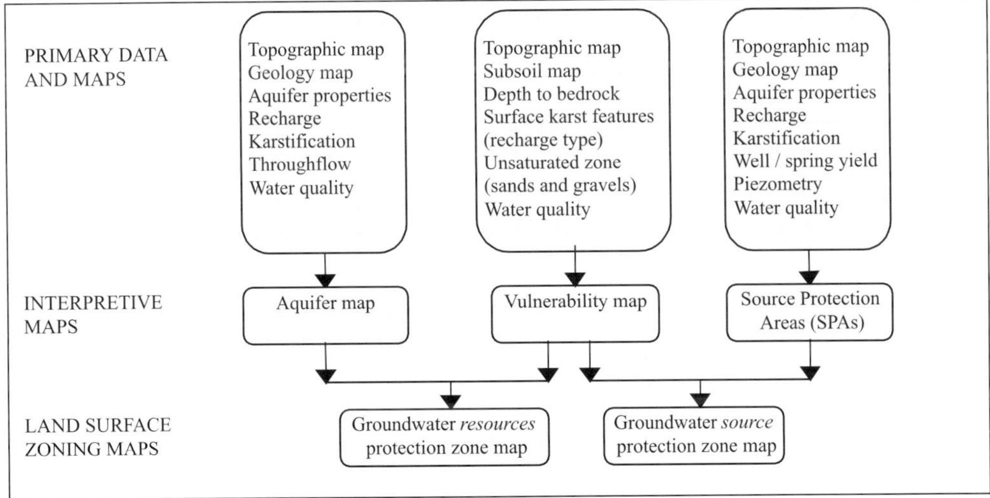

Fig. 2. Stages in land surface zoning and main data requirements.

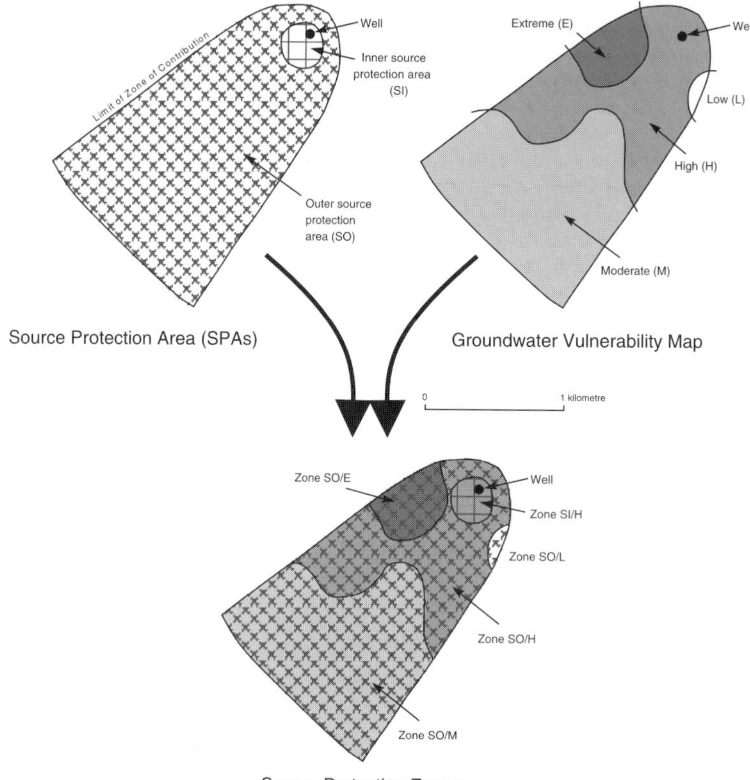

Fig. 3. Delineation of source protection zones around a public supply well by combining the source protection area map with the vulnerability map (after DoELG *et al.* 1999*a*).

Four levels of response (R) to the risk of a potentially polluting activity are determined:

- R1 Acceptable subject to normal good practice.
- $R2^{a,b,c}$ Acceptable in principle, subject to conditions in notes ab,c, etc. (The number and content of the notes vary depending on the zone and the activity).
- $R3^{m,n,o}$ Not acceptable in principle; some exceptions may be allowed subject to the conditions in notes m,n,o, etc.
- R4 Not acceptable.

Two groundwater protection responses have been published to date (DoELG *et al.* 1999*b*, 1999*c*). Table 4 shows a groundwater protection scheme matrix for the landspreading of organic

Table 4. *Groundwater protection response for landspreading of organic wastes (DoELG et al.* 1999*b)*

Vulnerability rating	Source protection		Resource protection Aquifer category					
			Regionally Important		Locally Important		Poor Aquifers	
	Inner	Outer	Rk	Rf/Rg	Lm/Lg	Ll	Pl	Pu
Extreme (E)	R4	R4	$R3^2$	$R3^2$	$R3^1$	$R3^1$	$R3^1$	$R3^1$
High (H)	R4	$R2^1$	R1	R1	R1	R1	R1	R1
Moderate (M)	$R3^3$	$R2^1$	R1	R1	R1	R1	R1	R1
Low (L)	$R3^3$	$R2^1$	R1	R1	R1	R1	R1	R1

agricultural wastes. In this table, the response R2[1], for example, indicates that the proposal is acceptable in this situation, subject to a maximum organic nitrogen load (including that deposited by grazing animals) of 170 kg/ha/a.

Limitations and key issues

A groundwater protection scheme must be sufficiently robust to be defensible when challenged. Yet every scheme incorporates some level of uncertainty (Committee on Techniques for Assessing Ground Water Vulnerability 1993). It is therefore important that the limitations of a scheme are clearly understood by its users. It is also important that the scheme should be sufficiently adaptable so that it can be updated and refined as further knowledge becomes available. Three key issues currently being addressed in Ireland are:

(a) the permeability classification of subsoils;
(b) the use of secondary indicators for mapping groundwater vulnerability;
(c) groundwater protection in karst areas

Subsoil permeability

The classification of groundwater vulnerability in Ireland relies partly on a qualitative assessment of whether the subsoil permeability is high, moderate or low (Table 2). For the scheme to be defensible, it is essential to have a systematic and consistent approach for assigning subsoil units to these permeability categories.

Grain size distribution is an important controlling influence on a sediment's permeability (e.g. Sperry & Peirce 1995). Even though this relationship is complex in fine-grained materials such as glacial tills (the most common subsoil in Ireland), it is important to try and describe the subsoils using criteria that bear some relationship to their permeability and hence to the vulnerability category.

On the current subsoil maps used for groundwater protection schemes in Ireland, the classification of till units is based mainly on the lithology of the stones (between 5 and 10 mm in size) within the till (Warren 1991); field observations regarding texture are also recorded. The broad permeability categories are then assigned to the mapped units. For example, granitic tills (tills containing a higher proportion of granite stones than any other lithology) are generally assumed to be sandy, based on their weathering characteristics and an assessment of the field observations, and are assigned a moderate permeability rating. This classification assumes that the mapped till units are relatively homogeneous in texture. However, in some situations this is clearly not the case, as shown by the large

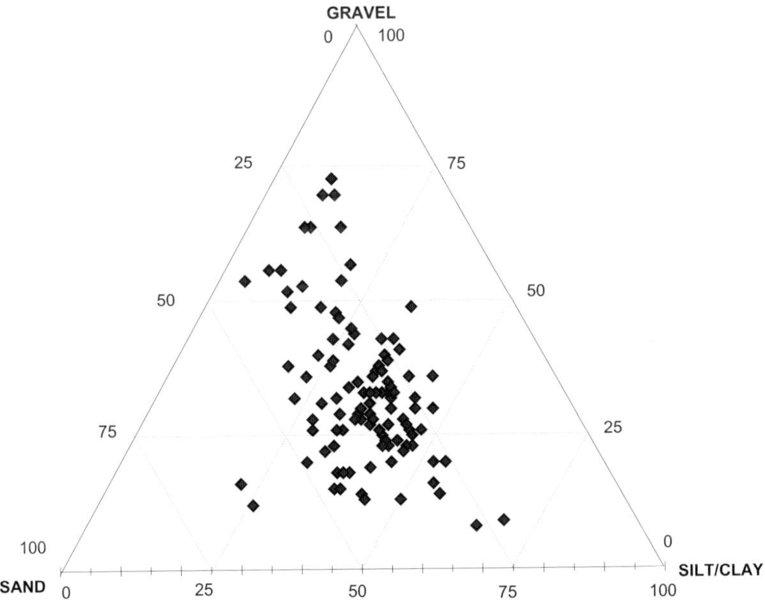

Fig. 4. Ternary diagram showing the range of particle sizes within the limestone till unit in County Clare (Swartz *et al.* 1999).

range in grain sizes for the limestone till unit illustrated in Fig. 4.

For groundwater protection schemes, the GSI is now adopting a revised approach to assessing the permeability of subsoils based on the British Standard (BS) 5930 system (1981), as updated in Norbury (1998). With this approach, textural descriptions of fine–grained materials are based on characteristics of behaviour such as dilatency and plasticity. BS 5930 also includes a description of other factors that may have a bearing on permeability, including density, compactness and the presence of discontinuities, and in general provides a systematic and consistent approach to describing subsoils. Using this system, materials described as 'CLAY' generally have a low permeability whereas those described as 'SILT' fall in the moderate category (Swartz 2000). The BS 5930 system has the added advantage that it is familiar to geologists and civil engineers who are the main users of the groundwater protection schemes in Ireland.

Secondary indicators of groundwater vulnerability

In addition to a direct evaluation of groundwater vulnerability based on the characteristics of the subsoil, it is useful to consider secondary indicators in any vulnerability mapping exercise. The following secondary indicators are being incorporated into the groundwater vulnerability mapping procedures used by the GSI:

- groundwater recharge;
- natural drainage density;
- artificial drainage density;
- vegetation characteristics.

For a particular aquifer, it is assumed that groundwater recharge is an indirect measure of topsoil and subsoil permeability, and hence of groundwater vulnerability. In qualitative terms, a high recharge rate implies that water can move relatively easily through the topsoil and subsoil to the aquifer, and hence indicates high aquifer vulnerability. Similarly, a low recharge rate indicates low aquifer vulnerability. Recharge can be assessed using a soil moisture balance, river baseflow estimate or other standard method. To be useful, this type of comparative recharge analysis should be applied to relatively small catchments that have similar topographical, rainfall and aquifer characteristics, and where the main control on groundwater vulnerability is due to the variation in the topsoil and subsoil. Further limitations are that water balance estimates of recharge tend to be very approximate and, in Ireland at any rate, there are relatively few long-term stream gauge records in small catchments available for analysing baseflow.

A survey of the density of natural (streams) and artificial (ditches that run along field boundaries) drainage can give a general indication of groundwater recharge potential, and hence of groundwater vulnerability. In areas with a low recharge potential, for example, there is proportionally less infiltration and more surface runoff, and this will be reflected in the drainage density. Both the natural and artificial drainage densities are measured in terms of length of channel per unit area. The main difference between the two approaches is in their scale of application: stream density assessments are useful on a catchment or regional scale, whereas artificial drainage density is more suitable for the field to catchment scale (Lee 2000). A limitation with both these approaches is that they cannot be applied to flat, low-lying areas with shallow water tables, where the drainage density would not reflect the recharge potential of the underlying material.

Floral species flourish under uniform habitat conditions (White & Doyle 1982), one of which is soil moisture content. Higher soil moisture content occurs where downward drainage is impeded by the nature of the topsoil or underlying material (subsoil or bedrock), or where the water table is situated close to the surface. Different plant species are associated with different soil moisture contents. For example, rushes and marsh thistles tend to be associated with wet, poorly drained conditions whereas meadow thistles and nettles usually thrive in drier areas. Vegetation surveys in association with drainage assessments have been found to give consistent and useful results (Lee 2000).

Karst

The degree of solution/karstification of the Carboniferous limestones of Ireland – which underlie almost half of the land surface – is highly variable. Yet solution has impacted on all the limestones to some degree and therefore all can be said to be karstified. Karst areas are characterized by rapid infiltration via features such as sinkholes, and by rapid groundwater movement within subsurface conduits. Hence karst groundwater, and wells and springs in karst aquifers, are often at particular risk from potentially polluting developments. Special measures are therefore required for protecting groundwater in such areas. All four elements of the protection scheme – vulnerability, source protection areas, aquifer categorization and groundwater protection responses – take account of karst. In contrast, in Britain and Northern Ireland, where karst limestones underlie a much lower proportion of the land surface, karstification is not currently a major factor in the groundwater protection policies.

In mapping vulnerability in Ireland, rapid recharge through karst features such as swallow holes, sinking streams and collapse features, is taken into account. Groundwater is classed as 'extremely' vulnerable within 30 m of karstic features (including the area of loss of losing or sinking streams) and within 10 m on either side of losing streams upflow of the area of loss. The distances can be varied depending on the circumstances – for instance, they are increased where overland surface runoff is likely.

In producing aquifer maps, the distinction is drawn between limestones with and without a high degree of karstification. Where the karstification is slight, the limestones are classed as 'fissured', although some karst features may occur. Aquifers in which karst features are more significant are classed as 'karstified' (Table 3).

There is normally no difference in the general principles used in delineating source protection areas in karst and non-karst areas. However, as travel times are rapid in karst limestones, sources are given a greater degree of protection by extending the boundary of the inner protection area for all wells and springs to coincide with the boundary of the zone of contribution. In the case of large karst springs containing a high proportion of surface water from sinking streams, the methodology is adapted to take account of the local hydrogeological situation (Deakin this volume).

Groundwater protection responses take account of the degree of risk posed by potentially polluting activities. As groundwater in karst areas is at greater risk than in fissured or granular aquifers, restrictions on most developments are greater in karst areas than elsewhere. For example, landspreading of piggery wastes is prohibited within 30 m of karst features.

Conclusions

The groundwater protection scheme in Ireland has been prepared using many of the same general hydrogeological principles that underlie groundwater protection policies in other countries, including the policy currently applied in England and Wales. The main features of the Irish scheme are vulnerability mapping, delineation of source protection areas, classification of aquifers, and preparation of responses to various polluting activities. The scheme, however, also incorporates specific hydrogeological criteria and pollution risks that are especially relevant to the Irish situation. These include the important role of the subsoil in groundwater protection, the relatively large number of point sources of pollution compared to diffuse sources, and the widespread occurrence of high vulnerability karstified limestone aquifers.

The groundwater protection scheme is intended to give guidance to developers and planners for minimizing the risks to groundwater from potentially polluting developments. In aiming to give the best guidance, the scheme must be sufficiently flexible to take account of new information and improved understandings of hydrogeology as these become available. A particular area under review is the preparation of groundwater vulnerability maps. Vulnerability is currently classified using qualitative assessments of permeability based on subsoil texture. However, many of the mapped subsoil units are relatively heterogeneous. A new system for assessing the permeability of subsoils, based on the British Standard BS 5930, is therefore being introduced for the purposes of groundwater vulnerability mapping. The subsoil units classified using this approach are more homogeneous in texture, and hence in permeability.

A further initiative in vulnerability mapping in Ireland is the use of groundwater recharge, drainage density and vegetation as secondary indicators of groundwater vulnerability. These parameters can give an indication of the relative ease with which water moves through the topsoil and subsoil, and hence of the groundwater vulnerability. They can provide information at the field to catchment scale, and therefore complement the site-specific data on subsoil characteristics and permeabilities provided by borehole investigations. It is intended to prepare a vulnerability mapping protocol that will incorporate these secondary indicators with the new subsoil classification system.

This paper is published with the permission of Dr Peadar McArdle, Director, Geological Survey of Ireland. The authors would like to acknowledge the research work of Monica Lee and Melissa Swartz on groundwater vulnerability mapping.

References

British Standards Institution. 1981. *Code of practice for site investigations: BS 5930: 1981.* BSI, London.

BURGESS, D. B. & FLETCHER, S. W. 1998. Methods used to delineate groundwater source protection zones in England and Wales. *In*: ROBINS, N. S. (ed.) *Groundwater pollution, aquifer recharge and vulnerability*. Geological Society, London, Special Publications, **130**, 199–210.

COMMITTEE ON TECHNIQUES FOR ASSESSING GROUND WATER VULNERABILITY. 1993. *Ground water vulnerability assessment: predicting relative contamination under conditions of uncertainty*. National Research Council, National Academy Press, Washington D.C.

CREIGHTON, J. R., DALY, D. & REILLY, T. A. 1979. *The geology and hydrogeology of County Dublin, with particular reference to the location of waste disposal sites*. Geological Survey of Ireland, Dublin, Ireland.

DALY, D. 1993. Groundwater resources in Ireland. *In*: MOLLAN, C. (ed.) *Water of Life*. Royal Dublin Society Seminar Proceedings, **5**, 37–46.

——1995. *Groundwater protection schemes in Ireland: a proposed approach – draft*. Geological Survey of Ireland, Dublin.

—— & WARREN, W. P. 1998. Mapping groundwater vulnerability: the Irish perspective. *In*: ROBINS, N. S. (ed.) *Groundwater pollution, aquifer recharge and vulnerability*. Geological Society, London, Special Publications, **130**, 179–190.

DEAKIN, J. 2000. Groundwater protection zone delineation at a large karst spring in western Ireland. *This volume*.

DEPARTMENT OF THE ENVIRONMENT. 1988. *Assessment of groundwater quality in England and Wales*. HMSO, London.

DEPARTMENT OF THE ENVIRONMENT AND LOCAL GOVERNMENT, ENVIRONMENTAL PROTECTION AGENCY & GEOLOGICAL SURVEY OF IRELAND. 1999*a*. *A scheme for the protection of groundwater*. Geological Survey of Ireland, Dublin.

——, —— & ——1999*b*. *Groundwater protection responses to the landspreading of organic wastes*. Geological Survey of Ireland, Dublin.

——, —— & ——1999*c*. *Groundwater protection responses for landfills*. Geological Survey of Ireland, Dublin.

DEPARTMENT OF THE ENVIRONMENT FOR NORTHERN IRELAND. 1994. *Groundwater vulnerability map for Northern Ireland*. Environment Service, Department of the Environment for Northern Ireland.

——1999. *Policy and practice for the protection of groundwater in Northern Ireland: a consultation document*. Environment Service, Department of the Environment for Northern Ireland.

DOWNING, R. A. 1993. Groundwater resources, their development and management in the UK: an historical perspective. *Quarterly Journal of Engineering Geology*, **26**, 335–358.

ENVIRONMENT AGENCY. 1997. *Groundwater pollution: evaluation of the extent and character of groundwater pollution from point sources in England and Wales*. Summary Report, Environment Agency, Bristol.

——1998. *Policy and practice for the protection of groundwater*. HMSO, London.

ENVIRONMENTAL PROTECTION AGENCY. 1999. *Water quality in Ireland 1995–1997*. Environmental Protection Agency, Wexford.

EUROPEAN COMMISSION. 1980. Directive 80/68/EEC: Council Directive of 17th December 1979 on the protection of groundwater against pollution from certain dangerous substances. *Official Journal of the European Communities*, **L 20**.

——1996. Proposal for Council Directive establishing a framework for Community action in the field of water policy. *Official Journal of the European Communities*, **C 184**.

K. T. CULLEN & CO. 1994. *Trace organic contaminants in Irish groundwaters*. Report prepared under the European Union STRIDE Programme, Environmental Subprogramme Measure 3.

LEE, M. 2000. *Surface hydrology and landuse as secondary indicators of groundwater recharge and vulnerability*. Unpublished MSc thesis, Trinity College Dublin.

MISSTEAR, B. D. R., JOHNSTON, P. & DALY, D. 1998. The risk concept as a basis of a national groundwater protection scheme in Ireland. *In*: BRAHANA, J. V., ECKSTEIN, Y., ONGLEY, L. K., SCHNEIDER, R. & MOORE, J. E. (eds) *Gambling with groundwater – physical, chemical and biological aspects of aquifer–stream interactions*. American Institute of Hydrology, St. Paul, 213–218.

NATIONAL RIVERS AUTHORITY. 1992. *Policy and practice for the protection of groundwater*. NRA, Bristol.

——1995*a*. *Guide to groundwater vulnerability mapping in England and Wales*. HMSO, London.

——1995*b*. *Guide to groundwater protection zones in England and Wales*. HMSO, London.

NORBURY, K. 1998. Rock and soil description and classification; the revised BS 5930, section 8. *In: Ground Investigation Practice in Ireland, Institution of Engineers of Ireland seminar*, Dublin, 2 April 1998.

OFFICE FÈDÈRAL DE LA PROTECTION DE L'ENVIRONMENT. 1982. *Instructions pratiques pour la détermination des secteurs de protection des eaux, des zones et des périmètres de protection des eaux souterrainnes*. Office fédéral de la protection de l'environment, Switzerland.

PAGE, D. & KEYES, J. 1999. Nitrates in water – the Offaly experience. *In*: *19th Annual Groundwater Seminar, International Association of Hydrogeologists (Irish Group)*, Portlaoise, 20–21 April 1999.

PALMER, R. C. & LEWIS, M. A. 1998. Assessment of groundwater vulnerability in England and Wales. *In*: ROBINS, N. S. (ed.) *Groundwater pollution, aquifer recharge and vulnerability*. Geological Society, London, Special Publications, **130** 191–198.

ROBINS, N. S., ADAMS, B., FOSTER, S. S. D. & PALMER, R. C. 1994. Groundwater vulnerability mapping: the British perspective. *Hydrogéologie*, **3**, 35–42.

SCOTTISH ENVIRONMENT PROTECTION AGENCY. 1997. *Groundwater protection policy for Scotland*. Scottish Environment Protection Agency, Glasgow.

SELBY, K. H. & SKINNER, A. C. 1979. Aquifer protection in the Severn–Trent Region: policy and practice. *Water Pollution Control*, **78**, 254–269.

SPERRY, J. M. & PEIRCE, J. J. 1995. A model for estimating the hydraulic conductivity of granular material based on grain shape, grain size, and porosity. *Ground Water*, **33**, 892–898.

SWARTZ, M. 2000. *Assessing the permeability of Irish subsoils*. Unpublished MSc thesis, Trinity College Dublin.

SWARTZ, M., MISSTEAR, B. D. R., FARRELL, E. R. & DALY, D. 1999. The role of subsoil permeability in groundwater vulnerability assessment. *In*: FENDEKOVÁ, M. & FENDEK, M. (eds) *Hydrogeology and land use management: Proc. XXIX International Association of Hydrogeologists Congress*, Bratislava, 6–10 September 1999, 443–448.

US ENVIRONMENTAL PROTECTION AGENCY. 1994. *Handbook: ground water and wellhead protection.*, U.S.E.P.A., Washington DC.

VAN WAEGENINGH, H. G. 1985. Protection of groundwater quality. *In*: MATTHES, G. *et al.* (ed.) *Theoretical background, hydrogeology and practice of groundwater protection zones*. IAH International Contributions in Hydrogeology, Vol. 6 Heise, Hannover, 159–166.

WARREN, W. 1991. Till lithology in Ireland. *In*: EHLERS, J. *et al.* (eds) *Glacial deposits in great Britain and Ireland*. A. A. Balkema, Rotterdam, 415–420.

WHITE, J. & DOYLE, G. 1982. The vegetation of Ireland; a catalogue raisonné. *In*: WHITE, J. (ed.) *Studies on Irish vegetation*. Royal Dublin Society, 289–357.

WRIGHT, G. 1999. How many wells are there in Ireland? *The GSI Groundwater Newsletter*, **35**.

WRIGHT, G. 2000. QSC graphs – an aid to classification of data-poor aquifers in Ireland. *This volume*.

Groundwater protection in Scotland

I. A. FOX

SEPA Borders Office, Burnbrae, Mossilee Road, Galashiels TD1 1NF, UK
(e-mail: ian.fox@sepa.org.uk)

Abstract: The way in which groundwater protection policy is implemented in Scotland is briefly reviewed, before the current status (July 1999) is examined. Imminent changing circumstances, particularly the advent of the Scottish Parliament and the implications of the forthcoming Water Framework Directive are discussed and related to a review of current and proposed groundwater research and development projects which should take forward the protection of Scottish groundwaters in the new millennium.

Setting the scene

Only 3% of Scotland's Public Water Supplies are currently derived from groundwater sources. This statistic explains, perhaps, why hitherto Scottish groundwaters have often been considered as 'out of sight–out of mind'. Yet, groundwater in Scotland is very much a 'hidden resource'. It is estimated that there may be 20 000 or so private potable supplies in Scotland which exploit groundwater storage and in recent years there has been a significant increase in the number of boreholes being drilled by private sector interests (industry and agriculture) as the annual charges for 'mains' supplies have increased dramatically.

Perhaps more important than the widespread exploitation of Scottish groundwaters is the fact that all surface waters derive, to a greater or lesser extent, from groundwater storage – in many Scottish watercourses, summer dry weather flows are entirely dependent on groundwater, supplemented in certain catchments by effluent discharges and 'compensation' water releases from water-supply and hydro-electricity generating reservoirs in their headwaters.

Against this background, a need has been identified to develop policies for the protection of groundwater. Various EC Directives have reinforced this need.

The purpose of this paper is to trace the history and development of groundwater protection in Scotland and to summarize the current (July 1999) position. 1999 has seen rapid developments in the Scottish groundwater 'field' and it is likely that these will continue into the new Millennium – several changing circumstances are reviewed, including the establishment of several research and development projects.

History and development of protection policies

Groundwater Protection Strategy for Scotland

In the early 1990s a working group comprising representatives of the River Purification Authorities (RPA), the Scottish Office (Environment Department) and selected other interests, was established. The group's deliberations included a review of the National Rivers Authority's Groundwater Protection Policy for England & Wales. The working group's work culminated in the publication by the Association of Directors and River Inspectors in Scotland (ADRIS) of a 'Groundwater Protection Strategy for Scotland' (ADRIS 1995).

This strategy drew heavily on the equivalent policy in England and Wales – the working group took the view that if the latter was appropriately robust enough to be implemented throughout England and Wales, it should be capable of being similarly attractive in the Scottish context. However, there was a requirement for the Scottish strategy to reflect the different institutional and legislative frameworks which existed north of the Border – in particular, the absence of a comprehensive form of abstraction control in Scotland and to reflect significant differences in the dominant hydrogeological regimes north and south of the Border.

From: ROBINS, N. S. & MISSTEAR, B. D. R. (eds) *Groundwater in the Celtic Regions: Studies in Hard Rock and Quaternary Hydrogeology*. Geological Society, London, Special Publications, **182**, 67–70. 1-86239-077-0/00/ $15.00

The Scottish strategy document included a national *Groundwater Vulnerability Map* (scale – 1:625 000); an implicit acknowledgment that the strategy was founded on the two basic concepts of risk and vulnerability – just as was the English/Welsh equivalent. Similar strategic statements included the need to identify resource and source protection zones.

Groundwater Protection Policy for Scotland

In 1997, the newly formed Scottish Environment Protection Agency effectively 'inherited' the ADRIS strategy when it published its Groundwater Protection Policy for Scotland (SEPA 1997). This took into account several new (i.e. April 1996) Scottish legislation, including the establishment of SEPA, the creation of the 3 Scottish Water Authorities and the formation of a 'single-tier' of 32 Scottish Councils.

EC Directives

The Scottish strategy and policy both acknowledged needs to comply with several EC Directives, notably the Groundwater Directive, the Nitrate Directive and the Drinking Water Directive – all of these required domestic (i.e. UK) legislation to be implemented.

The private water supplies (Scotland) Regulations 1992

The strategy and policy took cognisance of regulations in terms of private supplies, where new legal provisions took effect in Scotland in April 1992, when Local Authorities were given duties to check private supplies by:

(a) seeking information on the number, type, ownership and use of private water supplies for 'potable' use;
(b) classifying supplies according to size and use;
(c) taking samples for testing quality;
(d) maintaining registers of all private potable supplies in their areas.

Exact numbers of these supplies (registers now being maintained by the 32 'unitary' Councils in Scotland) are not known. It is thought that there could be between 20 000 and 30 000 – the great majority of which will be using groundwater sources. It should be noted that the public registers do not hold information which identifies each supply by either type or location of source – as such, the registers in this form have a limited value in helping to quantify pressures on water resources.

Current position

Early in 1999, two significant developments in the implementation of Scotland's Groundwater Protection Policy took place:

Groundwater Regulations 1998

These regulations came into force in December 1998, so that Government could fully implement the EC Groundwater Directive. The regulations require that 'the disposal (and tipping for the purpose of disposal) of List I & II substances to land be subject to a formal system of authorizations'. Important elements of the procedures which SEPA will require to have undertaken include:

(a) prior investigation of land suitability for surface disposal of List I & II substances;
(b) post-authorization surveillance of groundwater (in particular, its quality);
(c) recovering costs for implementing the Regulations.

Nitrate Directive

By the end of 1999, Government is required to demonstrate that it is fully complying with the monitoring requirements of the Nitrates Directive – in Scotland, these had previously been interpreted as being needed in respect of public water supplies only. In future, monitoring will be required for all groundwaters which are used, or could be used, for any supply of potable water, including private water supplies. Additionally, future strategies will also address needs to monitor groundwaters which, although they may not be exploited for anthropogenic use, are important in governing the quality of surface waters, especially where risks from diffuse pollution sources exist. An indication of the potential impact of these developments is that only one public water supply source in Scotland currently exceeds the prescribed limit of nitrate concentration, whereas several hundred private supplies currently 'fail'.

Changing circumstances

A number of circumstances which could impact on the issue of groundwater protection in Scotland have been highlighted in recent years, and particularly in 1999 (as indicated above).

Additionally, there are a number of new changes imminent. The most relevant are:

Scottish Parliament

Scotland's environmental issues will be 'devolved' to the new Scottish Parliament which was officially opened on 1 July 1999, when the new Scottish Ministers took office. It is understood that the Scottish Parliament will have a Transport and Environment Committee which, among other things, would be responsible for scrutinizing proposed environmental legislation. (In future, references to *Scottish Office* will be replaced by the *Scottish Executive*.)

EU Water Framework Directive

The draft of a proposed Directive to *'establish a framework for the protection of inland surface water, transitional waters, coastal waters and groundwater ...'* has been available for over a year. Although no final agreement between Member States on the text itself had been reached at the time of writing this paper, it seems likely that the European Parliament will sanction the Directive, with implementation to commence, in the year 2000.

Details of the likely implications for Scotland have been covered elsewhere (Marsden & Fox 1998). In the context of groundwater protection, two elements of the draft framework are particularly relevant:

River Basin Management. Member States will be required to introduce a statutory form of catchment management. This structure is expected to facilitate the achievement of water quality objectives (for all controlled waters, including groundwaters) by specified dates in the new millennium.

'Good' water status. Member States will be required to identify waters that are less than 'good' status and to produce and implement plans for them to reach 'good' status. A significant element of these processes will be the need to quantify the anthropogenic impacts on all waters – these will include the introduction, where appropriate, of a system of licensing (or permits) for abstractions. Article 13 3(e) of the draft text (Programme of basic measures) includes the words:

> 'Controls over the abstraction of fresh surface water and groundwater, including a register of abstractors and a requirement for prior authorisation for abstraction except in areas where the member state concerned has demonstrated, and reported to the Commission, that abstraction has no significant impact on water status and that the total level of abstractions amounts to a small proportion of the available resources'.

Research and development

A significant number of projects have been completed or initiated which are likely to inform the debate on groundwater protection in Scotland. Various organizations have funded, or agreed to co-fund these and it is appropriate to summarize the relevant funders:

- Scotland & Northern Ireland Forum For Environmental Research (SNIFFER) (SNIFFER is a plc funded by its members who are SEPA, SOAEFD, SNH, Forestry Commission (FC) and the DoE Northern Ireland, Environment & Heritage Service.)
- SEPA
- Scottish Office (Agriculture, Environment & Fisheries Department) (SOAEFD)
- Water Authorities
- British Geological Survey (BGS)

Database

SEPA has funded a national groundwater database of sources – it was developed by BGS and uses a MS ACCESS platform, ie it is a 'relational' database which offers the capability of searching records in different 'fields' – these include location, aquifer, borehole construction, owner, pump-test analysis, water chemistry etc. At the time of writing, the database comprises some 2800 sources. SEPA intends to place contracts to update this on an on-going basis. SEPA's existing agreement with BGS requires that information held on the database not be disclosed to third parties; search requests are directed towards BGS.

Detailed vulnerability maps

SEPA has commissioned BGS to produce detailed groundwater vulnerability maps (scale – 1:100 000). Two are currently available (Fife and Dumfries) and a third (Strathmore) is scheduled for production in the near future. The intention is to produce further detailed maps, on a prioritized phased basis, as and when funds are available – maps for Moray and the Scottish Borders are likely to complete the first tranche.

Groundwater monitoring networks

Towards the end of 1998, BGS completed a project, funded by SEPA, which recommended the basis for a national groundwater monitoring network. This identified 'primary' and 'secondary' networks where both groundwater level and quality should be monitored – quite properly, these were concentrated on Scotland's *major aquifers*.

In June 1999, a jointly funded SEPA/SOAEFD project commenced which aimed to recommend detailed monitoring sites for *all groundwater bodies* in Scotland. This is due to be completed by December 1999. One of the project's deliverables will be the creation of a national database summarizing information held in the Local Authority public registers for private water supplies.

Water Framework Directive

It is likely that the Directive will drive a number of projects in the short to medium term. It is too early to be prescriptive as to what form these might take.

SNIFFER projects

Current SNIFFER funded (or co-funded) projects which are relevant include:

- Representation on the UK Groundwater Forum's Steering Committee
- Production of a booklet – 'GROUNDWATER – OUR HIDDEN ASSET'
- Video 'HOW OUR RIVERS WORK'
- The use of gases in tracing groundwater movement
- Dating of groundwater recharge by isotopes

The first three of the above are aimed towards elevating the profile of groundwater in Scotland via membership of the UK Groundwater Forum.

The latter two projects are specifically directed towards a perceived need in Scotland and Northern Ireland to better understand the behaviour of groundwater in 'fractured rock' – many aquifers in these parts of the 'Celtic fringe' may not be as typical of those which occur in 'lowland' UK, where groundwater movement occurs as intergranular flow and is relatively well understood, although of course, there are many fractured and karstic aquifers in lowland UK.

Summary and conclusions

The latter half of the 1990's has seen significant developments in strategies and measures to protect Scottish groundwaters. These have been largely driven by statutory requirements of SEPA (and its predecessors) to '*maintain and enhance the quality of all controlled waters*'. It is now acknowledged that groundwaters deserve and require the same level of protection as surface waters in Scotland. Past practices and the lack of development control on groundwater management have presented regulators with a legacy of both quality and quantity issues which remain to be resolved.

The UK regulators (SEPA in Scotland) are also required to implement legislation associated with various EC Directives – the forthcoming Water Framework Directive is likely to have at least as great an impact on groundwater management as any of the existing Directives. The draft Framework Directive makes it clear that surface and groundwaters should be managed as an entity and that water quality and water quantity issues should be integrated. A statutory means of doing this may appear through River Basin District Management Committees. It is difficult to envisage how, in respect of groundwater at least, any realistic management strategies can be developed in the absence of a form of abstraction control n Scotland.

The advent of the new Scottish Parliament, with its devolved environmental powers, offers an opportunity for new Scottish primary legislation which would further implement EC Directives. It is not clear as to how any specific Scottish legislation would be reconciled with that of the UK (as a Member State).

The views expressed in this paper are entirely those of the author and do not necessarily represent those of SEPA or any other organization. The author acknowledges the helpful comments and suggestions received from his SEPA colleagues and others during the preparation of this paper – these are too numerous to name on an individual basis.

References

ASSOCIATION OF DIRECTORS AND RIVER INSPECTORS OF SCOTLAND. 1995. Groundwater Protection Strategy for Scotland. ADRIS, Perth.

SCOTTISH ENVIRONMENT PROTECTION AGENCY. Policy No. 19. 1997. Groundwater Protection Policy for Scotland. SEPA, Stirling.

MARSDEN, M & FOX, I. A. 1998. Implications for Scotland of the Proposed Water Framework Directive. CIWEM Scottish Annual Symposium, unpublished.

Groundwater vulnerability mapping in Scotland: modifications to classification used in England and Wales

M. A. LEWIS[1], A. LILLY[2] & J. S. BELL[2]

[1] *British Geological Survey, Wallingford, Oxon OX10 8BB, UK*

[2] *Macaulay Land Use Research Institute, Aberdeen AB15 8QH, UK*

Abstract: The recently published groundwater vulnerability map of Fife is the first in a series of maps for the Scottish Environment Protection Agency (SEPA), which includes the unpublished maps of the areas around Dumfries and Strathmore. Based on the methodology used on the Environment Agency maps of England and Wales, the lithology and permeability of the geological formations, and the physical and chemical properties of the soils are classified to produce 15 groundwater vulnerability classes. However, the Scottish maps incorporate several modifications that improve their accuracy and usefulness. These are:

(a) the geological formations are classified solely on the basis of their permeability and do not also incorporate aquifer potential;
(b) the occurrence of low permeability drift deposits at the surface are shown over the whole of the map area instead of only where they overlie aquifers;
(c) areas where borehole data indicate that significant thicknesses of low permeability deposits are present in the drift sequence are shown. In these areas groundwater in the underlying solid rock formations may have a lower risk of contamination than indicated by the vulnerability zones. This is particularly useful information where the clay occurs beneath permeable drift deposits. The borehole distribution is shown to give an indication of the reliability of the boundaries;
(d) nitrate vulnerable zones are shown;
(e) the leaching potential classification of soils with organic surface layers has been improved.

Groundwater occurs within various types of aquifer. Abstractions from these aquifers provide water for potable supplies and domestic, industrial and agricultural use. Some highly permeable aquifers are very productive and of regional importance as sources for public water supply while less permeable formations are locally important. Groundwater also provides baseflow to surface water courses. It is typically of high quality and often requires little treatment before use. However, it is vulnerable to contamination from both diffuse and point source pollutants, from direct discharges into groundwater and indirect discharges into and onto land. Aquifer remediation is difficult, prolonged and expensive, and therefore, the prevention of pollution is important (Palmer *et al.* 1995). The purpose of a vulnerability map is to show, in broad terms, the vulnerability of groundwater to contamination.

The concept of vulnerability has been widely discussed by Robins *et al.* (1994) and Foster & Skinner (1995). Vrba & Zoporozec (1994) reviewed the many methods of assessing and mapping the vulnerability of groundwater. One of the best known is the DRASTIC classification (Aller *et al.* 1987). This weighs and rates seven parameters (Depth to water table, net Recharge, Aquifer media, Soil media, Topography, Impact of vadose (unsaturated) zone and hydraulic Conductivity of aquifers), which are combined to give an overall groundwater contamination potential. This is relatively complex and the large amount of data required is only available for a few UK aquifers. Other approaches include those typified by Vereecken *et al.* (1995) and by Christiaens *et al.* (1995) which involve the linkage of soil water simulation models with solute transport models and, in the case of Christiaens *et al.* (1995), with distributed mechanistic hydrological models. These models attempt to express numerically the physical and chemical laws which govern the flow of water and solutes through the soil/plant/atmosphere continuum; they are both data and computationally intensive. The method proposed by

From: Robins, N. S. & Misstear, B. D. R. (eds) *Groundwater in the Celtic Regions: Studies in Hard Rock and Quaternary Hydrogeology*. Geological Society, London, Special Publications, **182**, 71–79. 1-86239-077-0/00/ $15.00 © The Geological Society of London 2000.

Palmer *et al.* (1995) and Palmer & Lewis (1998) is more rule-based and subjective, but it utilizes expert knowledge to interpret the large body of soil and geological information available from national surveys. It is also more applicable to assessing the relatively small local aquifers, typical of the UK.

The recently published groundwater vulnerability map of Fife and surrounding area (SEPA 1998), at a scale of 1:100 000, was commissioned by the Scottish Environment Protection Agency (SEPA) and compiled by the British Geological Survey (BGS) and the Macaulay Land Use Research Institute (MLURI). The methodology was based on that used to produce the Environment Agency's groundwater vulnerability maps for England and Wales, the original model being modified to better represent Scottish soil and aquifer conditions. Additional information on the thickness of low permeability deposits in the drift sequence was also incorporated.

The Groundwater Protection Policy for Scotland (ADRIS 1995) incorporates the concept of groundwater vulnerability based on that developed by the former National Rivers Authority (now part of the Environment Agency), Soil Survey and Land Research Centre and British Geological Survey for England and Wales (NRA 1992). The Policy includes a generalized vulnerability map at a scale of 1:625 000 for the whole of Scotland (BGS/SSLRC/MLURI 1995), defines source protection zones and contains a set of policy statements on the principal activities posing a threat to groundwater.

The Groundwater Protection Policy states that the key factors in assessing the vulnerability of groundwater to surface derived pollutants are:

(a) the presence and nature of the overlying soil cover;
(b) the presence and nature of the drift deposits;
(c) the nature and hydrogeological characteristics of the underlying strata;
(d) the thickness of the unsaturated zone.

The groundwater vulnerability assessment used on the maps incorporates all but the last of these, by combining a geological classification (based on lithology and permeability) with a soil classification (based on soil physical and chemical properties) to produce 15 groundwater vulnerability classes. The thickness of the unsaturated zone is not included in the classification, as there is only sufficient good quality data available for the main aquifers. It also changes seasonally in response to recharge. However, in fissured aquifers, which are in the majority in Scotland where the rocks are generally old and highly indurated, the thickness of the unsaturated zone does not significantly affect the time taken for recharge to reach the water table, and non-degradable pollutants are unaffected by it.

The vulnerability assessment used on the maps, classifies the physical and chemical properties of the soil and geological horizons of the unsaturated zone to assess the ease with which a pollutant released at the surface would be likely to reach the underlying groundwater body (Palmer *et al.* 1995). The maps only indicate the vulnerability of groundwater to diffuse (e.g. agricultural fertilizers, pesticides and sludge spreading) and point (e.g. chemical spillages) source pollutants applied at the ground surface. They assume that the soil is intact and has the potential to ameliorate the effect of any potential pollutant on the underlying groundwater (Palmer & Lewis 1998).

Land use is also a determinant in the risk of groundwater being contaminated by diffuse and point sources of pollution. However, land use is transitory and is not easily incorporated into a regional vulnerability assessment. Thus the maps show the potential risk onto which current land use can be overlain.

Geological classification

Geological formations have a range of hydraulic properties with flow through them being classified as fracture, intergranular or a combination of the two. The most permeable formations have the greatest capacity to transmit contaminated recharge. The groundwater vulnerability classification is therefore based on the permeability characteristics of the unsaturated zone of a formation.

In England and Wales the geological strata are classified on a combination of their aquifer potential and permeability as Major Aquifers (highly permeable), Minor Aquifers (variably permeable) and Non-Aquifers (negligibly permeable) (Palmer *et al.* 1995). This led to the karstic Carboniferous Limestone of the Mendips, south Wales and the Peak District being classified as a Major Aquifer, while that of the Pennines was classed as a Minor Aquifer, as their resource potentials are different, although their vulnerabilities are identical and, locally, extremely high. In Scotland the geological classification was based purely on permeability with the strata subdivided into:

(a) highly permeable;
(b) moderately permeable;
(c) weakly permeable.

Highly permeable formations are defined as those usually with a known or probable presence of significant fracturing. Some intergranular flow occurs in these formations, but fracture flow predominates. The aquifers may be highly productive and able to support large abstractions for public supply and other uses. Nationally they include the Permian basal breccias, the Permian and Triassic sandstones and the Upper Devonian sandstones. In the area bounded by the Fife map (SEPA 1998), part of the Carboniferous (Lower Limestone Group and Calciferous Sandstone Measures) and the Lower Devonian sandstones are also included.

Moderately permeable formations are defined as fractured or potentially fractured rocks that do not have a high primary permeability and other formations of variable permeability including unconsolidated deposits. Although these formations seldom produce large quantities of water for abstraction, they are important both for local supplies and in supplying baseflow to rivers. Nationally they include the Pliocene gravels, most of the Carboniferous sediments, the Middle Devonian and the Durness and Dalradian limestones. They also include permeable drift deposits such as blown sand, alluvium and river terrace deposits, raised beach deposits and glaciofluvial sand and gravel deposits.

Weakly permeable formations are generally of low permeability and do not normally contain groundwater in exploitable quantities. However, some groundwater flow does take place with formations often yielding water in sufficient quantities for domestic use and providing baseflow to rivers. This needs to be considered in assessing the risk associated with persistent pollutants. Examples on a national scale are the Silurian, Ordovician, Cambrian, Precambrian and intrusive igneous rocks. In the area bounded by the Fife map they also include the Lower Devonian mudstones and extrusives.

As the most productive aquifers in the United Kingdom are generally pre-Quaternary in age, the aquifer classification is applied initially to the uppermost solid formation present, whether it is at outcrop or subcrop. If this formation is classed as highly or moderately permeable, this is the classification shown on the map. However, Quaternary drift deposits overlying the consolidated rocks can be substantial in thickness. They are often variable in composition, changing from highly permeable glaciofluvial outwash gravels to low permeability clays over short distances, both laterally and vertically. Therefore, where the uppermost pre-Quaternary formation is classed as weakly permeable, if permeable Pleistocene or Recent deposits are present at the ground surface, they are shown instead (Fig. 1).

Where low permeability Quaternary deposits (such as peat, intertidal and saltmarsh deposits, lacustrine and glaciolacustrine silts and clays, and till) are present at the surface, their extent is

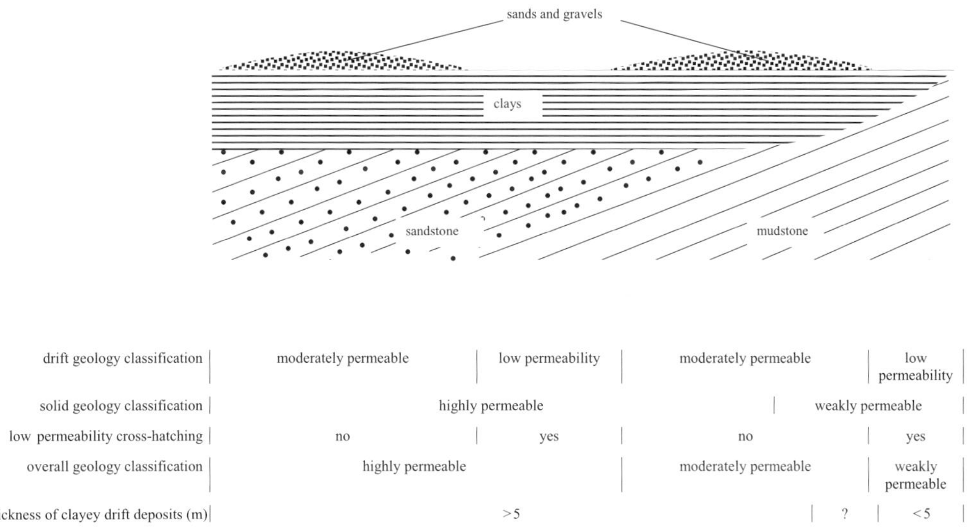

Fig. 1. Schematic cross-section indicating how the geological information is compiled onto the various layers input into the vulnerability map.

shown by cross-hatching. This is shown on the Scottish maps irrespective of the underlying geological classification. On the English and Welsh maps low permeability stipple was only shown over Major and Minor Aquifers. The Scottish vulnerability maps therefore better reflect the geological mapping from which they were compiled, and give additional information on the vulnerability of sites at the boundary between highly or moderately permeable and weakly permeable formations.

The geological classification is based solely on the permeability of the uppermost drift and the uppermost pre-Quaternary deposits, with any drift deposits not present at the ground surface ignored. Neither the overall thickness of the drift deposits nor of the underlying formations are shown. Hence on the English maps, Major Aquifers overlain by thick clays within the drift sequence and at little risk of contamination were portrayed as vulnerable, except in the north of the country where they were locally downgraded from Major Aquifers to Minor or Non Aquifers. But these adjustments to the generalized vulnerability classification were only made in areas where prior work had identified thick clay deposits as being present. In other areas where thick clays are known to be present (e.g. in Lincolnshire and East Anglia), but detailed thickness data were either unavailable or could not be interpreted within the time scale or budget of the project, aquifers were not downgraded. This was not a very satisfactory method of dealing with the low permeability deposits within the drift sequence, particularly as two different methods were used in the north-east and north-west of the country. Additionally in the areas where aquifers were downgraded, it is not possible to identify the sub-drift crop of the aquifers.

On the Scottish maps the same geological classification was used throughout, including where thick clayey deposits are present. However, the map also indicates areas where a total of more than five metres of clayey deposits have been proved or could possibly be present in the drift sequence. In these areas, groundwater in the underlying solid formations is less likely to be contaminated than indicated by the vulnerability zones. These clayey deposits may or may not be present at the ground surface; hence there is not necessarily a correlation between cross-hatched areas representing low permeability drift at the surface and areas with a significant thickness of clayey material in the drift sequence (Fig. 1). However it must be borne in mind that in areas where the moderately permeable classification represents a drift deposit, any clayey deposits present will be below this aquifer and not affect its vulnerability to surface pollution.

An example of where the contours provide additional information occurs on the Fife map (SEPA 1998), along the coast between the Tay and Eden estuaries, where the geological succession comprises blown sand, underlain by post-glacial raised beach deposits and late glacial marine clays on Devonian sandstones. The sandstones are classified as highly permeable and no low permeability cross-hatching is shown, as the clays never occur at the ground surface. However, the additional information that there are more than five metres of clayey deposits in the drift sequence, indicates that groundwater in the underlying highly permeable Devonian aquifer is afforded some considerable degree of protection. Therefore, showing where thick clayey deposits are present provides significant additional information, particularly in areas where they occur beneath permeable drift deposits.

Due to the complexity of the vulnerability linework, only the boundaries between the clay thickness areas are shown on the face of the Fife map, with the annotated version shown as an inset. The inset also indicates the borehole distribution used to derive the boundaries, hence giving an indication of their reliability. This additional analysis was only possible within the constraints of the project as the records were already digitized. The method was particularly successful in the Fife area where there are a large number of boreholes through the drift sequence (Fig. 2). In areas with less borehole data, the distribution of boreholes proving more or less than five metres of clayey deposits in the drift sequence provides additional information to the general vulnerability assessment, even where it is not possible to delineate areas with differing clay thicknesses (for example Dumfries (SEPA 1999) and Strathmore (SEPA 2000)). Overall, even where only the borehole locations are shown, this method is thought to be an improvement on the assessment used in England and Wales. Additionally the method can be used throughout the country, irrespective of the amount of borehole data available.

Only a few nitrate vulnerable zones (equivalent to the source protection zones in England and Wales) have been defined in Scotland. It is therefore possible to show their locations on the face of the map, thus bringing together source protection, site specific data and the overall aquifer vulnerability classification on one map. This was not possible on the English and Welsh maps due to the large number of source protection zones that have been defined.

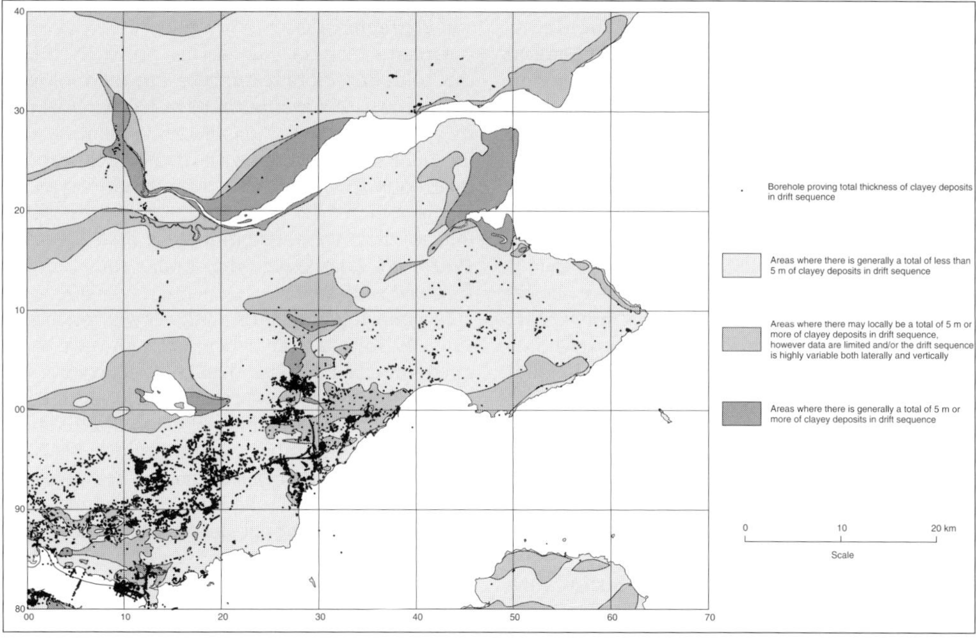

Fig. 2. Map showing thickness of clayey deposits within the drift sequence in Fife and surrounding area (after SEPA 1998).

Soil classification

Although the soil can offer a degree of protection to groundwater from contamination, through its ability to buffer potential pollutants chemically and biologically and to act as a physical barrier, these attributes vary considerably between different soil types. The chemical buffering capacity of the soil depends on the ability of the soil to bind potential contaminants which, in turn, depends on the thickness and arrangement of the soil horizons, and the amount and types of clay and organic matter. Plant and microbial metabolism can also help in the degradation process.

The rate at which a contaminant moves through the soil depends on the hydraulic conductivity of the matrix, the presence of preferential flow pathways, the moisture content and the sorption capacity of the soil. The slower this rate of movement, the greater the chance the pollutant will be degraded in the soil. Therefore, the role of the physical nature of the soil in determining the vulnerability of groundwaters to pollution cannot be over stressed. Almost all soils have some degree of structural development resulting in a wide range of pore sizes, including macropores. These can route solutes quickly to depth, bypassing the more chemically and biologically active parts of the soil. The smaller pores retain water for longer allowing chemical degradation of pollutants to occur. The rate of water movement through the soil varies seasonally giving a temporal aspect to the protection it affords (Lilly & Matthews 1994). About 80% of Scottish soils are waterlogged for some period during the year. Where this is the result of dense, slowly permeable soil horizons restricting the downward movement of water, the underlying groundwaters will be given some degree of protection, but where this is the result of fluctuating groundwater levels in permeable materials, there is an enhanced risk of contamination.

The complexity of the depositional environment during and since the last glaciation, and the influence of the various soil-forming factors over time, has resulted in a complex pattern of soil distribution throughout Scotland. This complexity is reflected in the soil maps and, as a consequence, any one aquifer is likely to be overlain by many different soil types. For the purposes of groundwater vulnerability mapping, the wide range of soils occurring in an area have been grouped into three Soil Leaching Potential classes and seven subclasses (three high, two intermediate and one low, plus an additional class where the soil type cannot be verified with any certainty). Each class is based on soil physical properties affecting the downward passage of water and on soil chemical properties affecting the attenuation of contaminants. These include: texture (in particular the amount and type of clay), structure, organic matter content, soil water regime and the presence of distinctive layers such as raw peaty topsoil and rock or gravel at shallow depth (Palmer & Lewis 1998).

In Scotland, over 50% of the land area is covered by soils with organic surface layers that equate with those described as 'raw peaty topsoils' in England and Wales (Palmer & Lewis 1998). These layers comprise humified plant remains and are amorphous, often waterlogged and poorly structured. They are found in both organic soils (peats) and in organo-mineral soils, which have a thin organic surface layer (<50 cm thick) overlying mineral material. In these latter soils the organic surface layer is described as 'peaty' (for example peaty gleys, peaty podzols and peaty rankers). Figure 3 shows the distribution of soils with these layers. As a consequence of the widespread distribution of soils with these peaty layers, the underlying subsoils exhibit a wide range of pedological conditions, which affect their ability to protect the underlying groundwater. The Soil Leaching Potential classification described by Palmer & Lewis (1998) does not adequately encompass this wide range of soil types. To ensure that all soils were successfully integrated into the existing classification, cognisance was taken of the information contained within the Hydrology of Soil Types (HOST) classification developed by Boorman, Hollis & Lilly (1995) which is a UK-wide soil hydrological classification. HOST describes the dominant pathways of water movement through both soil and substrate and takes account of the effects of seasonal waterlogging on both the flow pathways and the rate of flow.

The distinctive nature of these peaty layers provides some opportunity for the attenuation of pollutants due to the presence of negatively charged humic particles. Although these particles have a polarized charge, the cation exchange capacity is less than an equivalent volume of clay. However, the soils still have a large capacity to adsorb certain potential contaminants.

Palmer & Lewis (1998) describe only three soil types which have (or had) a waterlogged organic surface layer: upland blanket peats, undrained lowland peats and drained lowland peats. However, in Scotland there are more soil types with peaty surface layers. The subdivision of these additional soils into *Soil Leaching Potential* classes is based primarily on the subsoil characteristics.

Upland blanket peat forms primarily due to the cold, wet climate suppressing the biological breakdown of organic matter. The large capacity of these soils to adsorb potential contaminants, combined with their thickness and inherently low hydraulic conductivity (typically in the region of $6 \times 10^{-10}\,\mathrm{m\,s^{-1}}$ (Hobbs 1986)), means they were given a *Low Soil Leaching Potential (L)*, as were the similar soils found in England and Wales.

Organo-mineral soils with slowly permeable subsoils, for example, where the soils have developed on glacial lodgement till or have a thin but continuous iron-pan, are also deemed to have a *Low Soil Leaching Potential (L)*. This is due to the adsorptive capacity of the peat and the presence of a physical barrier to downward water movement. However, seasonal waterlogging within the upper layers may mean that contaminants move more rapidly in a lateral direction.

Soils with a thin (<50 cm thick) organic peaty layer which directly overlies solid rock (peaty rankers), have only a limited attenuation capacity and have been placed in the *Intermediate Soil Leaching Potential (I2)* class. These soils can transmit non- or weakly-adsorbed diffuse contaminants and liquid discharges due to their shallow nature, but the high levels of organic matter mean that they are unlikely to transmit adsorbed contaminants.

In low-lying areas the underlying subsoils of the organo-mineral soils are often wet and, in many cases, are likely to be in hydraulic continuity with underlying groundwater (for example, peaty gleys). The soils which have formed in highly permeable, gravelly drifts are often considerably drier in the upper layers but have very limited attenuation capacities and potential pollutants could move rapidly down to deeper groundwater, once they have penetrated the peaty topsoil. Both of these types of organo-mineral soils are deemed to have a *High Soil Leaching Potential (H1)*.

In peat deposits that have developed in shallow, waterlogged, semi-confined topographic depressions and valleys allowing limited surface outflow, the presence of a fluctuating shallow groundwater table affects the peat soils. Although often thick and, therefore having some attenuation capacity, these soils are also considered to have a *High Soil Leaching Potential (H1)* and equate with the undrained lowland peat soils described by Palmer & Lewis (1998).

Raised mosses up to 8 metres thick in north-east Scotland (Glentworth & Muir 1963) have

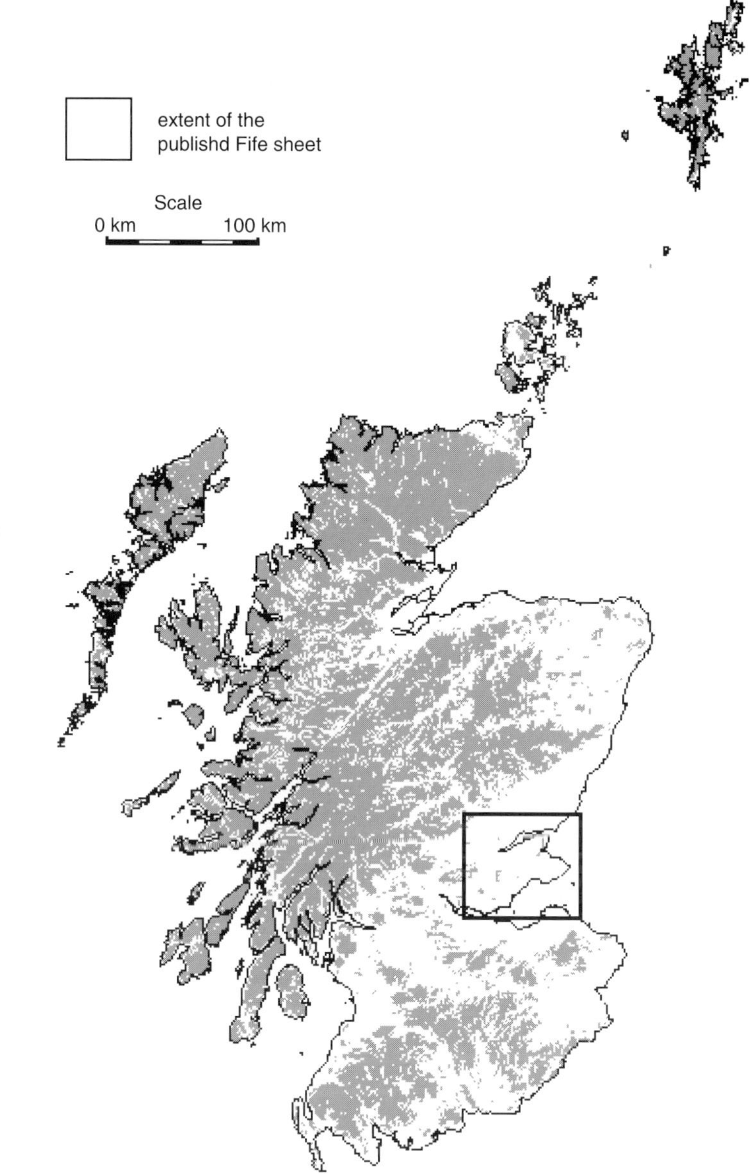

Fig. 3. Distribution of organic and organo-mineral soils in Scotland.

formed in confined depressions. Although many of these sites may have begun as shallow lakes which were gradually infilled, the thickness of organic matter and lack of hydraulic continuity with groundwater means that these specific forms of lowland peat deposits have been classed as having a *Low Soil Leaching Potential (L)*. This represents a change from the classification applied in England and Wales. In Scotland, the climate mainly precludes the use of these mosses for any agricultural activity apart from grazing livestock; however, peat extraction does occur and where the deposit becomes shallow or cut-over, there is a greater risk that a potential contaminant can penetrate to the underlying, but confined, water table.

The revised Soil Leaching Potential Classification used on the Scottish vulnerability maps and incorporating the above modifications for soils with peaty topsoils is shown in Table 1.

Table 1. *Soil Leaching Potential Classification*

High Soil Leaching Potential (H): soils with little ability to attenuate diffuse source contaminants. Non-adsorbed diffuse contaminants and liquid discharges have the potential to move rapidly to underlying strata or to shallow groundwater. Four subclasses are recognized:

H1 Soils that readily transmit liquid discharges because they are either shallow or susceptible to rapid by-pass flow directly to rock, gravel or groundwater.

H2 Soils with a low attenuation capacity due to low clay and/or organic matter contents, and with the ability to drain rapidly but have limited potential for by-pass flow.

H3 Soils with a moderate adsorption capacity due to the presence of organic matter and/or clay. Non-adsorbed contaminants and liquid discharges can be readily transmitted as these soils overlie rock or gravel at relatively shallow depths.

HU Soils over current and restored mineral workings and in urban areas are often disturbed or absent and the interpretation is based on fewer observations than elsewhere. A worst case vulnerability classification (equivalent to H1) is therefore assumed for these areas, until proved otherwise.

Intermediate Soil Leaching Potential (I): soils with a moderate ability to attenuate diffuse source contaminants but it is possible that some non-adsorbed diffuse-source contaminants and liquid discharges could penetrate the soil layer. Two subclasses are recognized:

I1 Soils with a moderate ability to attenuate a wide range of potential contaminants due to their thickness, moderate levels of both clay and organic matter. These soils have only a limited potential for by-pass flow.

I2 Soils that can possibly transmit non- or weakly-adsorbed diffuse contaminants and liquid discharges, but are unlikely to transmit adsorbed contaminants. These soils have a high topsoil organic matter content but relatively porous subsoils. In some cases the soils are shallow.

Low Soil Leaching Potential (L): soils in which contaminants are unlikely to penetrate the soil layer due to both the presence of a slowly permeable horizon and the ability of the soil to attenuate contaminants. Water and contaminant movement is, therefore, largely horizontal and the lateral flow from these soils may contribute to groundwater recharge elsewhere in the catchment.

Groundwater vulnerability mapping

The seven-fold Soil Leaching Potential Classification was applied to all soils above highly permeable and moderately permeable formations. The classification was not applied to weakly permeable formations as they contain little groundwater. This gave a total of fifteen groundwater vulnerability classes. The national map (BGS/SSLRC/MLURI 1995) only shows the Soil Leaching Potential Classification of soils overlying highly permeable formations. The classification was applied regardless of whether low permeability drift deposits (denoted on the map by cross hatching) are present at the surface or not, as the lithology of the whole drift sequence is not always known. Where the drift is thin and permeable, the soil classification assumes greater significance. Where thick, low permeability deposits are present at depth within the drift sequence, but are not the soil parent material, the soil classification becomes less significant and groundwater should be less vulnerable to contamination than shown by the vulnerability map.

The extent and geographical distribution of the fifteen groundwater vulnerability classes combined with whether low permeability deposits are present at the surface and/or are of a significant thickness, provide a broad assessment of groundwater vulnerability which is important for the protection and management of aquifers. The approach and classifications used to produce the groundwater vulnerability maps can also be used to assess specific land use practices, proposed developments and land use changes over aquifers, where these could affect groundwater quality. The maps are designed for use by planners, developers, consultants and regulators to provide guidance on possible constraints on potential development. More detailed site specific assessments of vulnerability will be required where development may have an impact on groundwater quality.

The Scottish vulnerability maps, similar to those in England and Wales, are a compromise between the representation of natural complexity and the simplified interpretation required at the given scale. This places limitations on the resolution and precision of generalized map information. Where the soil and/or underlying formations have been disturbed or removed, for example during mineral extraction, the vulnerability class could change and site specific data must be used to determine the vulnerability of the groundwater.

Conclusions

The methodology developed for Scotland to produce groundwater vulnerability maps builds on and improves the methodology developed for, and used to produce, similar maps in England and Wales by portraying additional data on the thickness of the overlying low permeability drift deposits and by extending the Soil Leaching Potential Classification. The maps do not include depth to water, which, together with a full assessment of the permeability and lithology of the drift sequence, would be additional useful features that in the future could be incorporated into large scale groundwater vulnerability maps. However, at present there are insufficient data for these to be included in any national classification, as they are only available for a few small areas where detailed investigations have been carried out.

There are significant differences in the proportions of soil types found in Scotland compared with those of England and Wales. Hence the Soil Leaching Potential Classification described by Palmer & Lewis (1998) did not take account of some of the soil types which are of significant extent in Scotland, but are of limited significance in England. Thus the leaching potential of soils with 'raw' peaty layers was developed further.

The concepts presented in this paper were developed for the groundwater vulnerability maps described, made under contract to the Scottish Environment Protection Agency. They are based on those used originally by the Environment Agency in England and Wales. Part of the work was funded by the Scottish Executive Rural Affairs Department. The paper is published with the permission of the Director of the British Geological Survey (NERC)

References

ADRIS. 1995. *Groundwater Protection Strategy for Scotland.* Association of Directors and River Inspectors of Scotland, Perth.

ALLER, L., BENNETT, T., LEHR, J. H., PETTY, R. J. & HACKETT, G. 1987. *DRASTIC: A standardized system for evaluating groundwater pollution potential using hydrogeologic settings.* US Environment Protection Agency, Ada, Oklahoma, **EPA/600/2-87/035.**

BGS/SSLRC/MLURI. 1995. Groundwater vulnerability Map of Scotland. 1:625 000. British Geological Survey/Soil Survey and Land Research Centre/Macaulay Land Use Research Instiituite, Edinburgh.

BOORMAN, D. B., HOLLIS, J. M. & LILLY, A. 1995. Hydrology of soil types: a hydrologically-based classification of the soils of the United Kingdom. Institute of Hydrology Report **No. 126**.

CHRISTIAENS, K., XEVI, E., MALLANTS, D. & FEYEN, J. 1995. Basin-scale applications of the integrated WAVE-MIKE SHE model. *In: Water and matter transport in soils at various scales.* ILWM, Katholieke Universiteit Leuven, Workshop Proceedings.

FOSTER, S. S. D. & SKINNER, A. C. 1995. Groundwater protection: the science and practice of land surface zoning. *IAHS Conference on Groundwater Quality: Remediation and Protection*, Prague, Czech Republic, 15–18 May 1995.

GLENTWORTH, R. & MUIR, J. W. 1963. *The soils of the country round Aberdeen, Inverurie and Fraserburgh. (Sheets 77, 76, 87 and 97). Soil Memoir. Soil Survey of Scotland.* HMSO, Edinburgh.

HOBBS, N. B. 1986. Mire morphology and the properties and behaviour of some British and foreign peats. *Quarterly Journal of Engineering Geology*, **19**, 7–80.

LILLY, A. & MATTHEWS, K. B. 1994. A Soil Wetness Class map for Scotland: new assessments of soil and climate data for land evaluation. *Geoforum*, **25**, 371–379.

NRA. 1992. *Policy and Practice for the Protection of Groundwater.* National Rivers Authority, Bristol.

PALMER, R. C., HOLMAN, I. P., ROBINS, N. S. & LEWIS, M. A. 1995. Guide to groundwater vulnerability mapping in England and Wales. National Rivers Authority, HMSO, London

—— & LEWIS, M. A. 1998. Assessment of groundwater vulnerability in England and Wales. *In:* ROBINS, N. S. (ed.) *Groundwater Pollution, aquifer recharge and vulnerability.* Geological Society, London, Special Publications, **130**, 191–198.

ROBINS, N. S., ADAMS, B. FOSTER, S. S. D. & PALMER, R. C. 1994. Groundwater vulnerability mapping: the British perspective. *Hydrogéologie*, **3**, 35–42.

SEPA. 1998. Groundwater vulnerability of Fife and surrounding area, 1:100,000. Scottish Environment Protection Agency, Perth.

——1999. Groundwater vulnerability of Dumfries and surrounding area, 1:100,000. Scottish Environment Protection Agency (unpublished map), Perth.

——2000. Groundwater vulnerability of Strathmore and surrounding area, 1:100,000. Scottish Environment Protection Agency (unpublished map), Perth.

VEREECKEN, H., DÖRING, U., KIM, D. J., FEYEN, J., MOUVET, C., MOREAU, C., BURAUEL, P. & HÜCKER, K. 1995. The fate and mobility of pesticides at laboratory, lysimeter and field scale: A European project. BCPC Monograph 62: *Pesticide movement to water.* 147–154.

VRBA, J. & ZOPOROZEC, A. (eds) 1994. Guidebook on Mapping Groundwater Vulnerability, *International Contributions to Hydrogeology*, **16**, International Association of Hydrogeologists.

Interdependence of groundwater and surface water in lowland karst areas of western Ireland: management issues arising from water and contaminant transfers

CATHERINE COXON[1] & DAVID DREW [2]

[1] *Department of Geology, Trinity College Dublin, Ireland (e-mail: cecoxon@tcd.ie)*
[2] *Department of Geography, Trinity College Dublin, Ireland*

Abstract: In lowland karsts, both surface and groundwater systems are often present. This is the case over large areas of limestone in the west of Ireland where gaining and losing streams and seasonal lakes (turloughs) are common and where much of the surface river system consists of artificial channels. A case study from County Clare illustrates the problems involved in delineating catchment areas for springs which have partially contributing surface stream sources. Case studies from counties Galway and Clare, with complex surface water–groundwater interactions, exemplify problems that may arise with water quality and quantity. Difficulties in defining realistic protection areas for groundwater resources and sources are discussed.

The interaction between surface water and groundwater is a fundamental component of the hydrological cycle. The fact that groundwater discharge is the main component of river base-flow, and the differences in the storage and transmission of surface water and groundwater, have encouraged the establishment of conjunctive management schemes in many suitable environments, as documented in Misstear (1999), Winter *et al.* (1998) and Rippon & Wyness (1994), for example.

In karstic environments, the exchange of water between the surface and subsurface systems is commonly very pronounced. Two-way movement, related to varying head conditions, may occur via seasonally losing or gaining streams or via estavelles (features which can function as sinks or springs depending on head conditions). The surface–subsurface interaction is less significant on upland plateau karsts where surface water is a rarity, temporally and spatially, but of great importance in lowland karsts where rivers and lakes may co-exist with a subterranean drainage system. Such areas include much of central Florida, karst areas of Puerto Rico, Cuba and the Yucatan Peninsula in Mexico, and extensive karsts in China, Vietnam and Malaysia.

In Ireland approximately 50% of the country is underlain by karstified or partly karstified limestone of Carboniferous age, much the greater part of which is lowland. The limestone is also the main groundwater source and resource in the country. Particularly in the most highly karstified western part of the limestone lowland, the existence of an efficient underground karst drainage system together with a surface drainage system leads to complex interactions between the two. The hydrological interactions result in a close interrelationship between groundwater and surface water quality, with surface water contaminants entering the aquifer via losing streams or swallow holes, and with contaminated groundwater passing rapidly through the aquifer by conduit flow and re-emerging to contaminate surface waters. In this paper these interactions and the implications for effective water management are illustrated using examples from the lowland karstified limestone aquifers of western Ireland.

Geology of the study area

The region and specific study areas described in this paper (Fig. 1) consist of a lowland (5–50 m above sea level) covering approximately 6000 km^2 in counties Galway, Clare and Mayo. The bedrock is Lower Carboniferous Limestone. Over much of the area, it corresponds to the Burren Formation: a pure bioclastic calcarenite of Asbian age which lacks primary permeability, but is well-bedded and jointed so has extensive secondary permeability and is susceptible to

From: ROBINS, N. S. & MISSTEAR, B. D. R. (eds) *Groundwater in the Celtic Regions: Studies in Hard Rock and Quaternary Hydrogeology*. Geological Society, London, Special Publications, **182**, 81–88. 1-86239-077-0/00/ $15.00

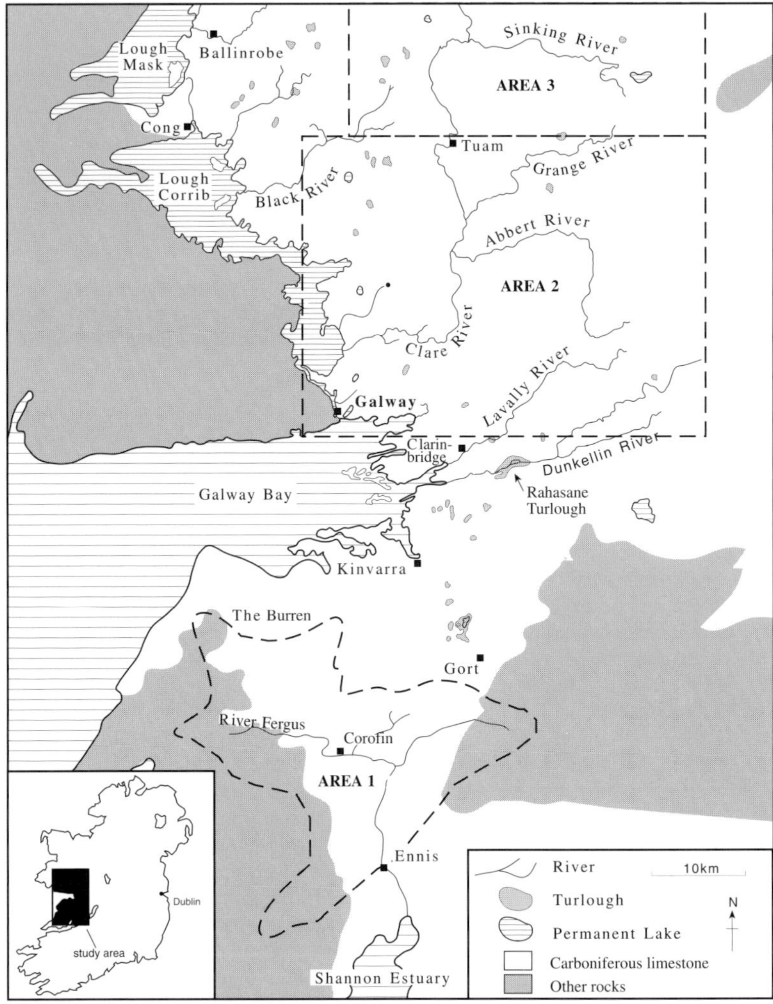

Fig. 1. The location of the study areas.

karstification. Impure limestones of lower permeability are also found, particularly to the east of the area in Fig. 1. The limestone plain narrows to the south, where it is bordered by Upper Carboniferous shales and sandstones to the west and the Devonian sandstone of the Slieve Aughty mountains to the east.

The bedrock is overlain by Quaternary deposits of very variable composition, including glacial till, fluvioglacial sand and gravel and areas of post-glacial basin peat and marl. The thickness of the deposits is highly variable, although generally it extends to several metres in the eastern part of the lowland, thinning to less than three metres in the centre and west of the area. Here, the Quaternary cover becomes more patchy and in places the bedrock is directly overlain by rendzina soil (Drew & Daly 1993).

Hydrological issues

Interchanges of water

The relationship between surface and groundwater hydrology is influenced by variations in the bedrock and Quaternary geology. Over much of the area shown in Fig. 1 where the pure limestone coincides with a thin Quaternary cover, the two hydrologic systems are completely intertwined. A characteristic feature of the lowlands, the turlough, illustrates this interlinkage. Turloughs are seasonal lakes which generally fill via springs or estavelles (sometimes in combination with surface inflows) in a matter of days each autumn, and then empty to groundwater, sometimes via diffuse seepage but more usually via discrete swallow holes or estavelles, over several weeks in the spring or early summer. They are generally associated with lines of higher permeability in the karst aquifer, with turlough water passing to springs at rates of the order of 100 metres per hour (Coxon & Drew 1986).

The natural interrelationship between surface and groundwater systems has been altered in much of the lowland by arterial drainage schemes carried out since the mid-nineteenth century. This has not merely consisted of the modification of existing river systems, but in many instances has involved the linking of lakes and turloughs to create wholly new river systems (Coxon & Drew 1986). However, under low stage conditions groundwater flows revert to what was presumably their natural condition prior to drainage and the artificial channels cease to link to groundwater and become dry.

Area 1, the Fergus catchment, is described at greater length elsewhere (Coxon 1995; Coxon & Drew 1998). The tortuous river course southwards across the lowland towards Ennis (Fig. 2) includes some artificial stretches created by drainage engineers in the nineteenth century. However, a significant proportion of the river flow does not follow the whole of this aboveground course, but takes a series of underground short-cuts from swallow holes in the river bed and lake shores to springs which feed the river further down-catchment (Fig. 2b). Water tracing experiments using optical brightener have demonstrated that underground velocities along these very localised routes vary from c. 20 m/hr along west–east routes to c. 100–200 m/hr along north–south routes. These rapid velocities indicate a high degree of karstification, particularly given the low hydraulic gradients (of the order of 1 m/km).

Further north, water tracing experiments using sodium fluorescein and optical brightener have shown that groundwater flow from east to west cuts across surface catchment boundaries (Drew 1992; Drew & Daly 1993). In Area 2 (Fig. 3), water passes underground from the Abbert river (a tributary of the Clare river) out of the Clare surface catchment to springs at Aucloggeen which flow directly into Lough Corrib. In Area 3 to the north (shown in Fig. 4), not only are there underground linkages within the Clare river catchment (with water passing from a losing reach of the Sinking river catchment to a spring feeding the main Clare river), but there is also an interbasin transfer of water to the head-waters of the Black river

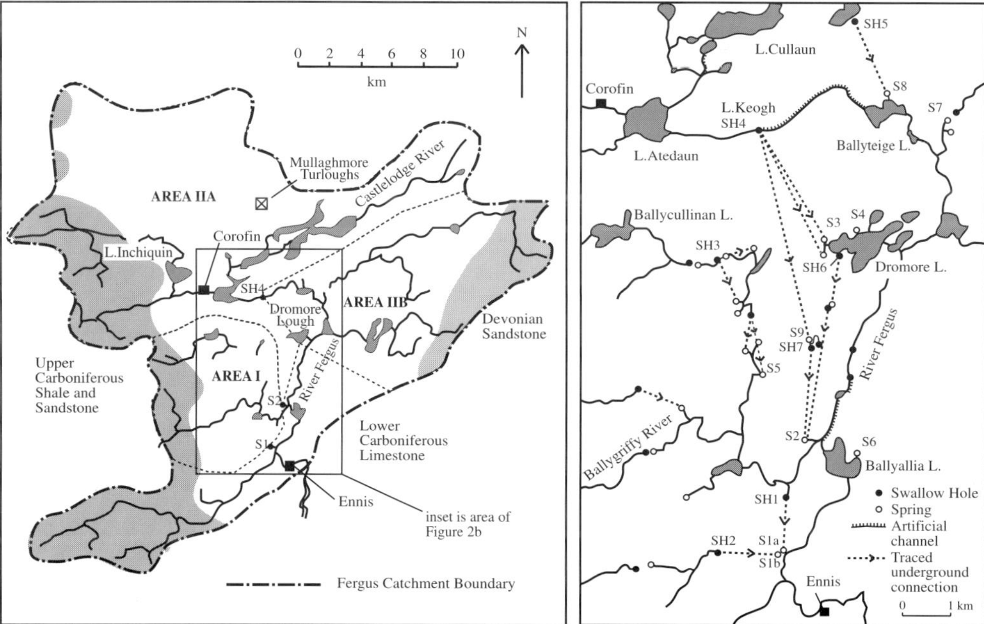

Fig. 2. (**a**) The Fergus catchment (Area 1 on Fig. 1), (**b**) Springs, swallow holes and underground connections.

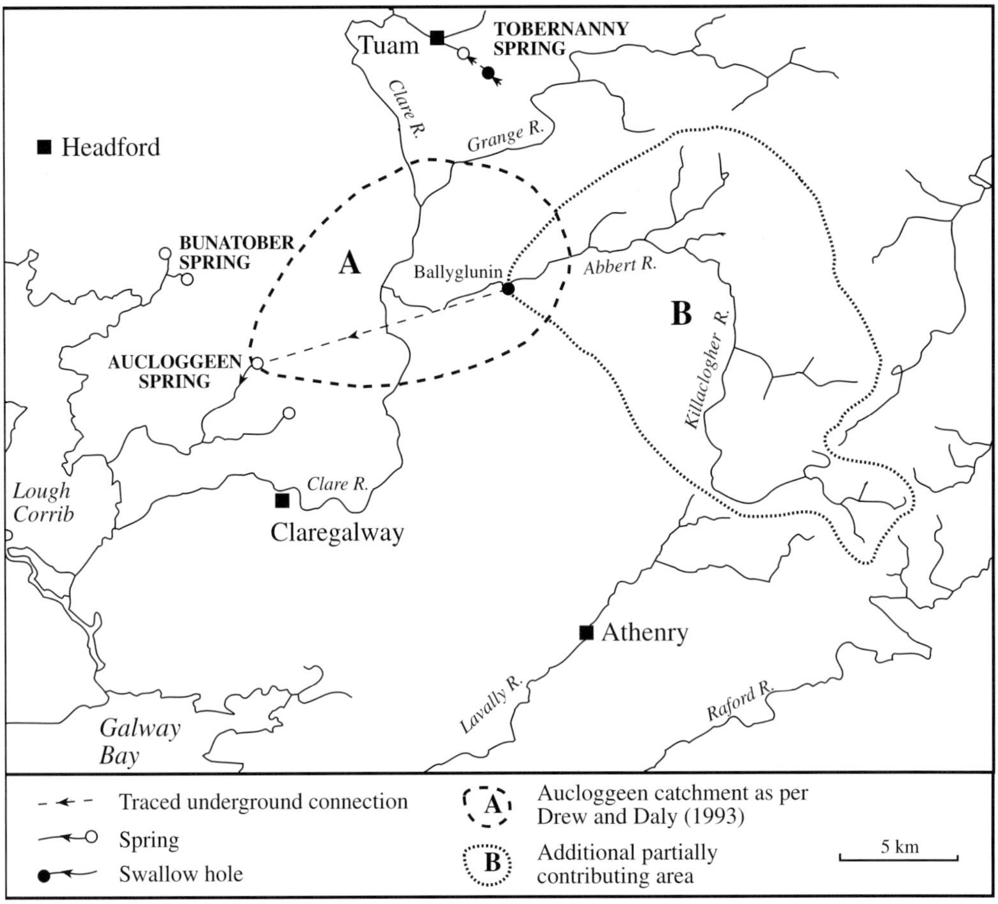

Fig. 3. Aucloggeen spring catchment (Area 2 on Fig. 1).

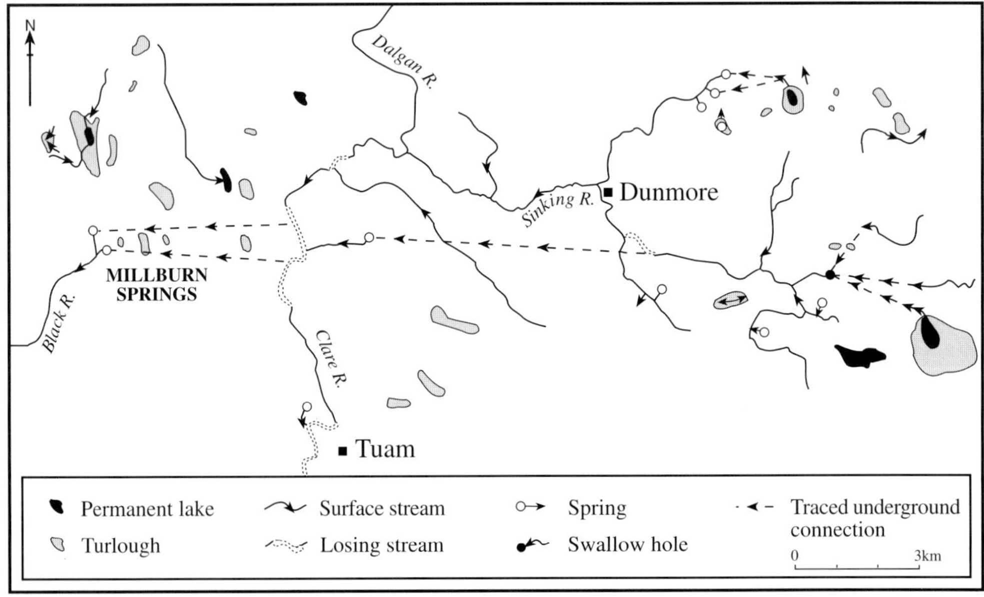

Fig. 4. Part of Millburn springs catchment (Area 3 on Fig. 1).

catchment. Again, these underground transfers are by rapid flow in karst conduits or solution-widened fissures; for example, water sinking at Ballyglunin passes to Aucloggeen springs at a velocity of 200 m/hr.

Difficulties in catchment delimitation

The close interconnection of surface water and groundwater means that lake and river catchments in this area cannot be defined solely on the basis of surface topographic divides, but must take groundwater inputs and outputs into consideration. Equally, the catchment areas of springs cannot be defined in conventional hydrogeological terms. The surface catchments of losing and entirely sinking rivers which contribute to spring flow must be taken into account. Karst groundwater flow may transfer water between surface catchments: a spring in one river catchment may receive water from a sinking stream in another catchment. Also, the catchment upstream of a swallow hole, or losing reach, may only contribute part of its flow to the spring, and the proportion of water passing to the spring may vary at different times of year and different water levels. This problem is illustrated below with reference to a number of case examples.

Drumcliff springs. Water tracing experiments using optical brightener, described in Coxon (1995), enabled the catchment boundary of Drumcliff springs (S1 in Fig. 2a) to be defined with a reasonable degree of certainty. It covers an area of approximately 60 km^2 (Area I in Fig. 2a). However, the area does not always drain solely to Drumcliff springs; at times of higher water levels, not all of the Ballygriffy river water sinks at SH1 (Fig. 2b) but some passes overground to the Fergus river. Also, at times of high water levels, some of the water sinking at SH2 (Fig. 2b) passes to a small ephemeral spring located 250 m to the south of Drumcliff springs. Thus Area I does not contribute all of its water to Drumcliff springs at all times.

Pouladower spring. This issue of catchment areas contributing only part of their flow to a spring is even more significant in the case of Pouladower spring (S2 in Fig. 2a). This spring is fed by Fergus river water sinking at SH4, 7 km to the north, which in turn is fed both by the Castlelodge river draining the lowland to the north, and by a large part of the Burren plateau to the north and north-west of Corofin (Fig. 2a). Thus the catchment area of Pouladower spring is of the order of 250 km^2 (Area IIA in Fig. 2a), yet the spring accounts for only part of the flow from this area, the remainder passing along an artificial surface channel to Ballyteige Lough (Fig. 2b). Furthermore, the channel between the Fergus river and Dromore Lough takes flow in either direction depending on relative water levels, and on occasions when water flows from the Fergus river into Dromore Lough, this means that a further 130 km^2 (Area IIB in Fig. 2a) could contribute part of its flow to Pouladower via the swallow holes in Dromore Lough.

Aucloggeen spring. The catchment of Aucloggeen spring as delimited in Drew & Daly (1993) is shown as Area A in Fig. 3. This catchment boundary was based on a combination of tracing test results and a calculation of the recharge area required to generate the spring flow. However, the fact that Abbert river water leaking into Ballyglunin Cave passes to the springs means that the Abbert river catchment upstream of Ballyglunin (Area B in Fig. 3) can also be regarded as part of the Aucloggeen catchment. However, again this is a partially contributing area only; a varying proportion of the flow of the Abbert river continues on its surface course to join the Clare river.

Millburn springs. The springs at Millburn form the headwaters of the Black river, which flows south-west to Lough Corrib (Fig. 1). Water sinking from the Clare river has been traced to these springs (Fig. 4). Thus the spring catchment includes a very extensive partially contributing area upstream of this point. Part of this area is shown in Fig. 4: water from the Sinking river reaches this losing zone of the Clare river by both surface and underground routes. In addition, the catchment extends northwards to encompass the catchment of the Dalgan river, which joins with the Sinking river to form the Clare river. In total, approximately 400 km^2 drains to the losing reach feeding Millburn springs. Flow measurements along the reach have shown that a varying amount of the river water sinks underground; in the summer, the river loses approximately 30% of its flow in this reach.

Karst water resource management issues

The lack of a clear distinction between surface and groundwater hydrologic systems in these western Irish karst lowlands has implications for water resource management. Not only does the water freely exchange between surface and underground, but the associated contaminants are also transferred between the two systems. Surface water contaminants may enter the aquifer readily via sinking streams, while the rapid transfer through the aquifer by conduit flow and the consequent lack of attenuation means that groundwater contaminants may in turn be discharged into surface waters. Thus the management of groundwater resources and groundwater supplies must take account of the close interaction with surface water, and equally the management of surface water bodies must take account of the fact that they may be largely, or entirely, groundwater-fed.

Surface water management: lakes and turloughs

Over much of the western Irish limestone lowlands, surface water bodies are fed largely, or entirely, by groundwater. As a result, both water levels and quality can only be managed on the basis of a proper understanding of the groundwater hydrology and quality.

The importance of groundwater hydrology to management of surface waters is most critical in the case of turloughs, given that these seasonal water bodies have no natural surface outlets and drain entirely to groundwater, and the vast majority also fill from groundwater. Turloughs are of considerable ecological importance, due to their unique flood-tolerant grassland communities, their distinctive freshwater invertebrate fauna and their role as winter feeding areas for ducks, geese and waders. They have been identified as priority habitats under the E.C. Habitats Directive (92/43/EEC) and 43 turloughs have been proposed as Special Areas of Conservation under this directive. The greatest threat to turloughs is posed by drainage schemes; as mentioned above, schemes carried out from the mid-nineteenth century onwards have frequently involved the linking of turloughs into an artificial surface drainage network. In a national survey of turloughs (Coxon 1987), it was found that a third of sites no longer flood to any significant degree due to such drainage schemes. The most extensive remaining area of naturally functioning turloughs is the Gort–Kinvarra area (Fig. 1). However, this area suffered extensive flooding during the winter of 1994/95, beyond the normal turlough flooding limits, inundating agricultural land, dwellings and roads, and resulting in local pressure for a drainage scheme. A detailed hydrological study of the area was carried out (Southern Water Global 1998), but it has not yet proved possible to resolve the conflict between flood prevention demands and the maintenance of turloughs of international ecological importance.

The water chemistry of lakes and turloughs is highly dependent on the surface water-groundwater relationships. In the Gort–Kinvarra area (Fig. 1), lakes fed in whole, or in part, by water originating on non-limestone bedrock, either by surface input or via karst conduits, have inflows that are sediment rich, acidic and variable in temperature in comparison with lakes fed by shallow epikarst water or karst springs (Southern Water Global 1998). The trophic status of groundwater-fed lakes and turloughs is controlled by the nutrient content of the inflowing groundwater. Phosphorus concentrations in groundwater in this region approach and sometimes exceed the threshold at which eutrophication is triggered (Kilroy *et al.* 1999). Nutrient inputs to lakes in this region, which constitute an important tourism and fishery resource, have given rise to concern, with some lakes already suffering periodic algal blooms. The issue of potential nutrient enrichment has also arisen in relation to turloughs in the Mullaghmore area, within the Burren National Park (Fig. 1). A visitor centre was proposed for this locality, and detailed studies of hydrology and water chemistry were undertaken because of concerns that nutrients released into groundwater from the proposed effluent treatment plant might enter the turloughs and change their trophic status. This issue has not yet been resolved, but if any effluent discharges are permitted at the site, they are likely to be severely constrained and closely monitored.

Dealing with rapidly fluctuating quality in groundwater supplies

The short underground residence times and lack of filtration associated with karst aquifers means that groundwater supplies in the area, particularly springs fed by point recharge from swallow holes, are very prone to pulses of contamination. Even where the point recharge forms a small proportion of the spring flow and so is relatively unimportant in quantitative terms, it may have a detrimental influence on the spring water quality. The rapid passage of contaminants through the conduit flow system means that such supplies may have more in common with surface water abstractions than with conventional groundwater sources, in terms of management strategies.

Irish groundwater source protection protocols (DoELG *et al.* 1999) involve the designation of an inner source protection area bounded by a 100-day travel time within the saturated zone. However, this is not feasible at many water abstraction sites across the western limestone lowlands, where rapid travel times along preferential flow routes through the aquifer are the norm. For example, Aucloggeen springs (Fig. 3), which supply water to several hundreds of people in an area of scattered rural population, are reached by water from the Abbert river within 48 hours, while at Drumcliff springs, which provide the water supply for the town of Ennis and its environs, there is a time-lag of only 7–9 hours between swallow hole SH1 and the springs (Fig. 2b). Short underground residence times mean that very little time is available for remedial action to avoid contamination of drinking water supplies, and an early warning system to enable shutdown of the supply may be required.

The water quality problems that arise are often of natural rather than anthropogenic origin. For example, in a number of springs and boreholes in the lower Fergus catchment, the breaching of drinking water standards for iron, turbidity and colour appears to be due to high iron concentrations in swallow hole recharge waters derived from the shale and overlying blanket peat to the east of the area. Tobernanny spring, which provides the supply to the town of Tuam (Fig. 3), has similar problems. It suffers from severe discoloration and iron concentrations exceeding 5 mg/l under high water conditions (when peaty water sinking in a swallow hole contributes *c.* 50% of spring outflow), but it is of potable quality under medium to low flows (when the swallow hole water contribution drops to 1% of spring outflow). In some supplies, such problems have been resolved by water treatment, while in the case of the Tuam supply, water quality management has involved separating the polluted sources from the clean water, and also ceasing to use the Tobernanny source entirely when outflows exceed the threshold above which the contaminated source becomes a majority of the total spring flow (Drew 1992)

Suspended sediment, which is more usually an issue in surface waters, can give rise to problems in karst groundwater discharges. For example, under very high flow conditions Bunatober spring in County Galway (Fig. 3) contains high concentrations of suspended silt, which appears to originate in infilled karst depressions and caverns 5–8 km to the north. The spring supplies a salmon hatchery, and on one occasion a suspended sediment concentration of 5 mg/l caused a high fish mortality. Management of this problem has involved the installation of an early warning system, based on continuous monitoring of rainfall and groundwater levels (Drew 1992).

The greatest anthropogenic contamination problem in the western limestone lowlands is undoubtedly faecal microbial contamination (Aldwell *et al.* 1988). This problem is frequently associated with inadequate means of subsurface effluent disposal, but it is aggravated in karst areas by sinking contaminated surface waters which pass freely through the aquifer along conduits and solution-widened fissures. For example, in the lower Fergus catchment, bacterial quality of untreated groundwaters is generally unsatisfactory, and problems are particularly acute in the springs (Coxon & Drew 1998). While this problem can be tackled to some degree with proper groundwater protection planning, in such a vulnerable situation it will be necessary to combine catchment protection measures with treatment of the water supply to remove faecal microorganisms. With increased awareness of the problem in recent years, such treatment has become the norm in the case of public water supplies, but the numerous private domestic and farm groundwater supplies in the region do not generally undergo any treatment.

Dealing with partially contributing areas in groundwater source protection plans

The outer source protection area designated in the standard Irish groundwater source protection protocol (DoELG *et al.* 1999) has an outer limit corresponding to the boundary of the catchment or contributing area. Planning controls are implemented within this area with varying degrees of severity depending on zones of varying intrinsic vulnerability. However, delimiting the boundary of karstic spring catchments poses problems in many instances, as seen from the foregoing discussion of partially contributing areas with reference to four case examples. In the case of Aucloggeen springs, the partially contributing catchment of the Abbert river, i.e. an area of *c.* 160 km^2 (Area B in Fig. 3) needs to be taken into account in addition to the direct contributing area (Area A in Fig. 3). At Pouladower spring, an even more extensive area of *c.* 380 km^2 (Areas IIA and IIB in Fig. 2a) could potentially impact on the springs, although the spring flow accounts for only a small proportion of total recharge in this area.

It is unrealistic in planning terms to provide an equally high degree of protection across the whole of these extensive catchments, and it may not be scientifically justifiable. The upper reaches of the partially contributing catchments pose less of a threat to the spring than areas of similar intrinsic vulnerability close to the spring, not only because they are further away but also because only a proportion of their recharge passes to the spring. Nevertheless, they cannot be ignored in the source protection plan, given the potential for pollution arising there to reach the spring. A modified protocol has recently been developed for the Pouladower spring catchment, taking such issues on board (Deakin this volume).

Conclusions

Many lowland karst areas have a high degree of interconnection and interaction between surface and underground waters with the magnitude and direction of water exchanges varying over time. Human alterations to surface water systems commonly provoke immediate changes in the groundwater system, both in terms of water quality and quantity and vice-versa. The existence of partial contributing areas to springs from losing and sinking streams further complicates the situation. Although site specificity is the norm in karst systems, the case studies presented in this paper from western Ireland do exemplify some of the problems involved in developing a strategy for the management of water resources, and in particular for the determination of protection areas for water sources in a karstic environment.

The research work in the Fergus catchment was funded by Forbairt, and technical assistance was provided by J. Plettenberg, C. Dalton, Ennis Urban District Council and the Office of Public Works. The provision of information and access by local landowners is also gratefully acknowledged. The diagrams were drafted by K. Blackmore and S. McMorrow, Geography Department, Trinity College Dublin.

References

ALDWELL, C. R., THORN, R. & DALY, D. 1988. Point source pollution in karst areas in Ireland. *In*: DAOXIAN, Y. (ed.) *Proceedings of the 21st IAH Congress*. Geological Publishing House, Beijing, 1046–1052.

COXON, C. 1987. An examination of the characteristics of turloughs, using multivariate statistical techniques. *Irish Geography*, **20**, 24–42.

——1995. Groundwater vulnerability and protection issues in the lower Fergus catchment, Co. Clare. *In*: *Hydrogeological Aspects of Groundwater Protection in Karstic Areas (COST Action 65 Final Report)*. European Commission D-G XII, Luxembourg, 162–169.

—— & DREW, D. P. 1986. Groundwater flow in the lowland limestone aquifer of eastern Co. Galway and eastern Co. Mayo, western Ireland. *In*: PATERSON, K. & SWEETING, M. (eds) *New Directions in Karst*. GeoBooks, Norwich, 259–280.

—— & ——1998. Interaction of surface water and groundwater in Irish karst areas: implications for water-resource management. *In*: BRAHANA, J. V. *et al.* (eds) *Gambling with Groundwater – Physical, Chemical and Biological Aspects of Aquifer–Stream Relations*. American Institute of Hydrology, St. Paul, Minnesota, 161–168.

DEAKIN, J. 2000. Groundwater protection zone delineation at a large karst spring in western Ireland. *This volume*.

DEPARTMENT OF THE ENVIRONMENT AND LOCAL GOVERNMENT, ENVIRONMENTAL PROTECTION AGENCY & GEOLOGICAL SURVEY OF IRELAND. 1999. *Groundwater Protection Schemes*. Department of the Environment and Local Government, Environmental Protection Agency & Geological Survey of Ireland, Dublin.

DREW, D. P. 1992. Water resource management and water tracing in a lowland karst aquifer, western Ireland. *In*: HOETZL, H. & WERNER, A. (eds) *Tracer Hydrology*. Balkema, Rotterdam, 221–225.

—— & DALY, D. 1993. *Groundwater and karstification in mid-Galway, south Mayo and north Clare*. Geological Survey of Ireland Report Series **RS 93/3**.

KILROY, G., COXON, C., ALLOTT, N. & RYBACZUK, K. 1999. The contribution of groundwater phosphorus to surface water eutrophication. *In*: *Proceedings of the 19th Annual Groundwater Seminar, International Association of Hydrogeologists (Irish Group)*, Surface Water and Groundwater: A Combined Resource, Portlaoise, **11**, 7pp.

MISSTEAR, B. D. R. 1999. Groundwater and surface water: exploiting a combined resource. *In*: *Proceedings of the 19th Annual Groundwater Seminar, International Association of Hydrogeologists (Irish Group)*, Surface Water and Groundwater: A Combined Resource, Portlaoise, **5**, 8pp.

RIPPON, P. W. & WYNESS, A. J. 1994. Integrated catchment modelling as a water resources management tool. *Journal of the Chartered Institution of Water and Environmental Management*, **8**, 661–679.

SOUTHERN WATER GLOBAL. 1998. *An investigation of the flooding problems in the Gort–Ardrahan area of south Galway*. Report to the Office of Public Works, Dublin.

WINTER, T. C., HARVEY, J. W., FRANKE, O. L. & ALLEY, W. M. 1998. *Ground Water and Surface Water: A Single Resource*. US Geological Survey Circular 1139, Denver, Colorado.

Groundwater protection zone delineation at a large karst spring in western Ireland

J. DEAKIN

Groundwater Section, Geological Survey of Ireland, Beggars Bush, Haddington Road, Dublin 4, Ireland

Abstract: Pouladower Spring is a large karst spring in County Clare, Ireland which is being considered for use as a public supply. Groundwater protection zones have been delineated as a water quality management strategy for the spring. The Irish national groundwater protection scheme methodology is adapted to take account of the hydrological and hydrogeological complexities of the karst regime. The catchment area for the spring is large (approximately 380 km^2) and comprises the zones of contribution for two major outlets of water, the spring and the River Fergus. The actual zone of contribution to the spring varies with different water level conditions and the risk to the source from any point within the catchment, at any given time, is less than that for a conventional groundwater source. The catchment area is highly vulnerable, but dilution and sedimentation occurring in the lakes up gradient of the source, the high throughput, and the contribution from fissures outside the main flow conduits have helped maintain good water quality at the spring. The source is considered to be a combination of both groundwater and surface water as they are intricately inter-linked throughout the catchment. An Inner Protection Area is delineated which does not provide the 100-days travel time to the source required by the national scheme, as this would be impractically large and over-protective. Rather, it delineates the area of highest hydrogeological risk to the source and should allow the local authority sufficient time to act in the event of an accidental spill. A certain degree of microbial contamination is inevitable in a karst regime and treatment is essential, as it would be for a surface water source. The remainder of the catchment is classed as an Outer Protection Area. These protection areas are then combined with the vulnerability in a GIS to give groundwater protection zones which will be used by the planners, in conjunction with groundwater protection responses, to control potentially contaminating activities.

Groundwater is an important natural resource which supplies some 20–25% of drinking water in Ireland and is important in maintaining wetlands and river flows through dry periods (Department of the Environment and Local Government *et al.* 1999*a*). Groundwater protection is a key environmental issue facing European Union (EU) member states with the proposed EU Framework Directive on Water (European Commission 1996) which states that deterioration of groundwater quality must be prevented.

Almost 50% of the land surface of Ireland is underlain by Lower Carboniferous limestones. The degree of karstification of these rocks is variable, ranging from extensive to minimal, but solution has impacted on all of them to a certain degree. Protecting groundwater in karstified areas is widely recognized as being particularly difficult owing to the degree of variability and the unpredictable nature of groundwater flow. Groundwater velocities are high, there is low storage, rapid response to recharge, and flow is concentrated through solutionally-enlarged conduits and cave systems which are difficult to isolate. In many of the extensively karstified areas in the west of Ireland, there is often little protective soil and subsoil cover (i.e. unconsolidated materials above the bedrock) where attenuation and filtration can take place. There is also a high degree of hydraulic interconnection between surface water and groundwater. Once contaminants gain access to the rock, they can make their way through the aquifer over large distances in a much shorter time than in other aquifer types.

In Europe there is increasing recognition of the distinctive nature of karst aquifers, and the complexities involved in trying to protect them. The EU, through the Co-operation in Science and Technology Programme (COST), has funded the exchange of information and experience in karst groundwater protection. The final report of the first initiative (COST 65)

From: ROBINS, N. S. & MISSTEAR, B. D. R. (eds) *Groundwater in the Celtic Regions: Studies in Hard Rock and Quaternary Hydrogeology*. Geological Society, London, Special Publications, **182**, 89–98. 1-86239-077-0/00/ $15.00

recommended a three-fold approach: (a) define wellhead protection areas in the immediate vicinity of the source; (b) define the catchment area; and (c) define inner protection zones based on time of travel to the source. If travel times are sufficiently rapid so that the entire catchment falls within the time required to ensure bacterial decay, subdivision may be made on the basis of vulnerability (European Commission 1995). The second ongoing European initiative (COST 620) is looking specifically at vulnerability mapping for the protection of karst aquifers.

In Europe, there are generally two main approaches to protecting groundwater in karst areas (European Commission 1995). The first is to delineate the entire spring catchment area, or groundwater drainage basin/watershed, as the basis for the main protection zone. For example, in Croatia, this included a lake catchment area with associated surface water tributaries. In Slovakia, the approach is to delineate the total area of carbonate rocks. Some countries then delineate sub-zones based on travel time to the source. Belgium, Croatia and Spain, for example, all delineate a 24 hour zone within which groundwater will take 24 hours or less to reach the spring. Some countries produce maps using specific index approaches where points are scored to indicate degrees of risk to the source within a number of different categories. Switzerland uses EPIK, where points are assigned in each of four categories: Epikarst development, thickness of Protective cover, type of Infiltration, and Karst development; and three resulting protection zones are delineated (Doerfliger & Zwahlen 1997). The second common approach is to identify all potential sources of pollution within the catchment area and either remove them, or install a comprehensive monitoring system for accidental spills so that treatment may be initiated at the source ahead of the contamination pulse.

As hydrogeological characteristics, data availability and potential hazards are variable across Europe, the methods outlined above are often specific to the country of origin. The Irish national groundwater protection scheme takes account of the wide variations in hydrogeological conditions found in Ireland and is not, as in some countries, specific to karst (Daly & Drew 1999). This paper presents a case study of source protection delineation around a large karst spring which is being considered for use as public supply in the west of Ireland. Following consultation with the local authority, the national standard source protection methodology has been adapted to provide a more pragmatic, practical approach that takes account of the karst hydrological and hydrogeological characteristics of the area, while providing reasonable protection to the spring.

Irish national groundwater protection scheme

The Irish national groundwater protection scheme was published in 1999 by the Department of the Environment and Local Government (DoELG), the Environmental Protection Agency and the Geological Survey of Ireland (DoELG *et al.* 1999*a*). The scheme uses a risk assessment and risk management approach, based on the hazard-pathway-receptor model, which has two separate, but inter-linked objectives: protection of individual sources and protection of groundwater resources in general (Misstear & Daly this volume).

Within the framework of the source protection methodology where the target is the well or abstraction point, the *pathway* has both vertical and horizontal components and is categorized using land surface zoning as follows:

(a) the vulnerability (vertical component) shows the degree of protection afforded to groundwater by the overlying subsoils by way of four categories: Extreme, High, Moderate and Low (Misstear & Daly this volume); and
(b) the Inner and Outer Protection Areas represent the horizontal time of travel of groundwater to the source (<100-days time of travel and the entire zone of contribution, respectively).

These maps are then combined using a GIS to produce a land use planning map which subdivides the land surface into a potential maximum of eight zones highlighting the varying degrees of risk to the source. The land use planning maps are used by the local authorities, in conjunction with response matrices (codes of practice) for each potentially polluting activity, as a framework to assist them in making decisions on the location, nature and control of potentially contaminating developments.

The protection scheme methodology is the same for both karst and non-karst sources. As groundwater velocities are usually very high in karst regimes, the Inner and Outer Protection Areas often constitute the same zone and the vulnerability determines the relative risk to the source. In especially large complex catchment areas, such as that for Pouladower Spring in County Clare, this can result in relatively stringent management practices over large areas which may not always be practical or defensible for reasons which will be outlined below. This methodology has been adapted at Pouladower to take account of the karst hydrological and hydrogeological characteristics of the area, while still providing protection for the spring and time for the local authority to react in the event of an accidental spill. The hydrogeology of the area is discussed followed by an explanation of the adaptations made to the national methodology.

Location and site description

Pouladower Spring is located in the Fergus River catchment in County Clare, approximately 3 km north of Ennis (Fig. 1). It is being considered as a backup supply to reduce the dependency on another nearby large karst spring currently supplying the town, and the local authority is keen to delineate protection zones around it and to investigate the likely risks to the source.

Spring flow measurements record a variation in discharge from 10 000 to 62 000 m^3/d although the high flow is difficult to ascertain as the area is liable to frequent flooding during periods of heavy rain (K. T. Cullen & Co. 1996). These variations are typical of a large karst spring. There are other springs in the area which exhibit the same characteristics, with mean flows of an order of magnitude greater than base flows, and peak discharges which may be as much as three orders of magnitude greater (Drew 1990).

Geological and hydrogeological setting

The bedrock geology has shaped the topography into four geographical regions which are also distinct hydrogeological units (Fig. 1). In the upland areas to the west and east, the rocks mainly comprise relatively low permeability shales, siltstones and sandstones which are classed as Poor and Locally Important aquifers. In the vicinity of the Fergus River, the land lies at an elevation of less than 30 m above Ordnance Datum (OD) and comprises a relatively flat, low-lying plain, which is underlain by clean Lower Carboniferous limestones. These rocks extend northwards to the Burren Plateau (280 m OD) and are classified throughout the area as a Regionally Important Karst aquifer (Deakin & Daly 1999).

Bedrock generally dips 10–15° to the west and has undergone various degrees of structural

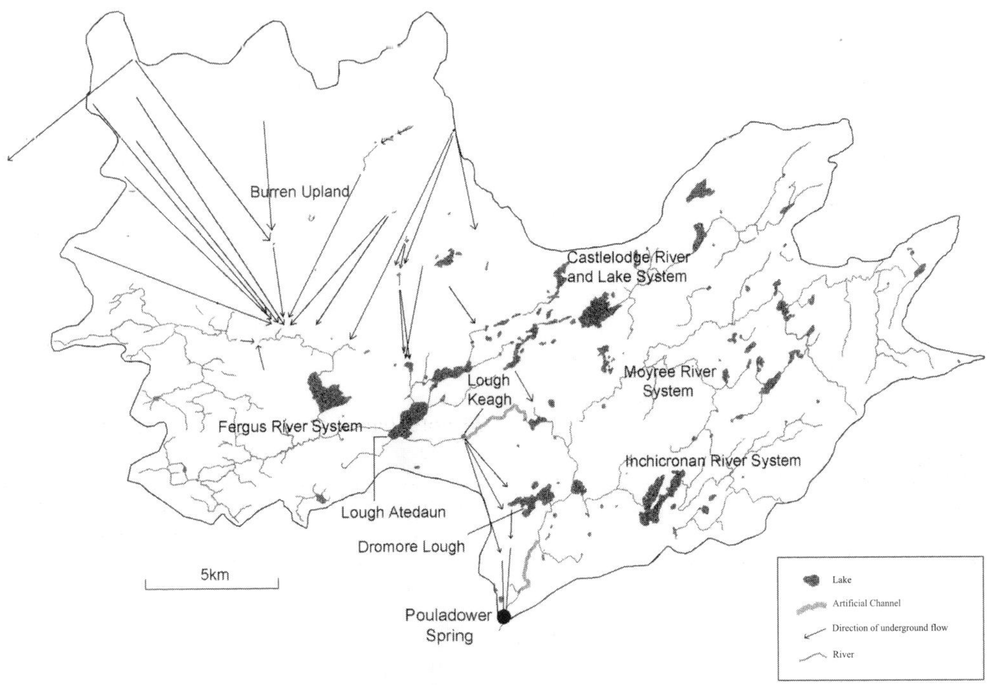

Fig. 1. Location map for Pouladower Spring catchment area.

deformation which has provided the secondary permeability facilitating groundwater flow. The major fault trends are north-south and east-west, and there is a local area of intensive north-south faulting in the low-lying clean Visean limestones north of the spring. Groundwater therefore flows mainly in these directions along associated joints and fissures (Deakin 1999).

The subsoils in the area are predominantly tills which have been directly influenced by the underlying bedrock. Where the subsoils are relatively thick, the clean limestones have given rise to moderately permeable silty, sandy and gravelly tills. The lower permeability rocks have a higher shale and mudstone content which has broken down to more clay-rich tills with low permeabilities. Where the tills are relatively shallow and are not as well developed, or where they were deposited in meltout conditions, they have a higher proportion of coarse-grained loose material and therefore a moderate permeability (Deakin & Daly 1999). Rock is present at or near the ground surface over much of the area although drumlins are common where subsoils may reach a thickness of more than 30 m. The degree of protection afforded by the subsoils is therefore highly variable throughout the area.

Surface water and groundwater are closely inter-linked throughout the Fergus River catchment (Cox & Drew this volume). Many of the surface water courses sink underground before rising to the surface further down-gradient and then sinking again (Fig. 1). A number of water traces have been conducted which prove underground connections between various parts of the Fergus River system (Drew 1988; Drew *et al.* 1995). Two stretches of artificial channel now ensure a direct link from source to sea but before they were constructed, some parts of the river system had no surface water outlet. There are also known links between many of the numerous swallow holes and caves to surface water courses and springs. Depending on rainfall and water levels throughout the catchment, rivers may be either gaining or losing, and flooding is common.

It is probable that a high proportion of flow to springs is in direct-route, underground, solutionally enlarged conduits in the limestones, with a somewhat lesser contribution from the smaller, more diffuse network of fissures and smaller conduits in the surrounding rock. The proportion of flow travelling through large conduits will vary with different water levels: there is likely to be more flow in the diffuse fissures at lower water levels. During medium to high flow, tracing has showed groundwater velocities to be high (90–250 m/h) (Drew *et al.* 1995).

Specific protection issues

The aims and objectives of delineating protection zones for karst sources are similar, though not identical, to those for granular and fissured aquifers, and can be summarized as follows:

(a) Protection from contamination of groundwater in the zone of contribution to the abstraction source. Two areas are delineated for microbial (Inner Protection Area) and chemical (Outer Protection Area) protection with the inner area carrying more stringent controls than the outer area.
(b) Provision of enough time to react to unexpected contamination incidents within the above areas.

Three main issues arise at Pouladower Spring which warrant special consideration: (1) the catchment is extremely large and is only partially contributing to the spring; (2) the lakes and the large throughput help maintain good water quality; and (3) the source is a combined groundwater and surface water source and a protection strategy must therefore take this into account.

The large, partially contributing catchment area

Underground links have been traced to the source, at medium to high water levels, from sinks present in Dromore Lough, and (indirectly) from Lough Keagh (Drew *et al.* 1995). The catchment area for the spring must therefore incorporate the catchment areas to Lough Keagh and Dromore Lough, and this includes essentially the entire Fergus River catchment (Fig. 1).

Two tributaries, the Moyree and Inchicronan River systems, join the River Fergus upstream of Dromore Lough. The flow between Dromore Lough and the River Fergus is via a small channel in which the flow regularly reverses depending on different hydrological conditions and the height of the water in the lake (Coxon & Drew 1998). This is not a seasonal variation but rather can change on a daily basis (Enda Mooney, Office of Public Works, pers. comm.). Therefore at times when the flow is from east to west, i.e. from the River Fergus into the lake, the tributaries and their catchments may also contribute to the spring. When flow is from the lake into the river, contribution from these areas is unlikely.

The catchment area is much larger (*c.* 380 km^2) than the area required to supply the source, perhaps more than 10 times larger at some water level conditions. It incorporates the intricately inter-linked catchment areas for two major outlets of water from the system, Pouladower Spring and the River Fergus. It is impossible to isolate the actual area contributing to the spring as it will vary with different water level conditions. At high water levels, for example, smaller sub-catchments, which would normally be isolated from the main flow to Pouladower, may overflow into the main zone of contribution and contribute to the spring. Groundwater may also leave the system via other conduits connected to the main river channel. As it is possible that contaminants reaching groundwater or surface water anywhere within the catchment may influence water quality at Pouladower, the entire area must be considered in protecting the source. In terms of risk management, however, the risk to the source from a given point in the catchment is significantly less than it would be to a source in a non-karstified aquifer where the pathways were not so variable. This large catchment area also extends into a neighbouring local authority area which poses additonal concerns with regard to exercising stringent developmental controls.

The influence of the lakes and high throughput

There is an extensive northeast–southwest band of lakes present in the Castlelodge River system further to the north, which separates the upland and lowland karst limestone areas (Fig. 1). The lakes are a particularly important factor within the catchment area as they constitute a substantial body of water where significant dilution and sediment deposition is occurring. There is also an exceptionally high throughput from the large catchment area (up to 380 km^2) through the relatively narrow discharge area in the vicinity of the spring and the river.

Water quality at Pouladower Spring is relatively good with very low levels of chemical contaminant indicators such as chloride, nitrate and potassium (Deakin 1999). Faecal coliforms (*E. coli*) are often present but this is typical in karst areas where groundwater travel times are often much less than the 100 days allowed for bacteria and viruses to die off (DoELG *et al.* 1999*a*). In terms of drinking water standards, dilution has a much greater impact on chemical contamination in karst areas than on bacterial contamination, particularly faecal bacteria. This is because the organic wastes generated in a catchment such as Pouladower (mainly farmyard wastes and septic tank effluent) contain very large numbers of faecal bacteria, for instance >100/100 ml for septic tank effluent, while the drinking water standards require a count of 0 per 100 ml of water. In contrast, for most chemical contaminants, far less dilution is required, firstly because the initial concentrations are lower relative to faecal bacteria (e.g. 50 mg/l N for septic tank effluent) and secondly because the drinking water limits are seldom 0 (e.g. 50 mg/l NO_3 (11.3 mg/l N) for nitrate).

Overall, the water quality at Pouladower Spring is better and generally more stable than in the nearby Fergus River and in other large karst springs in the catchment (Cronin & Deakin 1999). The dilution and sedimentation in the lakes, together with the input of groundwater from fissures, may account for the relatively stable, good quality groundwater which is discharging at the spring.

While it is accepted that 'dilution is not the solution to pollution', it is recognized that it is a significant factor in lowering chemical contaminant concentrations in the rapid conduit flow component of an extensive karst hydrogeological regime, such as the Pouladower catchment. It is far more significant in such a karst regime than it would be in granular or fissure flow aquifers, or at more tightly constrained karst springs.

The combined groundwater and surface water source

Pouladower Spring is considered as a combined surface water/groundwater source, derived largely from river waters sinking at Dromore and Lough Keagh, with a certain contribution from fissures in the surrounding rock. The river waters, however, are derived largely from groundwater. As the surface water and groundwater systems are so well interconnected throughout the catchment, they need to be considered together in protecting the source.

Microbial contamination in vulnerable karst aquifers is inevitable due to the rapid groundwater velocities which do not allow the bacteria and viruses time to die off. Treatment at surface water drinking sources is accepted as standard and is also needed at karst public supplies to eliminate microbial content. This is particularly true at Pouladower as groundwater and surface water have such a high degree of interconnectivity.

Source protection at Pouladower Spring

Pouladower has a very large potential catchment area, incorporating both groundwater and surface water, only part of which contributes to the source at any one time. An innovative approach to delineating protection zones is therefore required which retains a pragmatic and defensible methodology that provides adequate protection for the source. Discussions were held with the local authority and the following approach, based on an adaptation of the national Groundwater Protection Scheme (DoELG *et al.* 1999*a*), was agreed to be more appropriate.

Delineating source protection zones addresses the *pathway* element of the risk assessment approach. As mentioned above, there are two subcomponents to the pathway: the horizontal component dictated by the time of travel to the source (the Inner and Outer Protection Areas) and the vertical component which is given by the vulnerability. The horizontal component was adapted at Pouladower to take account of the specific hydrogeological conditions in the catchment area.

Inner and outer protection areas

The Inner Protection Area is delineated to protect against contaminants which may have an immediate effect on the source, particularly microbial contamination, and it is normally defined by the area within which groundwater takes 100 days or less to reach the source. Based on the tracing carried out in the area by Coxon and Drew (Drew 1988; Drew *et al.* 1995) it is expected that groundwater from anywhere within the limestone areas may reach Pouladower within 100 days, depending on the water level conditions. If the standard Inner Protection Area was delineated it would result in an extremely large area (*c.* 250 km^2) which is impractical and unnecessary for the management of potentially contaminating activities for the following reasons:

(1) The delineated area includes the catchment areas for two major outlets of water from the system, Pouladower Spring and the River Fergus. It is impossible to isolate the actual zone of contribution to the spring at any given time as it will vary with different water level conditions. The overall risk to the source from any given point in the system is therefore significantly less than it would be in a non-karst catchment area, which is more tightly constrained.

(2) There is significant dilution occurring throughout the Pouladower system within the large catchment area, particularly in the lakes and also as a consequence of the high throughput flowing through a relatively small discharge area, which is not usually the case at other groundwater sources.

(3) At sources in other aquifer types, the Inner Protection Area is delineated to prevent microbial contamination occurring. A certain degree of microbial contamination of vulnerable karst sources such as Pouladower is inevitable due to the rapid groundwater velocities and the high interconnectivity with surface water. Treatment of the supply is essential, as it would be for a surface water source.

It was concluded that it is more appropriate to delineate an area of highest risk to the source as an Inner Protection Area, to assist the local authority in prioritizing its resources, e.g. monitoring, hazard mapping etc. Groundwater within this area of highest risk can reach the source in much less than 100 days – based on surface water velocities, groundwater tracing and known response times to rainfall events, perhaps within 2 days on occasion. However, in the event of an accidental spill, it should be possible to temporarily cease abstraction or increase treatment provided any such incident is reported immediately and/or there is an adequate monitoring network.

The Inner Protection Area, in this instance, therefore includes the southernmost part of the catchment closest to the source, from Pouladower Spring as far up-gradient as the band of lakes where most of the dilution is taking place (Fig. 2). With the rapid groundwater velocities in this high risk area, advance warning of a day or two should be possible, depending on proximity to the source and water level conditions. The area is classed as the Inner Protection Area so that it will be given the highest level of protection available. The terminology already defined in the groundwater protection scheme methodology was maintained as the framework and responses are already in place.

Outside the highest risk area, even though groundwater in the limestones may reach the source well within 100 days, there is considerable dilution of potential contaminants occurring due to the number of lakes in the system and the high groundwater throughput in the karst aquifer. There may also be significant delay in travel times, at certain water levels, as water moves through the lakes. A further consideration is that the actual zone of contribution to the source in this area is far less tightly constrained than in the highest risk area, and at some water levels potential contaminants may bypass the source altogether. It is considered that hazards in these areas pose a significantly lesser threat to the

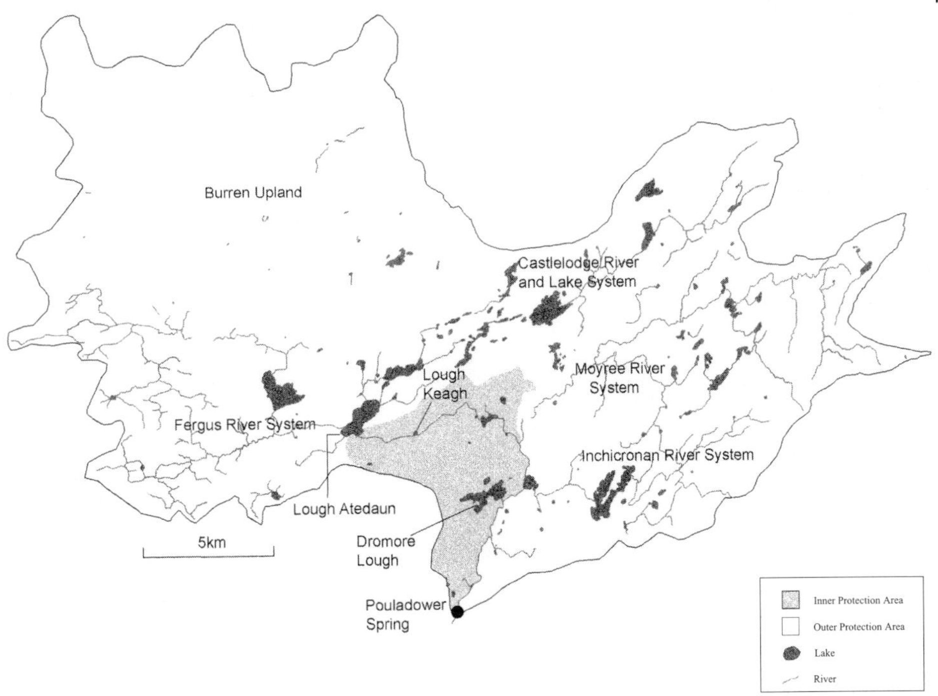

Fig. 2. Inner and Outer Protection Areas for Pouladower Spring.

quality at the source than hazards within the Inner Protection Area. Therefore the remainder of the catchment is classed as the Outer Protection Area and protection is provided by the appropriate responses. Part of this area is located outside the local authority jurisdiction, in the neighbouring county.

Vulnerability

The vulnerability of the groundwater (the vertical component of the pathway) depends on the thickness, type and permeability of the subsoils. In the Pouladower catchment, the subsoil thickness is predominantly less than 3 m and this gives rise to a vulnerability category of 'Extreme' over large areas (Fig. 3). As most surface water is connected to groundwater prior to reaching Pouladower Spring, an area of 'Extreme' vulnerability is delineated along all surface water channels throughout the catchment as a means of indicating the threat to the source from surface runoff of contaminants into streams. This area also comprises a 30 m buffer zone added to the normal water channel boundary on the karstified limestones, with a buffer zone of 10 m along channels on the non-karstified and lower permeability rocks. There are a number of karst features in the catchment area such as caves, swallow holes, turloughs and collapses that are all designated as points of 'Extreme' vulnerability on the vulnerability map as they provide direct access to groundwater for potential pollutants.

The remainder of the subsoils are divided into moderate and low permeability classes (there are no high permeability deposits) and are combined, using a GIS, with the thicknesses of the deposits to give the final vulnerability categories (Table 1). Depending on the depth to rock, the *vulnerability* of the moderately permeable deposits will range from 'High' (3–10 m thick) to 'Moderate' (>10 m thick), while the low permeability deposits range from 'High' vulnerability (3–5 m thick), through 'Moderate' vulnerability (5–10 m thick), to 'Low' vulnerability (>10 m thick).

Groundwater source protection zones

Combining the Inner and Outer Protection Areas with the vulnerability categories gives the following groundwater protection zones around Pouladower Spring which are shown on the groundwater source protection zone map (Fig. 4):

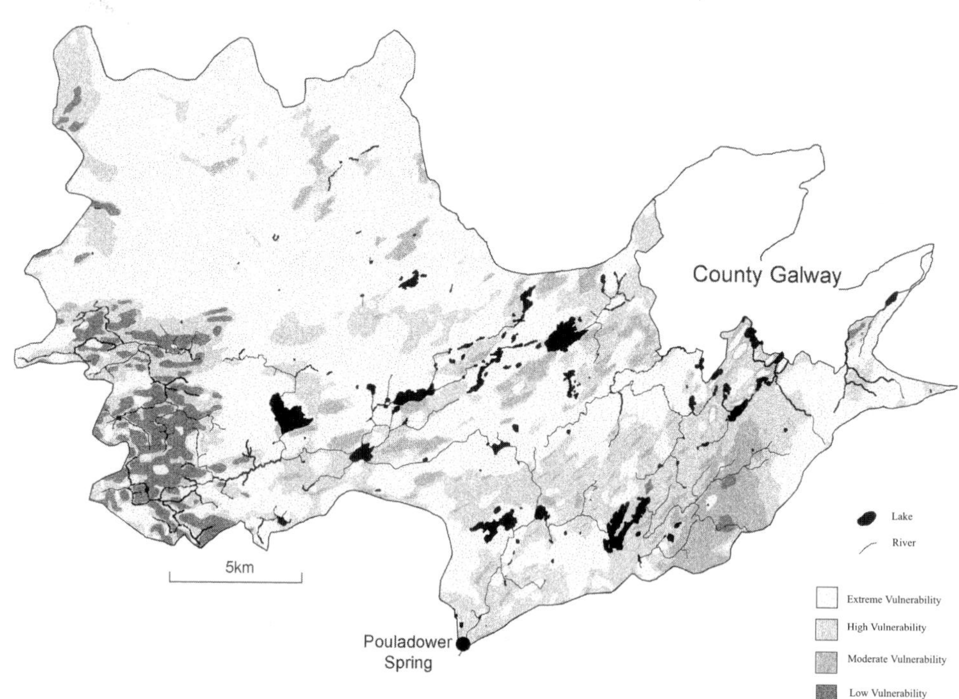

Fig. 3. Vulnerability of groundwater in the Pouladower Spring catchment area.

Table 1. *Geological and hydrogeological conditions determining vulnerability mapping categories*

Subsoil thickness*	Hydrogeological requirements				
	Diffuse recharge Subsoil permeability and type§			Point recharge†	Unsaturated zone‡
	High	Moderate	Low		
0–3 m	Extreme	Extreme	Extreme	Extreme (<30 m radius)	Extreme
3–5 m	High	High	High	N/A	High
5–10 m	High	High	Moderate	N/A	High
>10 m	High	Moderate	Low	N/A	High

* Release point of contaminants is assumed to be 1–2 m below ground surface.
† Includes swallow holes, losing streams, etc.
‡ Only relevant in sand/gravel aquifers.
§ Permeability classifications relate to the material characteristics as described by BS5930:1981 (British Standards Institution 1981).
Adapted from Deakin & Daly (1999).

SI/E Inner Protection Area with subsoils 1–3 m thick.
SI/H Inner Protection Area with subsoils 3–10 m thick.
SI/M Inner Protection Area with subsoils >10 m thick.
SO/E Outer Protection Area with subsoils 1–3 m thick.
SO/H Outer Protection Area with subsoils 3–10 m thick.
SO/M Outer Protection Area with moderate permeability subsoils >10 m thick.
SO/L Outer Protection Area with low permeability subsoils >10 m thick.

The appropriate groundwater protection response (code of practice) for various potentially

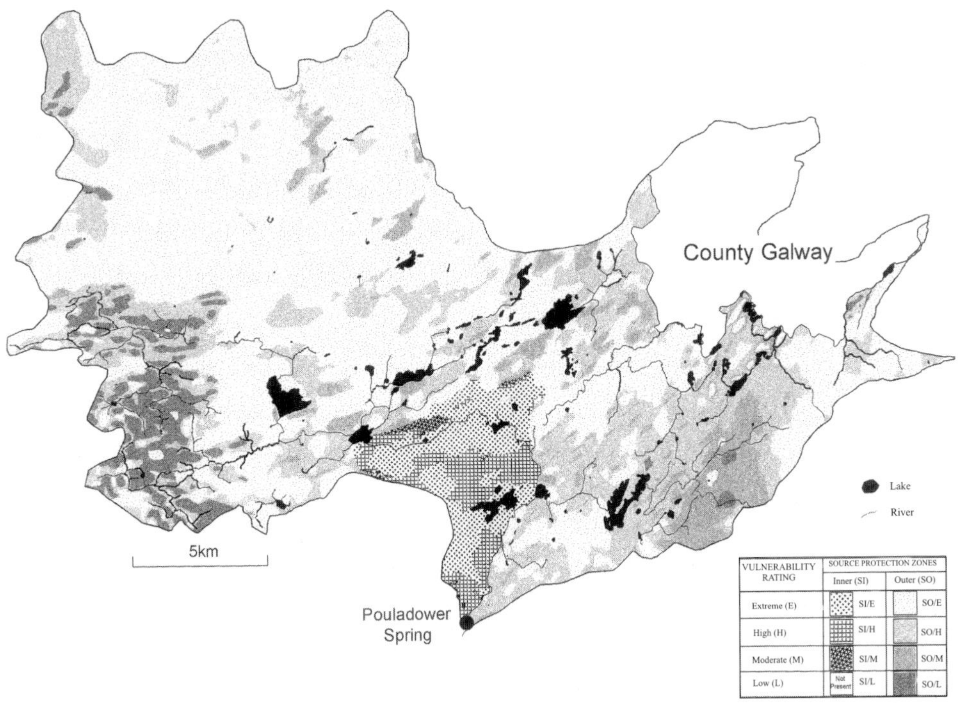

Fig. 4. Groundwater protection zones for Pouladower Spring.

contaminating activities is then consulted to obtain the degree of restriction necessary in each protection zone. Groundwater protection responses for landfills and landspreading of organic wastes have been developed by the DoELG, EPA and GSI to date (DoELG *et al.* 1999*b*, *c*) and there will be others to follow.

Conclusion

Pouladower spring is a large karst spring located in the Fergus River catchment, and is considered to be a combined surface water and groundwater source. It responds rapidly to recharge, is extremely vulnerable to contamination, yet has better and more stable water quality than the nearby river and other karst springs in the area. This good water quality is likely to be a consequence of dilution and sedimentation of potential contaminants occurring in a large band of lakes in the catchment, and the significant groundwater contribution to the source from smaller fissures outside the main conduits of flow.

An innovative approach to protecting the source is required which is pragmatic and defensible, yet provides reasonable protection for the source. Following consultation with the local authority, the national groundwater protection methodology has been adapted to take account of the local hydrological and hydrogeological characteristics. The Inner Protection Area, which is delineated to protect the source from microbial contamination, comprises the area of highest risk to the source between Pouladower Spring and the lakes. It does not, in this instance, include the 100-day travel time zone which would have included the entire limestone area (c. 250 km^2): this is considered to be impractical and unnecessary for the management of potentially contaminating activities for the following three reasons:

(1) The delineated area includes the catchment areas for two major outlets of water from the system, Pouladower Spring and the River Fergus. It is not feasible to define the actual zone of contribution to the spring at any given time as it will vary with different water level conditions.

(2) There is significant dilution occurring throughout the Pouladower catchment. Dilution takes place in the lakes and as a consequence of the high throughput flowing through a relatively small discharge area.

(3) The Inner Protection Area at other groundwater sources is delineated to prevent microbial contamination occurring. Pouladower is a combined groundwater/surface water

source and so a certain degree of microbial contamination is inevitable. Water treatment is therefore essential to maintain a good quality drinking water supply.

The Outer Protection Area, designed to protect from chemical contamination, encompasses the remainder of the catchment area. These protection areas are combined with the vulnerability classifications in a GIS to give the groundwater source protection zones.

The groundwater protection zones will be used by the local authority in conjunction with groundwater protection responses (codes of practice) for each potentially contaminating activity to highlight the development restrictions, design and construction conditions, and site investigation requirements which may be appropriate. The maps are produced with the best available geological information and are used as a guideline by the local authorities, but are not intended to be prescriptive. The onus is then on the developer to conduct site investigations and gather new information to prove the actual groundwater protection zone on site.

The national groundwater protection scheme methodology has been retained at Pouladower Spring, but has been adapted to take account of the local hydrogeological situation, in an acceptable, practical, sustainable way, while still providing protection for the source. This new approach will however, be reliant on having accidental spills identified as soon as they occur. Developing a programme of education and awareness within the highest risk area would be beneficial.

The author would like to acknowledge the work carried out by C. Coxon and D. Drew on the hydrogeology of this area. Assistance and advice from the following is also gratefully acknowledged: M. Burke and D. Timlin, Clare County Council; M. Keegan and A. Fanning, EPA; D. Daly, Groundwater Section, GSI. This paper is published with the permission of Dr. P. McArdle, Director, Geological Survey of Ireland.

References

British Standards Institution. 1981. *Code of Practice for Site Investigations*. British Standards Institution, London, BS 5930:1981.

Coxon, C. & Drew, D. 1998. Interaction of surface water and groundwater in Irish karst areas: implications for water resource management. *In*: Brahana, J. V., Eckstein, Y., Ongley, L. K., Schneider, R. & Moore, J. E. (eds) *Gambling with Groundwater – Physical, Chemical and Biological Aspects of Aquifer-Stream Interactions*. Proceedings of the XXVII IAH Congress, Las Vegas, 161–168.

—— & ——2000. Interdependence of groundwater and surface water in lowland karst areas of western Ireland: management issues arising from water and contaminant transfers. *This volume*.

Cronin, C. & Deakin, J. 1999. *An Assessment of Groundwater Quality in County Clare*. Geological Survey of Ireland Report to Clare County Council.

Daly, D. & Drew, D. 1999. Irish methodologies for karst aquifer protection. *In:* Beck, Pettit & Herring (eds) *Hydrogeology and Engineering Geology of Sinkholes and Karst*. Balkema, Rotterdam, 267–272.

Deakin, J. 1999. *Pouladower Public Supply Groundwater Protection Zones*. Geological Survey of Ireland Report and Maps prepared for Clare County Council, Draft.

—— & Daly, D. 1999. *County Clare Groundwater Pro*tection Scheme. Joint Geological Survey of Ireland/Clare County Council Report, Draft, 73pp.

Department of the Environment and Local Government, Environmental Protection Agency & Geological Survey of Ireland. 1999*a*. *Groundwater Protection Schemes*. Joint Department of the Environment and Local Government, Environmental Protection Agency and Geological Survey of Ireland Report.

——, —— & ——1999*b*. *Groundwater Protection Responses for Landfills*. Joint Department of the Environment and Local Government, Environmental Protection Agency and Geological Survey of Ireland Report.

——, —— & ——1999*c*. *Groundwater Protection Responses for Landspreading of Organic Wastes*. Joint Department of the Environment and Local Government, Environmental Protection Agency and Geological Survey of Ireland Report.

Doerfliger, N. & Zwahlen, F. 1997. EPIK: a new method for outlining of protection areas in karstic environments. *In*: Gunay and Johnson (eds) *Karst Waters and Environmental Impacts*. Balkema, Rotterdam, 117–124.

Drew, D. 1988. The hydrology of the upper Fergus River catchment, Co. Clare. *Proceedings of the University of Bristol Spelaeological Society*, **18**(2), 265–277.

——1990. The hydrology of the Burren, Co. Clare. *Irish Geography*, **23**(2), 69–89.

——, Doak, M., Daly, D., Thorn, R., Warren, W. & Coxon, C. 1995. National report for Ireland. *In*: European Commission *Hydrogeological Aspects of Groundwater Protection in Karstic Areas*. COST Action 65 Final Report No. **EUR 16547**, 149–169.

European Commission. 1995. *Hydrogeological Aspects of Groundwater Protection in Karstic Areas*. COST Action 65, Final Report No. **EUR 16547**.

——1996. Proposal for Council Directive establishing a framework for Community Action in the field of water policy. *Official Journal of the European Communities*, **C184**.

K. T. Cullen & Co. 1996. *Investigations at Pouladower Spring*. Report to Ennis Urban District Council.

Misstear, B. & Daly, D. 2000. Groundwater protection in a Celtic region: the Irish example. *This volume*.

Groundwater resources and vulnerability in the Cretaceous Chalk of Northern Ireland

S. BARNES

Geology Department, The Queen's University of Belfast, Belfast, Northern Ireland BT7 1NN, UK
Present address: S.I.T.A., Narborough Lodge, Huncote Road, Narborough, Leicester LE9 5RQ, UK (e-mail: narborough_lodge@sita.fr)

Abstract: A confined Cretaceous Chalk aquifer underlies approximately one quarter of Northern Ireland, yet little is known about its groundwater resource potential. This issue has been addressed on the catchment scale by analysing spring discharge and hydrochemical fluctuations.

The Chalk springs are recharged by allogenic leakage and surface runoff from overlying Tertiary basalts. Sources connected to river-sinks show greater variation in flow and quality reflecting a much shorter residence time than those predominantly derived from the diffuse recharge. Discharge from the confined region becomes proportionally significant during prolonged dry spells, but is typically a minor component compared with groundwater circulation volumes in the unconfined region.

Spring flood recessions are rapid (recession coefficients up to 0.125 per day) and suggest that the Chalk has a high hydraulic conductivity and a low storage capacity. These characteristics together, with the essentially impermeable matrix, are consistent with an aquifer dominated by a dispersed fracture network.

Conceptual aquifer classification suggests that the outcrop region is a highly sensitive karst aquifer. The subcrop areas can only be exploited via boreholes and are likely to be less productive, although the water quality has been shown to be more stable and less vulnerable to contamination.

The Cretaceous Chalk or Ulster White Limestone Formation (UWLF) in Northern Ireland is an extremely heterogeneous and anisotropic aquifer. Groundwater flow direction is often not coincident with hydraulic gradient (Barnes 1999), and hydraulic conductivities derived from borehole tests range from values indicative of a shale to those well within the karstic limestone range (data courtesy of Haul Waste, in Kirk *et al.* 1994). As a consequence of this hydraulic variability, the traditional pumping test approach to determine quantitative aquifer properties is potentially erroneous, and confident regional extrapolation of the results impossible.

The behaviour of carbonate aquifers is fundamentally governed by groundwater recharge, storage and transmission characteristics (Smart & Hobbs 1986). In the absence of human influences they essentially control the spatial physical and chemical fluctuations of natural springs. Understanding what controls spring response is thus important to achieve an understanding of aquifer structure or properties, resource potential and vulnerability.

This paper presents data obtained from monitoring discharge and hydrochemical variations of springs draining UWLF groundwater catchment areas. The information has been utilized to contribute to a qualitative understanding of regional aquifer characteristics.

Geology and hydrogeology

The UWLF aquifer is confined at depth beneath Tertiary Basalt Lavas over about one quarter of Northern Ireland, but its unconfined outcrop area is restricted to approximately 80 km^2 around the periphery of the overlying igneous rock (Fig. 1). It has a low matrix porosity of 2.3–10.4% (Maliva & Dickson 1997) and conductivity of 10^{-5}–10^{-6} m d^{-1} (Robins 1996). Groundwater movement mainly relies on secondary fracture porosity.

A series of natural springs issues from the base of the UWLF outcrop at its junction with the underlying aquiclude rocks (typically the Mercia Mudstone Formation). In the Garron Peninsula

From: Robins, N. S. & Misstear, B. D. R. (eds) *Groundwater in the Celtic Regions: Studies in Hard Rock and Quaternary Hydrogeology*. Geological Society, London, Special Publications, **182**, 99–112. 1-86239-077-0/00/

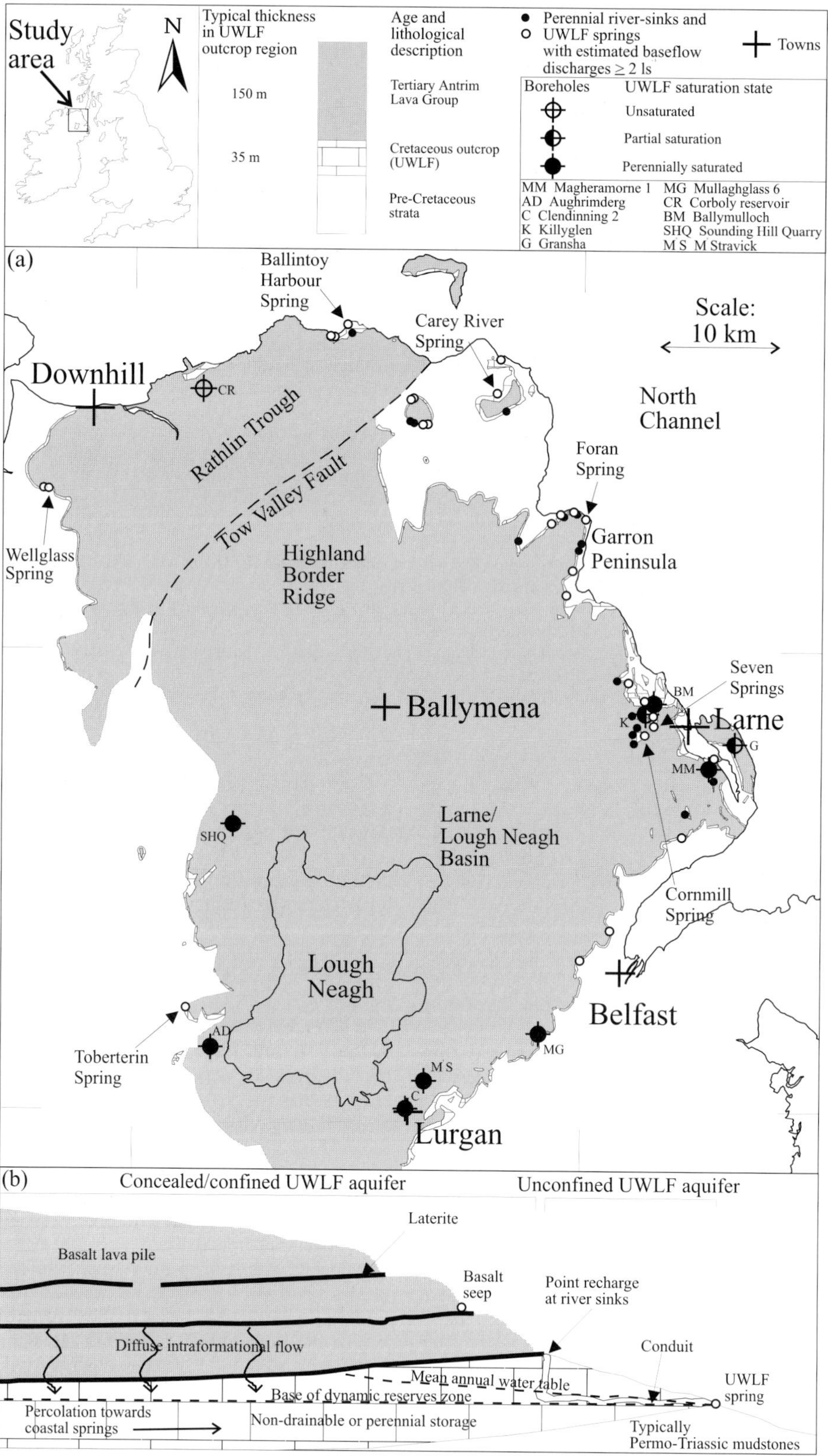

Study area
N
Typical thickness in UWLF outcrop region
150 m
35 m
Age and lithological description
Tertiary Antrim Lava Group
Cretaceous outcrop (UWLF)
Pre-Cretaceous strata
Perennial river-sinks and
UWLF springs with estimated baseflow discharges ≥ 2 ls
Towns
Boreholes
UWLF saturation state
Unsaturated
Partial saturation
Perennially saturated
MM Magheramorne 1
AD Aughrimderg
C Clendinning 2
K Killyglen
G Gransha
MG Mullaghglass 6
CR Corboly reservoir
BM Ballymulloch
SHQ Sounding Hill Quarry
M S M Stravick
(a)
Ballintoy Harbour Spring
Carey River Spring
Scale: 10 km
Downhill
CR
Rathlin Trough
Tow Valley Fault
North Channel
Foran Spring
Garron Peninsula
Wellglass Spring
Highland Border Ridge
Seven Springs
BM
Ballymena
K
Larne
G
MM
SHQ
Larne/ Lough Neagh Basin
Cornmill Spring
Lough Neagh
Belfast
Toberterin Spring
AD
MG
M S
C
Lurgan
(b)
Concealed/confined UWLF aquifer
Unconfined UWLF aquifer
Laterite
Basalt lava pile
Basalt seep
Point recharge at river sinks
Diffuse intraformational flow
Conduit
Mean annual water table
Base of dynamic reserves zone
UWLF spring
Percolation towards coastal springs
Non-drainable or perennial storage
Typically Permo-Triassic mudstones

Table 1. *List of hydrochemical determinants and the methods/procedures used for their analysis*

	Determinand	Method
Field Measurements	pH and Temperature	Temperature compensated Horiba pH meter
	Total alkalinity	HCl titration with BDH 4.5 indicator
	Electrical conductivity	Whatman conductivity meter
Laboratory Measurements	Cl^-, SO_4^{2-} and NO_3^-	Dionex ion chromatograph
	Ca^{2+}, Na^+, Mg^{2+} and K^+	Perkin Elmer atomic absorption spectrometer
	SiO_2	Spectrophotometric analysis of heteropoly blue method

area (Fig. 1a), the water emerging from these sources is predominantly supplied by allogenic river-sink water from surface catchments on top of the basalt plateau. A minor component also derives from basalt leakage water (Barnes & Worden 1998). The concentrated point recharge to the UWLF outcrop has been associated with conduit or karst development (Barnes 1999).

The UWLF is known to dip gently away from outcrop beneath the basalt cover (typically 5°). From this structure and background hydrogeological knowledge, a cross-section has been drawn to depict the conceptual hydraulic setting supplying UWLF spring sources (Fig. 1b).

Methods

Both discharge and hydrochemistry were monitored for a full annual cycle during part of 1995 and 1996 at seven different UWLF spring sites (Cornmill Spring, Seven Springs, Foran Spring, Carey River Spring, Ballintoy Harbour Spring, Wellglass Spring and Toberterin Spring; see Fig. 1a). Data were typically collected every three weeks. The sources chosen are among the most prolific from the UWLF aquifer, include existing public and private supply sites, and are thought to drain groundwater catchments encompassing a range of recharge and flow mechanisms (Barnes 1999).

Hydrology

Point spring discharge measurement. Crude discharge estimates from Carey River, Ballintoy Harbour and Toberterin Springs were achieved by multiplying the measured stream cross-section by the velocity of a surface float. Results were generally reproducible to within 20%. At Cornmill, Foran and Seven Springs point discharge data were obtained from a combination of dilution gauging and current metering approaches (Shaw 1994). A V-notch weir was utilized for discharge measurement at Wellglass Spring.

Spring hydrograph determination. At Cornmill, Foran and Seven Springs permanent gauging stations were constructed. Fixed regular man-made channels were utilized at each locality and a weir installed at the former site to aid accurate discharge measurement. Stilling wells with Munro horizontal float recorders were put in place at the edge of each channel for continuous stage measurement. Rating curves and subsequent rating equations were developed from the point discharge data.

Discharge recession analysis. Reliable hydrograph portions that include flood recession curves running into baseflow contributions only (i.e. exponential decay (Mangin 1975)) were plotted in a semi-logarithmic format. Superimposing these sections allows the construction of a master recession or depletion curve (Wilson 1994).

Water chemistry

At each locality an unfiltered sample was taken for immediate alkalinity titration and laboratory measurement of electrical conductivity. A 50 ml

Fig. 1. (**a**) Simplified solid geology map of the study area showing the outcrop pattern of the Cretaceous Ulster White Limestone Formation (UWLF) in Northern Ireland (based on the published hydrogeology map (BGS/DoENI 1994)). Piezometers/boreholes where the UWLF saturation state is known are presented together with the locality of the main river-sinks and spring sources associated with the aquifer. (**b**) Schematic summary cross-section through a typical UWLF outcrop area depicting the conceptual hydraulic background supplying UWLF springs. This model was constructed using knowledge of potential recharge pathways, and the geological structure and saturation state of the aquifer.

Table 2. *Summary of point discharge data from monitored UWLF springs*

Determinant	Statistics	Springs connected to surface river-sink					Diffusely recharged springs	
		Cornmill Spring $n=10$	Seven Springs $n=17$	Foran Spring $n=16$	Carey River Spring $n=19$	Ballintoy Harbour Spring $n=13$	Wellglass Spring $n=15$	Toberterin Spring $n=18$
Discharge ($l\,s^{-1}$)	Mean	108	55.1	51.8	26.3	11.3	29.8	8.00
	Range	14.1–444	27.3–112	5.72–132	15.0–50.0	7.00–18.0	7.00–55.0	8.00
	SD	153	29.0	34.8	9.93	3.32	16.1	

Toberterin Spring discharge variations were too subtle to detect with the technique used.
n = number of observations.

aliquot was also passed through a 0.2 µm cellulose nitrate filter into a clean plastic screw-top bottle. This sample was later analysed for dissolved constituents after a maximum of one week refrigerated storage (Table 1).

Results

Spring discharge fluctuations

UWLF springs known to be connected to surface swallets via conduit systems (Cornmill, Foran, Carey River, Ballintoy Harbour and Seven Springs) typically exhibited a large variation in point discharge observations (Table 2). At Cornmill Spring the peak value ($444\,l\,s^{-1}$) was approximately thirty times greater than that of the lowest ($14.1\,l\,s^{-1}$). Wellglass and in particular Toberterin Springs, which have no influence from river-sinks, produced more subdued discharge variations. Flow rates from the former source varied between 7 and $55\,l\,s^{-1}$, which is a similar range to the more conservative values from springs benefiting from point recharge. Toberterin Spring sustained yields of around $8\,l\,s^{-1}$.

Spring discharge recession characteristics. Semi-logarithmic plots of the master recession curves from Cornmill, Foran and Seven Springs reveal what can be interpreted as several linear segments to each recession (Fig. 2a–c). Each segment can be described by a simple exponential relationship (see equation (1)). Values for the recession coefficient (Maillet 1905), response time (Burdon & Papakis 1963), recession constant (Ford & Williams 1989) and the time to halving of baseflow (Martin 1973) have been computed (Fig. 2) for each of these sections using

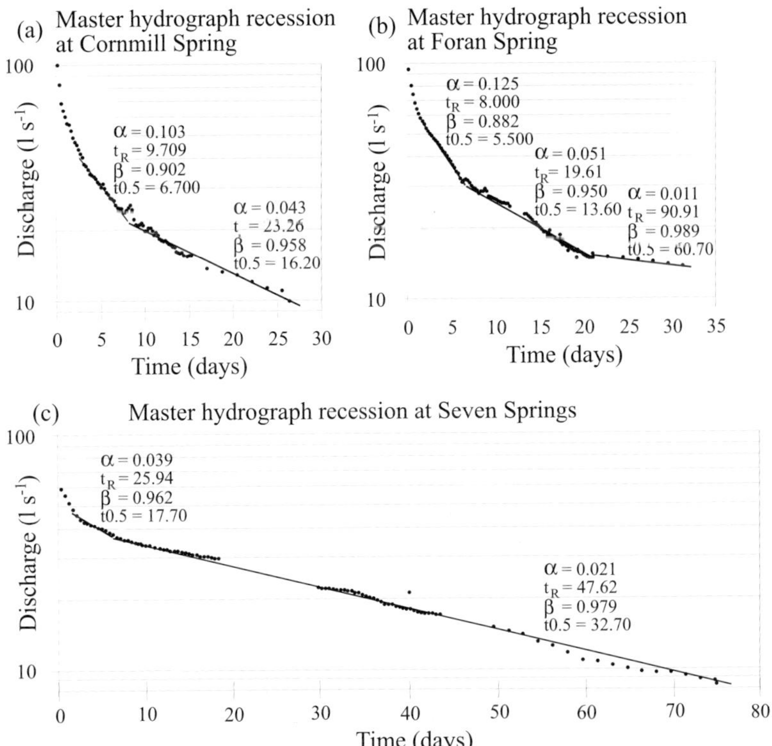

Fig. 2. Master recession curves for Cornmill, Foran and Seven Springs. The recessions were built up from numerous portions of hydrograph collected over a two year period. The initial non-linear segments represent quickflow recession and the subsequent linear sections are due to baseflow from storage within the aquifer. The steeper portions of these may reflect rapid drainage from the largest fissures or conduits, whereas the less steep linear sections can be interpreted as the slower drainage of smaller fissures in the UWLF. The characterizing parameters of recession coefficient (α), response time (t_R), recession constant (β) and the time to the halving of baseflow ($t0.5$) are presented for each linear section.

Table 3. *Reference to terms used in recession limb analysis equations (1) to (4)*

α	recession coefficient (day^{-1})
t_R	response time (days)
β	recession constant
$t0.5$	the time to halving of baseflow (days)
Q_t	discharge at time t ($m^3 s^{-1}$)
Q_0	discharge at time zero ($m^3 s^{-1}$)

equations (1) to (4), respectively, and the terms in Table 3.

$$Q_t = Q_0 e^{-\alpha t} \quad (1)$$

$$t_R = 1/\alpha \quad (2)$$

$$\beta = e^{-\alpha} \quad (3)$$

$$1/2 = \beta^{t0.5} \quad (4)$$

(where e is the base of natural logarithms)

Considering the data collectively, the recession coefficients (α) range from 0.125 to 0.011 per day, the response times (t_R) from 8 to 90 days, the recession constants (β) from 0.882 to 0.989 and the times to halving of baseflow ($t0.5$) from 5.5 to 60.7 days. It can be observed from the results and Fig. 2 that the data represent rapid baseflow recession. Seven Springs appears to recess least rapidly of the three sites.

Hydrochemical data

All water chemistry results were scrutinized for ion balance using PHREEQC speciation software (Parkhurst 1995). Samples that fall outside a 5% ion balance error margin have been rejected from the data sets. Details of the remaining data and their correlation with corresponding flow rate from individual UWLF springs are presented in Table 4. Springs connected to surface river-sinks typically produce a larger range of hydrochemical determinands than their diffuse fed counterparts, and their mean values tend to be lower. With the exception of dissolved CO_2, most correlation coefficients between the different parameters and discharge rates at the time of sampling are negative. Therefore, at lower discharges effectively all dissolved ions are more abundant.

Discussion

UWLF spring discharge fluctuations

Cornmill, Foran and Seven Springs produced the largest range of measured flow rates (up to thirty fold) (Table 2). This behaviour probably reflects the prevalence of point recharge and conduit systems supplying these sources. Continuous stage recordings from these sites suggest that an even greater variation in discharge may be occurring (up to fifty fold at Cornmill Spring). The lowest flow rates derived from the stage monitoring at Cornmill, Foran and Seven Springs are $8.6 l s^{-1}$, $5.7 l s^{-1}$ and $22.7 l s^{-1}$ respectively (between 1.9 and 16.2% of maximum). These data suggest that sustained drainable storage from the concealed portion of the UWLF aquifer is often the subordinate source of spring water supply.

Carey River and Ballintoy Harbour Springs also have access to point recharge, yet their discharge ranges were comparatively narrow. From field observation it would appear that in the former case this behaviour is a reflection of a partially blocked river-sink which regulates recharge rate, while the latter source has a comparatively small contributory surface catchment area.

The narrow range of measured discharges from Wellglass and in particular Toberterin Spring (Table 2) may reflect the lack of observed river-sink recharge associated with these sites, and probably a much less karstified or more subdued groundwater transit system as a result.

Indications of aquifer storage and flow characteristics from flood recessions. Hydrograph generation at Cornmill, Foran and Seven Springs allowed baseflow drainage characteristics to be investigated. All three sources produced a 'stepped' flood recession (Fig. 2a–c) which is often typical of karst aquifers and can be interpreted as drainage from a variety of void sizes (Milanovic 1981). The initial period of most rapid spring recession information complete in just a few days and could represent drainage from a main conduit system. The subsequent more subdued groundwater withdrawal rates can be interpreted as dewatering of smaller sets of contributory voids which would then be the main source of baseflow drainage.

Despite the above interpretation, Nutbrown & Downing (1976) have demonstrated that individual recession characteristics from even the simplest of groundwater systems represent a superposition of many exponential terms (similar to equation (1)). Some of these influences will necessarily be created by the dynamics of the groundwater system alone and are therefore independent of aquifer structure. The discharge recession information presented for the UWLF is an accumulation of many flood peak recessions observed during the monitoring period and is therefore thought to best represent a permanent structure to the aquifer rather than a dynamic feature of the groundwater. Nevertheless, given the complexity of the UWLF aquifer, a potential influence from continual river-sink recharge, and the current data available, it is considered most appropriate to examine and compare the overall drainage rates rather than the origin of individual recession segments.

Parameters α, t_R, β and $t0.5$ are all expressions of the recession curve slope or the rate of groundwater withdrawal. They therefore reflect aquifer properties; in particular effective porosity and transmissivity (Ford & Williams 1989). The α and t_R values obtained from UWLF springs are relatively large in comparison to those from other karst aquifers. The steeper portions of UWLF baseflow recession produce values closest to the 'fast-response springs' category (t_R range from 4.1 to 19 days) described by White (1988). Even the slower draining sections of the UWLF baseflow recession can either be classified as 'fast-response' or 'intermediate-response' springs (t_R range 65–86 days) using the same classification ranges. Correspondingly, the times taken for each portion of baseflow to halve ($t0.5$) tend to be comparatively small, and the data suggest the rapid drainage of relatively large voids (i.e. high hydraulic conductivity) with limited drainable storage to sustain spring water supply. These data are consistent with an aquifer dominated by a dispersed fracture/fissure network. At outcrop this regime will be punctuated by occasional conduits carrying swallet recharge to springs. This conceptual understanding of the aquifer's heterogeneity can also account for the fact that groundwater flow direction is often determined by fracture orientation, and not by hydraulic gradient (Barnes 1999).

Small amounts of recharge (via river-sinks) could be observed to persist during the lowest recorded UWLF spring-flow rates. This occurrence may help subdue the rapid rate of spring discharge decline (inferred rate of aquifer drainage), and thus may enhance the implied groundwater resource potential. However, with regard to the three UWLF springs with gauging stations, the smallest contributing river-sink catchment area (probably with the least potential for persistent recharge) was associated with the slowest hydrograph decline. Swallets supplying Cornmill, Foran and Seven Springs have catchment areas of approximately 7.7, 4.9 and 2.8 km^2, respectively, yet the longest sections of consistent recession rate from these sites produce $t0.5$ values of 16.2, 13.6 and 32.7 days, respectively. These data suggest that the rate of UWLF spring hydrograph decline is not significantly altered by small amounts of persistent recharge.

Stage measurements at Cornmill, Foran and Seven Springs suggest maximum discharges of 444 $l\,s^{-1}$, 141 $l\,s^{-1}$ and 140 $l\,s^{-1}$, respectively. The higher value is associated with the greater contributing swallet catchment area. Therefore, the portion of the UWLF aquifer supplying Cornmill Spring does not appear to suppress the transmission of increased runoff/recharge.

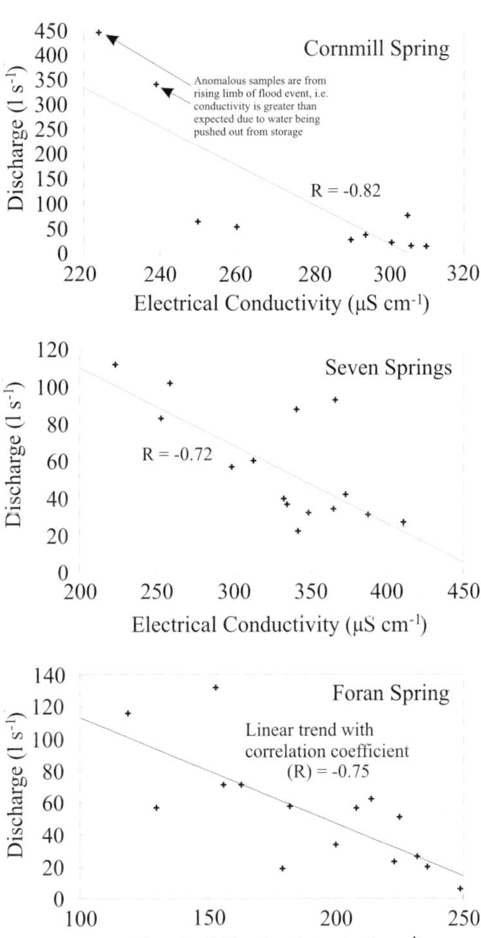

Fig. 3. Scatter plots showing the inverse correlations between UWLF spring discharge rate and electrical conductivity. The relationship is probably due to increased residence time and water rock contact ratios in both the recharging surface catchments and the UWLF aquifer at low flow rates. It is suggested that the quality of the relationship in part relies on the fact that most sampling took place during flood peak recessions, i.e. rising limbs tended to be very rapid and short lived.

Table 4. *Summary of chemical variations from monitored UWLF spring*

Determinand	Statistics	Springs connected to surface river-sink					Diffusely recharged springs	
		Cornmill Spring $n=10$	Seven Springs $n=17$	Foran Spring $n=16$	Carey River Spring $n=19$	Ballintoy Harbour Spring $n=13$	Wellglass Spring $n=15$	Toberterin Spring $n=18$
pH	Mean	6.82	6.71	7.09	6.61	6.72	6.84	6.69
	Range	6.16–7.51	6.10–7.36	6.29–7.83	6.08–7.29	6.03–7.43	6.08–7.42	6.22–7.24
	SD	0.55	0.48	0.59	0.39	0.47	0.50	0.36
	R	−0.46	−0.40	−0.56	−0.23	−0.39	−0.68	
Temp. (°C)	Mean	8.53	7.97	7.68	7.57	8.66	8.95	9.22
	Range	6.90–10.5	7.10–9.00	5.50–9.70	5.50–10.4	7.80–9.80	8.60–9.20	8.70–9.40
	SD	1.15	1.29	1.20	1.61	0.71	0.22	0.14
	R	−0.28	−0.42	0.08	0.21	0.05	**−0.78**	
Na^+ (mg l^{-1})	Mean	11.1	11.6	12.6	13.5	25.2	12.6	10.8
	Range	8.95–14.0	9.80–17.0	9.40–22.0	9.80–25.0	12.0–30.4	9.00–16.0	8.50–15.0
	SD	1.98	2.31	3.51	3.90	5.12	2.63	1.66
	R	−0.36	0.01	−0.41	−0.28	−0.19	−0.47	
K^+ (mg l^{-1})	Mean	1.26	0.44	0.34	0.70	1.76	0.48	1.28
	Range	0.60–2.10	0.30–0.70	0.22–0.45	0.20–1.70	1.30–2.20	0.26–0.76	0.90–1.50
	SD	0.56	0.09	0.06	0.30	0.31	0.16	0.20
	R	0.27	0.52	−0.37	−0.52	−0.23	0.50	
Mg^{2+} (mg l^{-1})	Mean	10.5	9.10	7.40	5.56	10.8	14.3	13.8
	Range	6.90–13.0	6.10–11.0	4.20–10.0	0.50–9.70	7.99–13.0	12.0–16.0	12.0–15.0
	SD	1.99	1.47	1.76	2.32	2.13	1.27	0.94
	R	**−0.80**	−0.56	**−0.81**	−0.68	−0.64	−0.49	
Ca^{2+} (mg l^{-1})	Mean	36.0	48.6	18.5	31.3	66.7	35.5	84.4
	Range	24.0–51.0	28.0–68.0	9.90–31.0	18.8–47.0	34.0–97.0	26.9–48.0	63.7–101
	SD	7.60	10.8	6.25	8.62	20.6	6.48	11.7
	R	−0.59	**−0.73**	**−0.73**	**−0.71**	−0.65	−0.18	
HCO_3^- (mg l^{-1})	Mean	133	169	82.3	106	221	143	277
	Range	83.3–178	90.3–208	34.5–136	52.5–223	121–285	113–171	272–285
	SD	28.1	31.6	25.5	39.9	52.4	19.4	4.08
	R	**−0.89**	**−0.83**	**−0.81**	−0.61	**−0.74**	**−0.86**	

Cl^- (mg l^{-1})	Mean	16.0	21.0	18.0	23.7	47.2	21.9	22.7
	Range	13.6–18.3	17.0–23.8	13.0–23.8	17.2–32.2	33.5–59.5	19.4–24.4	18.7–25.1
	SD	1.46	1.89	3.21	4.94	6.99	1.60	1.43
	R	−0.40	−0.07	−0.62	−0.54	−0.45	**−0.74**	
SO_4^{2-} (mg l^{-1})	Mean	10.4	8.91	7.21	6.64	16.1	8.63	11.0
	Range	7.99–15.4	7.13–13.0	6.02–9.23	5.09–11.9	13.8–24.0	7.07–15.0	10.3–12.8
	SD	2.27	1.35	0.96	1.49	3.05	2.02	0.85
	R	0.25	−0.30	−0.32	0.39	0.13	−0.64	
NO_3^- (mg l^{-1})	Mean	8.46	5.70	1.79	0.87	9.19	16.0	21.6
	Range	5.38–12.1	3.22–7.48	0.00–5.34	0.00–1.58	5.62–12.0	11.2–19.0	15.2–26.4
	SD	2.17	1.17	1.51	0.51	2.13	1.76	3.45
	R	0.36	−0.36	−0.36	−0.07	−0.46	0.12	
SiO_2 (mg l^{-1})	Mean	17.9	19.1	10.5	7.51	13.7	18.9	15.1
	Range	12.2–21.0	13.5–31.2	7.70–13.9	5.13–14.5	10.9–20.1	15.6–23.9	13.9–17.3
	SD	2.43	3.83	1.82	2.23	2.19	1.93	1.11
	R	**−0.7**	−0.56	−0.47	−0.43	−0.32	−0.34	
Electrical cond.	Mean	278	332	194	242	484	316	508
(μS cm^{-1})	Range	224–310	223–411	119–249	166–416	361–642	285–363	458–549
	SD	31.6	49.9	40.2	56.0	80.5	24.1	24.4
	R	**−0.82**	**−0.72**	**−0.75**	**−0.67**	−0.64	**−0.72**	
SIc (log scale)	Mean	−1.17	−1.08	−1.41	−1.57	−0.85	−1.13	−0.65
	Range	−1.95–−0.31	−2.02–−0.29	−2.58–−0.29	−2.32–−0.53	−2.00–0.05	−2.00–−0.45	−1.18–−0.18
	SD	0.63	0.56	0.77	0.54	0.62	0.59	0.38
	R	−0.63	-0.61	**−0.72**	−0.49	−0.58	−0.69	
PCO_2 (bars)	Mean	1.27	1.29	1.17	2.14	1.68	1.33	0.55
	Range	9.1E-3–3.75	1.3E-2–7.55	3.1E-3–7.48	2.4E-2–5.29	7.1E-3–8.97	1.3E-2–8.78	1.8E-2–1.88
	SD	1.43	1.86	1.92	2.61	2.85	2.26	0.67
	R	0.40	0.45	**0.78**	0.50	0.68	0.46	

The above data represent point hydrochemical observations. The number of data sets (*n*) for each determinand vary due to either equipment failure or poor ion balances.
R, correlation coefficient with discharge (exceptional correlations (≥0.7) are highlighted).
SIc, saturation state with respect to calcite (calculated using PHREEQC).

UWLF spring hydrochemical fluctuations

All of the springs that are directly connected to surface river-sinks produced chemographs that have broad seasonal trends, but analogous to their hydrographs they often fluctuated rapidly. This erratic hydrochemical behaviour can be explained by linking the determinand concentrations to precipitation or recharge events. Spring discharge rates typically have quite strong inverse relationships with the different solute concentrations. Correlation coefficients (R) of −0.64 or better are produced with electrical conductivity (Table 4). This relationship may result from simple dilution, increased runoff/flow-through velocity or the resulting residence/reaction time reduction associated with flooding in the surface catchment area. A similar association between dissolved constituents and discharge is also evident from Wellglass Spring which is without any obvious influence from point recharge. However, fluctuations of the various solutes from this source are much more subdued (electrical conductivity SD = 24.1) than even the most conservative of sources benefiting from point recharge (electrical conductivity SD = 31.6). The above observations suggest that Wellglass Spring is influenced to a limited degree by rapidly circulating precipitation in its UWLF outcrop region.

Significant changes in stream water chemistry can occur at similar discharges depending upon the stage of a flood event (Drever 1997). Such variation in this study is not particularly apparent from the often strong correlations between solute concentrations and discharge rate. This observation may be due to the fact that water sampling almost invariably coincided with discharge recession. All three spring gauging stations confirm that rising limbs of flood pulses typically had durations of several hours, and only the two highest discharge samples from Cornmill Spring can be confirmed as being from the ascending limb or crest of a discrete flood event. These samples have anomalously high conductivity values and may represent groundwater in storage being pushed out ahead of the influx (Fig. 3).

Wellglass and Toberterin Springs, which appear to be predominantly fed by diffuse recharge from the overlying basalts, exhibited comparatively suppressed hydrochemical fluctuations (lower electrical conductivity SDs, see Table 4). This behaviour is characteristic of springs with one recharge source and probably limited karstic groundwater transit conditions (Mazor 1991). Slower groundwater flow would promote relatively long residence times in the aquifer and thus a degree of water–rock equilibrium to be attained. It is likely the diffuse recharge envisaged for Wellglass and Toberterin Springs will also contribute to their less erratic hydrochemistry as opposed to a sporadic surface runoff component.

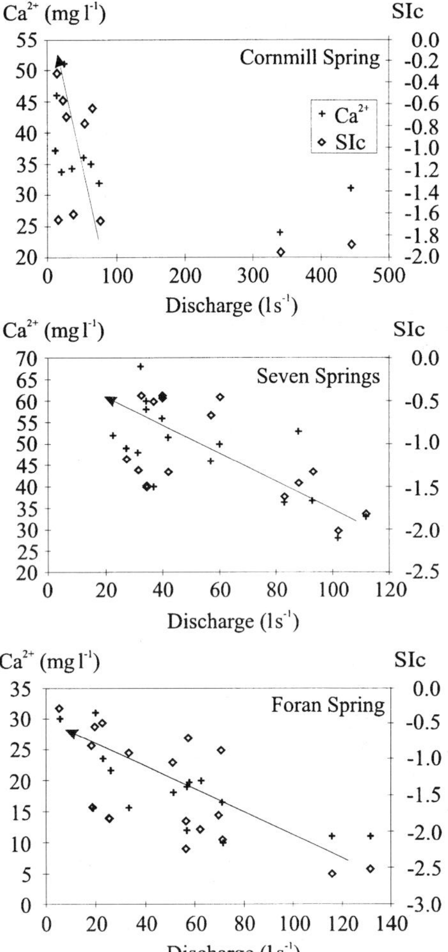

Fig. 4. Ca^{2+} concentration and saturation state with respect to calcite as a function of flow rate from three UWLF spring sources. The data form fairly linear relationships and suggest that as spring discharge recession advances the groundwater has attained a greater degree of equilibrium with the aquifer mineralogy. This behaviour is consistent with the drainage of progressively smaller void spaces where flow-through is less rapid.

Indications of UWLF recharge and storage characteristics. The carbonate reaction system. All of the monitored UWLF springs typically produce waters which are undersaturated with respect to calcite. As the aquifer is essentially pure $CaCO_3$,

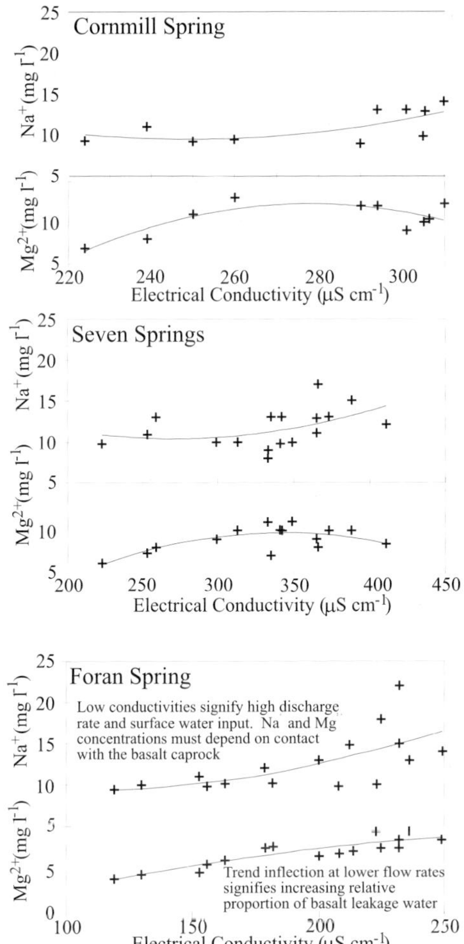

Fig. 5. Scatter plots showing the Na^+ and Mg^{2+} concentrations as a function of conductivity from three UWLF springs. Both ions appear to increase with total dissolved solids (as discharge decreases) until an advanced stage of recession, thereafter, Na^+ appears to increase at an accelerated rate and Mg^{2+} may actually decrease. These alterations are consistent with an increased component of basalt leakage water (Barnes & Worden 1998).

rock dissolution will therefore increase Ca^{2+} and alkalinity concentrations depending upon the reaction rate and groundwater residence time.

Figure 4 presents the calcium concentration and saturation state with respect to calcite (SIc) as a function of discharge rate from the three spring sources with gauging stations (i.e. most accurate discharge methods used). These data produce fairly linear inverse relationships and show that groundwater emerging at more advanced stages of recession has achieved a greater equilibrium with aquifer mineralogy. This behaviour is consistent with drainage of smaller void spaces at lower flow rates. Toberterin spring which appears to best represent drainage fed solely by diffuse basalt leakage recharge is always relatively close to saturation with respect to calcite and has comparatively elevated Ca^{2+} concentrations (Table 4). These characteristics represent the more consistent and advanced extent of water–rock interaction promoted by dispersed recharge and subsequently less karstic flow characteristics. This source in particular therefore provides an insight into the flow regime and groundwater quality below the basalt cover.

At all UWLF springs, maximum pH and minimum PCO_2 levels were attained during the lowest summer discharges. These data sets therefore produce negative and positive correlations with discharge rate respectively (Table 4). This behaviour is contrary to the annual soil zone CO_2 maximum, and at odds with the behaviour of numerous other carbonate springs (Shuster & White 1971; White 1988). It is inferred that after prolonged recession the influence of surface recharge is reduced and the proportion of evolved basalt leakage water emerging from the concealed portion of the aquifer becomes significant. The latter source will have suppressed PCO_2 levels because of the increased residence time and thus the potential for soil-zone derived CO_2 to de-gas or to be consumed by mineral weathering.

Ions outside the carbonate reaction system. As the UWLF aquifer is essentially monomineralic calcite, groundwater solutes outside the carbonate reaction system should be predominantly controlled by recharge hydrochemistry. These ions typically correlate with discharge with varying degrees of linearity. However, data from the three springs with permanent gauging facilities (Cornmill, Foran and Seven Springs) reveal that Na^+ and Mg^{2+} have a distinct non-linear relationship with electrical conductivity and discharge. Thus their hydrogeochemical evolution at different flow rates is not consistent with the bulk of other dissolved species (Fig. 5). At higher discharge rates and thus at lower conductivity values the Na^+ and Mg^{2+} concentrations can be observed to increase with conductivity. These larger discharges from the aforementioned sources are indicative of flooding from river-sink recharge. Therefore, Na^+ and Mg^{2+} must increase on top of the basalt due to the increasing significance of surface catchment alteration processes. At discharges of below about $40 \, l \, s^{-1}$ (as total dissolved solids increase further) the Na^+ and Mg^{2+} concentrations show a substantial deviation from that of other dissolved ions (conductivity). These non-linear trends cannot be due to evolution along a simple reaction/dilution pathway.

Waters which percolate through the basalt pile have been demonstrated to increase and decrease their Na^+ and Mg^{2+} concentrations, respectively (Barnes & Worden 1998). It is therefore apparent that during advanced base-flow conditions basalt leakage derived water is mixing with the surface water input in significant proportion to alter the total spring water chemistry. At high discharge rates the chemical signature of the basalt leakage component to recharge would be overwhelmed by the proportionately huge swallet input.

From Table 4 it is evident that the UWLF springs supplied by diffuse recharge have higher

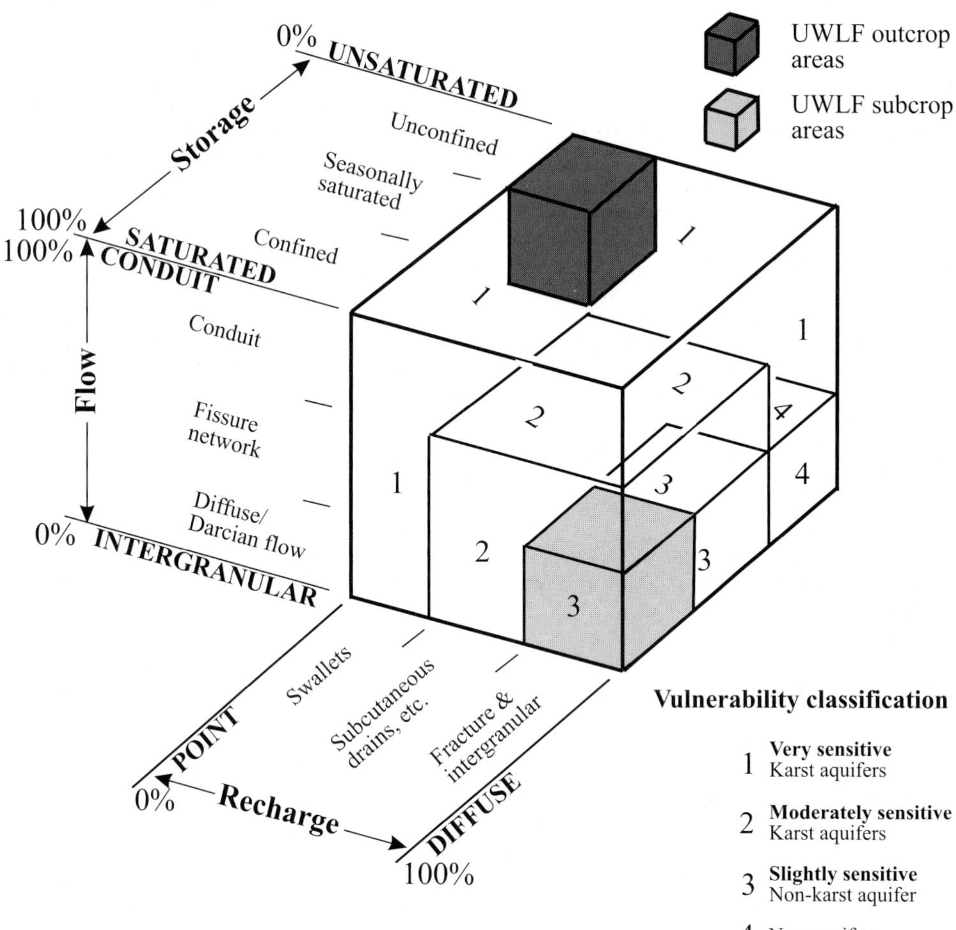

Fig. 6. Conceptual model for classification of carbonate aquifers involving end members on a percentage basis for storage, flow and recharge characteristics (modified after Quinlan *et al.* 1991). Boundaries between vulnerability fields are approximate and the volumes plotted in the cube for the UWLF are based on the dominant form of each of the three parameters. The UWLF at outcrop is seen to be only partially saturated, dominantly recharged by swallets and has conduit transmission characteristics. Conversely, the confined UWLF aquifer can only be directly recharged by diffuse basalt leakage water and appears to be largely dominated by more diffuse flow characteristics. From this classification it can be inferred that the best development strategy for UWLF groundwater resources would involve exploitation of conduit water sources at springs in outcrop areas. At subcrop UWLF groundwater can only be exploited via boreholes, although the water is likely to be of better quality and less vulnerable to contamination.

mean NO_3^- concentrations. This occurrence is despite percolation via the soil zone and the apparently less vulnerable status of the aquifer in these areas. It is proposed that the observed variation in NO_3^- between the different sources is a direct result of more intensive agricultural practices in the less karstic regions (i.e. the west of the study area where there are fewer river-sinks (Fig. 1a)) rather than a consequence of aquifer throughput characteristics. This conclusion suggests that denitrification is not particularly active in Toberterin and Wellglass Spring catchment areas.

Conclusions

Five of the seven largest UWLF springs are known to be directly connected by a conduit system to surface swallets. Substantial discharge fluctuations from these sources have demonstrated that river-sink recharge is often their predominant supply, and that sustained drainage from the large concealed region of the aquifer is proportionately small. Therefore, these springs largely represent drainage from the unconfined portion of the aquifer which is characterized by point recharge and conduit flow. This region of the aquifer has been independently classified in the conceptual summary model presented in Fig. 6 (modified after Quinlan *et al.* 1991). The categorization status suggests that the outcrop region is a highly sensitive karst aquifer and is best exploited by development of spring sources. The non-recommendation of boreholes is consistent with the haphazard nature of a dispersed fracture/fissure network. This flow system has been predicted by the rapid descent of UWLF spring flood recessions (high recession coefficients) which is characteristic of a high formational hydraulic conductivity and a low storage capacity.

At subcrop the UWLF can only be directly recharged by diffuse basalt leakage water. Springs which predominantly drain this portion of the aquifer (i.e. are without the benefit of river-sink recharge) tend to exhibit comparatively suppressed discharge and chemical fluctuations. Both of these characteristics are indicative of a non-karstic flow regime, and the confined aquifer has been correspondingly classified in Fig. 6. This portion of the aquifer can only be exploited via boreholes and is likely to be less productive, although the water quality has been shown to be more stable and less vulnerable to contamination.

The largest UWLF springs show signs of both river-sink and basalt leakage recharge. The minor leakage component is only apparent from hydrochemical variations at low flow rates. These alterations during advanced recession include elevated Na^+ and suppressed Mg^{2+} concentrations (relative to other dissolved species). At the same time Ca^{2+}, alkalinity and calcite saturation states increase in a manner reflective of the increased UWLF aquifer residence and thus dissolution time experienced by groundwaters recharged via the basalt. Minimum PCO_2 levels are attained during suppressed summer discharges, contrary to the annual soil zone CO_2 maximum. This phenomenon can also be explained by an increasingly significant component of groundwater flow from the confined UWLF aquifer.

The author wishes to thank the Department of Education for Northern Ireland for funding this research. J. Parnell (The University of Aberdeen) and A. Ruffell (The Queen's University of Belfast) are also acknowledged for their support and advice throughout the work. The Geological Survey of Northern Ireland and J. Marshall at The University of Liverpool deserve much credit for providing field equipment and access to hydrogeochemical software, respectively. Lastly, thanks are also due to the staff in the Biography Lab. at Queen's for carrying out most of the chemical analyses. The opinions expressed in this paper are those of the author alone.

References

BARNES, S. 1999. Karstic groundwater flow characteristics in the Cretaceous Chalk aquifer, Northern Ireland. *Quarterly Journal of Engineering Geology*, **32**, 55–68.

—— & WORDEN, R. H. 1998. Understanding groundwater sources and movement using water chemistry and tracers in a low matrix permeability terrain: the Cretaceous (Chalk) Ulster White Limestone Formation, Northern Ireland. *Applied Geochemistry*, **13**, 143–153.

BGS/DOENI. 1994. *Hydrogeological map of Northern Ireland. 1:250 000*. British Geological Survey, Keyworth.

BURDON, D. J. & PAPAKIS, N. 1963. *Handbook of Karst Hydrogeology*. United Nations, Athens.

DREVER, J. I. 1997. *The geochemistry of natural waters*. Prentice-Hall, Inc., Englewood Cliffs, N.J.

FORD, D. C. & WILLIAMS, P. W. 1989. *Karst Geomorphology and Hydrology*. Unwin Hyman Inc., London.

K.M.M. 1994. *Magheramorne landfill project hydrogeological report*. K.M.M., Consulting Engineers. Report No. **2924.01/G 01/94**.

MAILLET, E. 1905. *Essais d'hydraulique souterraine et fluviale*. Hermann, Paris.

MALIVA, R. G. & DICKSON, J. A. D. 1997. Ulster White Limestone Formation (Upper Cretaceous) of Northern Ireland: effects of basalt loading on chalk diagenesis. *Sedimentology*, **44**, 105–112.

MANGIN, A. 1975. *Contribution a l'etude hydrodynamique des aquiferes karstiques*. DES thesis, University of Dijon, France.

MARTIN, G. N. 1973. Characterization of simple exponential baseflow recessions. *Journal of Hydrology*, **1291**, 57–62.

MAZOR, E. 1991. *Applied chemical and isotopic groundwater hydrology*. Open University Press, Milton Keynes.

MILANOVIC, P. T. 1981. *Karst Hydrogeology*. Water Resources Publications, Colorado.

NUTBROWN, D. A. & DOWNING, R. A. 1976. Normal-mode analysis of the structure of baseflow-recession curves. *Journal of Hydrology*, **30**, 327–340.

PARKHURST, D. L. 1995. *User's guide to PHREEQC – A computer program for speciation, reaction-path, advective-transport, and inverse geochemical calculations*. US Geological Survey Water-Resources Investigations Report **95–4227**.

QUINLAN, J. F., SMART, P. L., SCHINDEL, G. M., ALEXANDER, E. C. JR., EDWARDS, A. J. & SMITH, A. R. 1991. *Proceedings of the Third Conference on Hydrogeology, Ecology, Monitoring, and Management of Ground Water in Karst Terranes*, Nashville, Tennessee, USA, 4–6 December, 573–635.

ROBINS, N. S. 1996. *Hydrogeology of Northern Ireland*. HMSO, London.

SHUSTER, E. T. & WHITE, W. B. 1971. Seasonal Fluctuations in the chemistry of limestone springs: A possible means for characterizing carbonate aquifers. *Journal of Hydrology*, **14**, 93–128.

SHAW, E. M. 1994. *Hydrology in Practice* (Third Edition). Chapman and Hall, London.

SMART, P. L. & HOBBS, S. L. 1986. Characteristics of carbonate aquifers: A conceptual base. Environmental Problems in Karst Terranes and Their Solutions Conference (1st, Bowling Green, Ky.). *In*: *Proceedings of the National Water Well Association*, Dublin, Ohio, 1–14.

WHITE, W. B. 1988. *Geomorphology and Hydrology of Karst Terrains*. Oxford University Press, Oxford.

WILSON, E. M. 1994. *Engineering Hydrology* (Fourth Edition). The Macmillan Press, Ltd., Houndmills, Basingstoke, Hampshire and London.

Influent rivers: a pollution threat to Schwyll Spring, South Wales?

STEVEN L. HOBBS

Enviros Aspinwall, 5 Chiltern Close, Cardiff CF14 5DL, UK
(e-mail: steve.hobbs@enviros.com)

Abstract: A number of influent rivers cross the southern outcrop of the Carboniferous Limestone in South Wales. They lose a proportion of their flow to groundwater via both discrete sinks and leaky river beds. These influent rivers represent a pollution threat to local springs that is dependent upon many factors including antecedent conditions, pollutant type, the location of pollutant loss and the nature and volume of influent flow. An attempt is made to define the catchment area of Schwyll Spring public water supply, near Bridgend in South Wales, prior to considering the potential for pollution from influent rivers.

Schwyll Spring is located near the town of Bridgend in South Wales. The discharge from the 13 individual springs at this location used to flow into the adjacent River Ewenny (Knox 1933) until 1872, when the water was first used for supply (Jones 1985). Early in the twentieth century a shaft was dug to the south-east of the springs to prevent the influx of poor quality river water to the springs during high water. This shaft encountered a natural chamber that has subsequently been entered by cave divers and which continues for over 440 m upstream (and is still open), with several routes going off the main passage.

Welsh Water has a licence to abstract up to 7955 $Ml\,a^{-1}$ from Schwyll Spring, although for operational reasons the spring is not currently being utilized. Water quality is generally good except following heavy rainfall when the water can run turbid and the spring is taken out of supply.

Physical setting

Regional setting

Bridgend is located close to the South Wales coastline in a relatively flat area with low hills. To the north is higher terrain associated with the western end of the South Wales valleys and the Brecon Beacons beyond. In simple terms the geology of South Wales comprises a broad east–west orientated syncline with Carboniferous Coal Measures strata forming the core some 25 km in width. A relatively narrow band of Carboniferous Limestone surrounds this core with Devonian Old Red Sandstone outcropping around this. Triassic and Jurrasic deposits, dominated by mudstones, overlie the limestone and sandstone in places along the coast.

Local setting

Schwyll Spring is located at about 5 m above Ordnance Datum (m AOD) on the south side of the River Ewenny (Fig. 1). The latter flows south-west for 1 km before becoming a tributary to the Ogmore River which discharges into the Bristol Channel some 3 km to the south-west of the spring. To the south the land rises steeply to Ogmore Down and Beacon Down which form a plateau surface at about 80 m AOD. The surface of this plateau is dissected by several dry valleys on its north-west and north-east sides. The plateau is cut by the River Alun, a small north-west flowing watercourse within a steep sided valley. This river is a tributary to the Ewenny upstream of the spring. To the west of the spring, on the far side of the Ewenny and Ogmore Rivers lies the sand dune system of Merthyr Mawr Warren. To the north and north-east the land rises gently towards the town of Bridgend and the M4 motorway (Fig. 1).

The geology in the vicinity of Schwyll Spring comprises Dinantian Carboniferous Limestone (on the southern side of the South Wales coalfield syncline) which is overlain in places by superficial deposits including wind blown sand, head and alluvium adjacent to the river. The limestone dips at 5° to 15° to the south and is cut by several faults, one of which, the Rhiw Fault, is probably partly responsible for the location of the spring. Some 1 km to the north and 3 km to the south and east of the spring the limestone is overlain by Lower Lias limestones and mudstones and Triassic mudstones. The latter

From: ROBINS, N. S. & MISSTEAR, B. D. R. (eds) *Groundwater in the Celtic Regions: Studies in Hard Rock and Quaternary Hydrogeology*. Geological Society, London, Special Publications, **182**, 113–121. 1-86239-077-0/00/

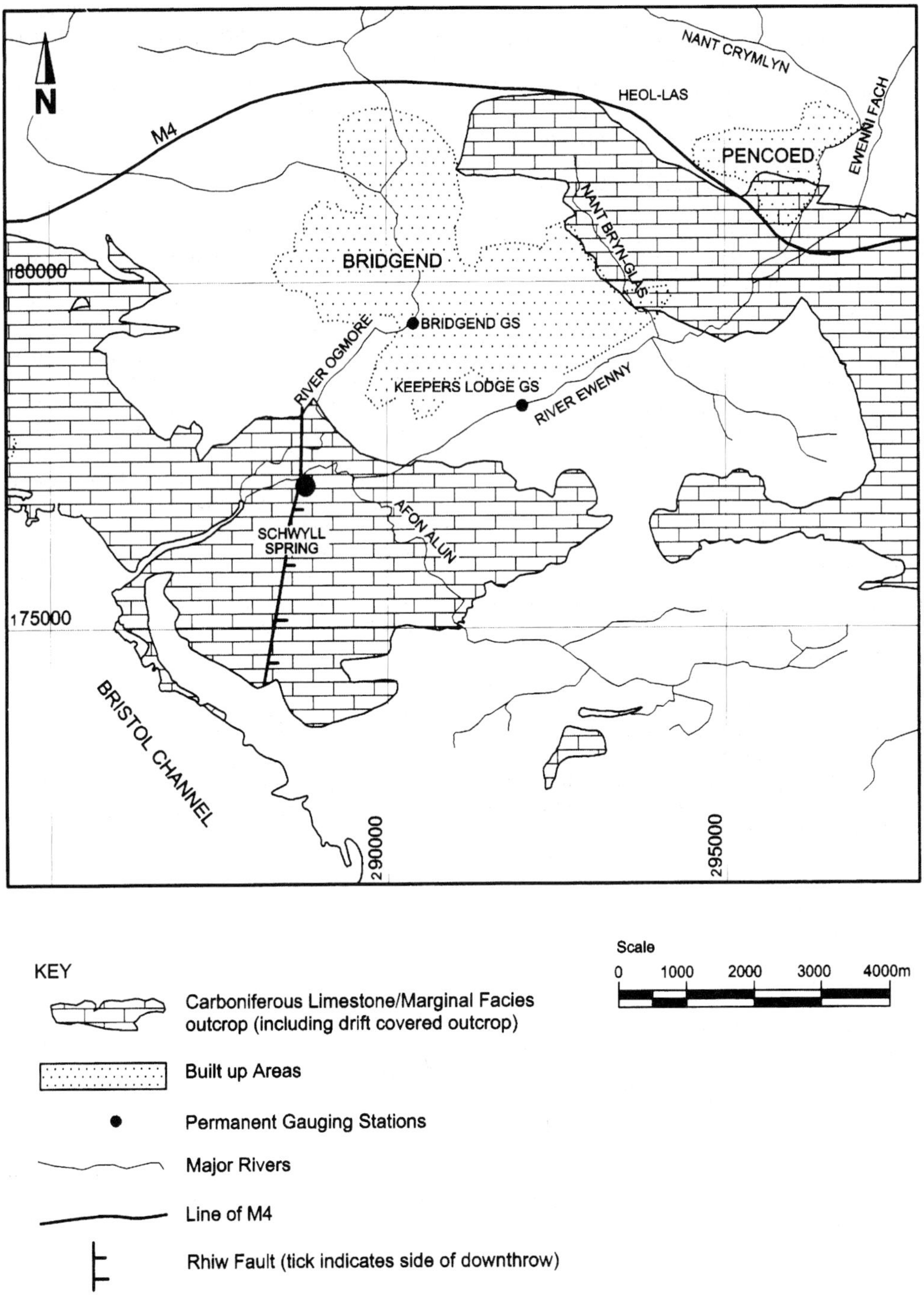

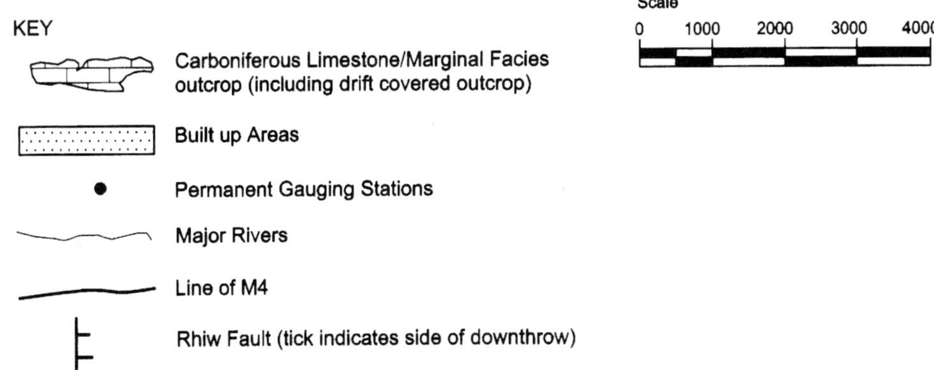

Fig. 1. Study area and outline of the solid geology.

include the conglomeratic marginal facies which can be in hydraulic continuity with the limestone.

Within the Ogmore catchment the rivers Ogmore, Ewenny and Alun all flow across the limestone outcrop along part of their length. In such areas loss of river water to groundwater takes place, perhaps the most striking local example of which is at Merthyr Mawr. Sinks on the bank of the River Ogmore take water that flows through a series of conduits before the water rises and discharges back into the river 700 m downstream. The conduits are intersected in places by shallow potholes that allow human access to the underground passages. At other locations, such as on the south side of Bridgend, loss of river water by diffuse bed seepage is more common (Welsh Water Authority, undated). In dry weather sections of the Afon Alun are completely lost to groundwater.

Spring catchment

Catchment area

The Carboniferous Limestone in the Celtic regions of Britain and Ireland has a low primary porosity and permeability. However, preferential dissolution of carbonate rocks along bedding planes, fractures and joints increases the porosity and can increase the permeability by many orders of magnitude. Spatial variations in dissolution can produce a heterogeneous aquifer in which the permeability at two locations a short distance apart can vary substantially (Ford & Williams 1989). The location and nature of these preferential flow routes are of great importance in establishing the catchment of a spring in such an aquifer.

As a starting point to determine the approximate spring catchment a simple water balance was calculated. This utilized the estimated spring discharge and the effective rainfall in the area to determine the outcrop necessary to support the spring. This calculation was at best an estimate as there is no continuous measurement of discharge at the spring and only occasional 'spot' measurements. These indicated an annual average discharge of the order of 12 300 Ml a^{-1} (Hobbs 1993). With an effective rainfall in the area of some 540 mm a^{-1} this is equivalent to a catchment of some 23 km^2. Based on the available outcrop area, the area calculated above and the estimated area for other abstractions in the locality (the Bridgend and Pwllwy boreholes), a conjectural catchment area has been formulated (Fig. 2).

Water tracing

The above calculation indicates an approximate catchment area for Schwyll Spring but does not allow an estimate of its extent. A number of attempts have been made to trace sinking streams and influent rivers to the spring (Knox 1933; Aldous 1986, 1988; Dixon *et al.* 1986; Williams & Brown 1989). These attempts confirmed that the spring is recharged by dispersed rainfall over the outcrop of the limestone in addition to concentrated recharge from a small number of stream sinks and from rivers influent to the Carboniferous Limestone. A summary of the traces is presented on Fig. 2 and detailed in Table 1. Travel times have not been calculated for traces with sporadic breakthrough as they can grossly over-estimate the flow velocity. Times of travel based on time to first arrival and time to peak can also over-estimate actual travel time, which should be based on time to centroid of breakthrough curves. Where tracer recovery is low such calculations cannot be made accurately.

Of most interest here are the traces attempted from the River Ogmore, River Ewenny, River Alun and the Merthyr Mawr sinks. Each of these will be considered briefly (see Hobbs 1993, for more details).

River Ogmore traces (traces 2 and 3). These were undertaken from two different sections of the Ogmore using bacteriophage. The tracers were injected upstream of where the river flows over limestone strata to ensure that the tracer could move to groundwater via any influent zones along the river. The traces proved a positive connection with Schwyll Spring, albeit with sporadic tracer breakthrough. The traces confirm that the Ogmore is influent where it crosses the Carboniferous Limestone.

River Ewenny traces (traces 5, 6 and 10). A series of traces were undertaken at different times with both fluorescent dye and bacteriophage. Both tracers were injected into a sink in the bed of the Ewenny; however, only the bacteriophage traces were positive at Schwyll Spring. A further bacteriophage trace was completed some 4 km downstream of the above sink. The tracer was injected into the river and was detected at the spring, albeit sporadically with low recovery.

River Alun traces (traces 7, 8 and 9). Bacteriophage injected into the river were not detected at Schwyll. However, both fluorescent dye and

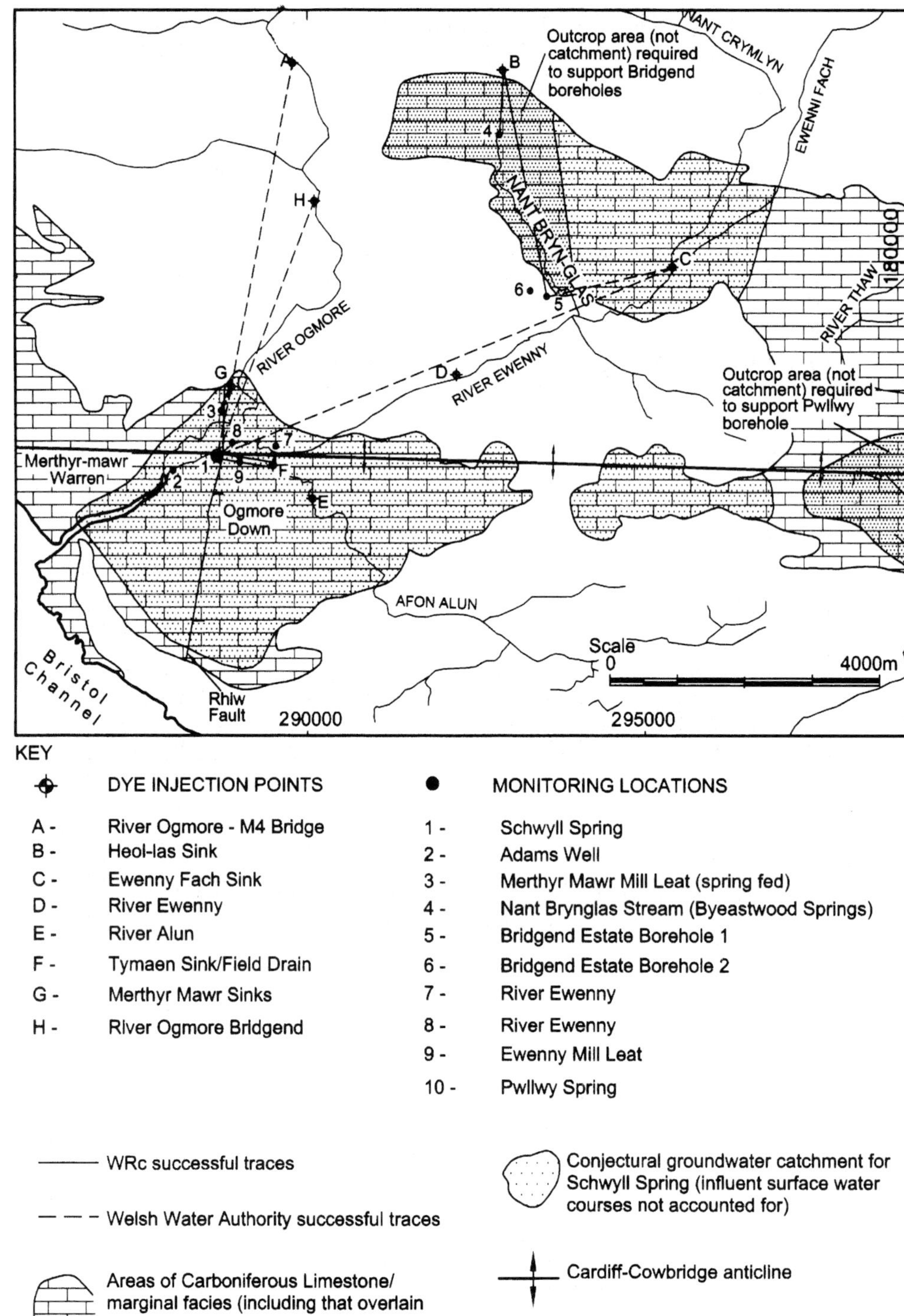

Fig. 2. Water tracing and conjectural catchment for Schwyll Spring.

Table 1. *Summary of water traces carried out in the Schwyll Spring area*

Trace no.	Completed by	Date	Type of tracer	Input location	Recovery location	Travel Time*		Flow Velocity†		% Tracer recovery	Comments
						First (hrs)	Peak (hrs)	First (m d)	Peak (m d)		
1	Knox	1920s	Flurorescein	Schwyll Spring Pumping Shaft (SS 8880 7706)	12 of the 13 other springs which rise in the vicinity of schwyll	N/D‡	N/D	N/D	N/D	N/D	No breakthrough at spring in bed of R. Ewenny
2	Welsh Water Authority	1986	Bacteriophage	River Ogmore, Bridgend (SS 9013 8065)	Schwyll Spring	–	–	7800	1500	N/D	Sporadic breakthrough
3	Welsh Water Authority	1989	Bacteriophage	River Ogmore, Bridgend (SS 8980 8283)	Schwyll Spring	–	–	6100	2600	N/D	Sporadic breakthrough
					Merthyr Mawr Mill Leat			–	–	N/D	
4	Welsh Water Authority	1989	Bacteriophage	Merthyr Mawr Sinks (SS 8901 7807)	Merthyr Mawr Mill Leat	–	–	3000	2000	<1%	Sporadic breakthrough
					Schwyll Spring			2400	1200	N/D	
5	Welsh Water Authority	1989	Bacteriophage	Ewenny Fach Sink (SS 9542 7990)	Schwyll Spring	–	–	8500	3400	N/D	Sporadic breakthrough
					WDA BH No. 1			900		N/D	
6	Welsh Water Authority	1989	Bacteriophage	River Ewenny (SS 9213 7822)	Schwyll Spring	–	–	2600	2200	0.01%	Sporadic breakthrough
7	Welsh Water Authority	1989	Bacteriophage	Afon Alun (SS 9007 7652)	No recovery	–	–	–	–	–	
8	Welsh Water Authority	1989	Bacteriophage	Field Drain (SS 8943 7705)	Schwyll Spring	1	3	15600	5200	<1%	
					River Ewenny						
9	Water Research Centre	1988	Fluroescein sodium	Tymaen Sinkhole (SS 8943 7705)	Schwyll Spring	4.5	5.5	3500	2800	20%	
					River Ewenny						
10	Water Research Centre	1988	Fluroescein sodium	Ewenny Fach Sink (SS 9542 7990)	No dye recovered	–	–	–	–	–	Test repeated twice, once using 7 kg, then 14 kg of Fluroescein dye
11	Water Research Centre	1988	Fluroescein sodium	Merthyr Mawr Sinks (SS 8901 7807)	Mill Leat Rising	4.75	6.5	2500	1800	57%	
					River Ogmore	5	7	–	–	17%	
12	Water Research Centre	1988	Fluroescein sodium	M4 sink at Heol Las (SS 9288 8267)	River Ewenny (via Byeastwood Springs)	9	11	–	–	63%	25 kg of fluroescein used, much lost to overflow culvert
					Bridgend Estate BH	21	31	3900	2600		

* Travel times have not been calculated for traces with sporadic tracer breakthrough as they can result in a gross overestimate of flow velocity.
† All flow velocities are based on straight line connections.
‡ N/D, data not available to allow calculations to be made.

phages injected to the Tymaen sinkhole adjacent to the lower stretch of the river, 0.5 km east of Schwyll, proved positive at the spring.

Merthyr Mawr trace (trace 11). Both fluorescent dye and bacteriophage traces were positive to the downstream risings on the Ogmore. Only the bacteriophages were also positively traced to Schwyll Spring.

The above traces indicate that influent river water comprises a component of the water discharged from Schwyll Spring. However, the results of the tracing are very variable with low recoveries suggesting that dilution is high and that there is not a direct open conduit (passage) link between the influent water courses and Schwyll.

Influent rivers

The work outlined indicates that some of the water rising at Schwyll Spring comprises influent river water. However, the tracing work was unable to estimate the proportion of the spring discharge that comprises influent river water. Two further means of estimating this have been examined, flow gauging and spring chemistry.

Flow gauging

Although there are four permanent gauging stations in the River Ogmore catchment only two are in the study area (Fig. 1) and neither of these are suitably located to determine the river loss to groundwater. A number of 'spot' gauging exercises have been undertaken but these were largely random and discontinuous in nature. A total of 15 sites have been monitored since 1965 but only on one occasion are the data suitable for examining river losses. A survey of 5 points on the River Ogmore carried out on 11 August 1986 indicates that 135 $Ml\,d^{-1}$ (20% of the river flow) of water entered the ground between Pen y Cae bridge over the M4 motorway and the gauging station near the Swing Bridge in Bridgend town (Fig. 1). This is six times the licensed abstraction rate (22 $Ml\,d^{-1}$) and four times the estimated discharge (34 $Ml\,d^{-1}$) from Schwyll Spring. In contrast Glamorgan River Authority (1971) estimated a 4 $Ml\,d^{-1}$ loss from the River Ogmore in the Bridgend area based on analysis of residual flow diagrams.

Water chemistry

A brief examination of the water quality at Schwyll Spring was made to determine if any clear indication of river water recharge could be gleaned. However, excepting the turbidity pulses, there was insufficient information to allow for an estimate of the importance of river water recharge. This may be because river water is a very small contributor to spring discharge, in which case dilution of river water in groundwater would mean that no effect would be observed at Schwyll. Alternatively, it may be that travel times are sufficiently slow for chemical changes to take place in the river water (by ionic exchange) so that its chemistry is not significantly different to that of the groundwater by the time it resurges at Schwyll.

Discussion

Of the traces to Schwyll Spring only those from the Tymaen sink, 0.5 km away, were particularly successful, and even then the maximum tracer recovery was only 20%. The remainder of the traces indicate that connections exist between the influent rivers and Schwyll Spring although there is not a direct open conduit (passage) link between the influent rivers and the spring.

The assessment of water chemistry is of little assistance in estimating the proportion of influent river water appearing at Schwyll whilst the spot gauging does not indicate a consistent volume of influent water, even if substantial seasonal fluctuations are taken into account. The gauging carried out in 1986 was completed in summer when groundwater levels were likely to have been low, so that river leakage would have been high. At such times dispersed river bed loss is maximized in addition to that which can enter stream sinks as concentrated recharge. The water lost to the ground may have reappeared further downstream or at one of the other springs which are present in the area in addition to Schwyll.

In addition to the tracer tests, gauging and water analysis, the potential catchment area has been assessed based on 'first principles'. The approximate spring discharge is known, the effective rainfall in the area is known as is the available outcrop area. Calculations based on these figures indicate that there is an excess of aquifer available to support Schwyll Spring flow and other known discharge points (Bridgend and Pwllwy boreholes – Fig. 2.). The small number of known discharge points does, however, have to be treated with caution. Substantial submarine springs are known to exist in the Bristol Channel – that from the Great Spring which was intersected when boring the Severn

Railway Tunnel encountered a flow estimated as 27 000 Ml a^{-1}.

The data indicate that river water from the Ogmore catchment does enter the ground at a number of locations. Some of these result in recharge to the Carboniferous Limestone with flow eventually reaching Schwyll Spring. However, the sporadic tracer recovery suggests that the route to the spring is neither direct nor particularly rapid, with the exception of the Tymaen Sink/Field Drain traces to Schwyll and the Mertyr Mawr Sink to Mill Leat dye trace. The results do not indicate a well developed conduit system from the influent rivers to the spring.

Potential for pollution

Even though the results of the above investigations do not allow the exact proportion of influent river water to be determined the data do have implications for pollution prevention at the spring. The potential impact of influent river water upon Schwyll Spring is difficult to predict as it will be dependent upon a number of factors as follows:

(a) antecedent conditions including rainfall, groundwater levels and river discharge;
(b) the type of pollutants, especially whether they float, sink, remain in suspension or are dissolved;
(c) the location within the catchment where the pollutant is lost;
(d) the nature of the connection between the point of loss and the spring.

The importance of antecedent conditions is that they influence the amount of water that may move from surface to groundwater. With low groundwater levels, and especially if river levels are high (following a summer thunderstorm for example), then the amount of influent water may be expected to be greater than when groundwater levels are high and river flow is moderate. River discharge *per se* is also important. High discharges mean that there is a greater volume of water available for dilution should a pollution incident occur. However, the converse can also be true if the pollutant is natural such as suspended sediment (see below).

The type of pollutant is critical. Where the pollutant is spilt or leaks into a river and is a light non aqueous phase liquid (LNAPL), such as diesel for example, then the pollution potential to groundwater is reduced as the diesel would float and only that part moving into solution would enter the groundwater system. However, if river flow is very low there may be potential for direct movement of LNAPLs to groundwater via exposed fractures and fissures in the river bed. If the pollution spill occurred in the upper reaches of the catchment this gives the relevant authorities time to deploy downstream booms or other means to reduce movement of floating free products. This is clearly not the case where river levels are low or where the majority of the river water sinks at a discrete point. At such a time even the floating material will be lost to groundwater. Where dense non aqueous phase liquids (DNAPL) pollute a river these will sink and may take time to move downstream (depending upon river discharge). This allows the relevant authority time for mitigation measures, although these are more complex to carry out successfully than with LNAPLs.

The greatest threat to groundwater from influent rivers is from pollutants that move into solution. These can travel at the same speed as the river water and can become dispersed so that the opportunities for mitigation are fewer. However, through the process of dispersion the impact can be reduced. Pollutants can also move in suspension and cause a problem, although flow must be turbulent to support these particles and in terms of groundwater, pollution is only of relevance to those systems with solutionally–enlarged flow paths.

The location of a pollutant loss within a catchment is an important influence upon the potential impact. Losses in the upper reaches of a catchment mean that a significant period of time may elapse before the pollutant reaches an influent point lower in the catchment. This may allow time for the relevant authorities to undertake mitigating action. Longer travel times/distances also provide additional opportunities for dilution, dispersion and attenuation as the pollutant plume moves downstream.

The nature of the connection between an influent reach and the spring is a major control on the type of incident that may occur at the spring. Losses close to the spring, or where there are high velocity, well defined flow paths (e.g. from Tymaen sinkhole), can result in relatively short duration high peak impacts. On the other hand, cases where the losses are more distant and the flow paths less direct can result in a pollutant plume of longer duration but with a peak with lower concentration.

Potential pollutants may be either natural or from human activities. The most common natural pollutant, and one that regularly affects Schwyll, is turbidity. This is probably caused by suspended sediment–laden runoff entering a discrete stream sinkhole (such as Tymaen sink) following rainstorms, rather than from influent rivers. The elevated turbidity pulses, which result in the spring being taken out of supply, suggest a rapid local connection between this sink and the spring.

At Schwyll the main known potential sources of spring contamination are:

- Tymaen sink – a discrete sinking stream which has been traced to Schwyll;
- Merthyr Mawr sinks – discrete stream sinks on the bank of the River Ogmore which have been traced (albeit with sporadic breakthrough) to Schwyll;
- influent sections of the rivers Ewenny, Alun and Ogmore.

Of these, Tymaen sink is probably partly responsible for the turbidity pulses at Schwyll. If they become a significant problem then it may be that management options (eg. stream diversion) could be examined to control the entry of turbid water to the sink. Although Tymaen sink is very close to Schwyll, such that dilution and dispersion of any human induced pollutant would be limited, the stream is small and from a rural area. It is therefore under a relatively low threat from such pollutants. There may, of course, be other discrete sinks such as that at Tymaen that have not yet been located and which contribute to the suspended sediment pulses observed at Schwyll.

The traces from Merthyr Mawr sinks were not strongly positive, so that any pollutant is unlikely to move rapidly or without significant dilution to Schwyll. The pollutant threat to the spring in this situation is considered to be small.

The influent rivers probably represent a greater threat to Schwyll than the discrete sources. In theory the whole catchment upstream of the influent sections of river could contribute to the spring. Within this catchment there are many industrial discharges that may cause pollution if a leak occurs or if there is a failure in their systems that results in an unauthorized effluent discharge.

In addition to the risk of pollution from industrial discharges there is also the potential for contaminants from road accidents to enter rivers in the catchment. This is particularly so in the vicinity of the motorway which passes some 5.5 km to the north of the spring (Fig. 1). One such accident that occurred in 1985 resulted in a significant loss of oil to the River Ewenny. However, a large proportion of this oil was trapped by booms placed across the river by Welsh Water and there was no detectable increase in oil concentrations at Schwyll.

Conclusions

The information collected as part of this investigation helps the Environment Agency to assess planning applications for new developments within the catchment area for Schwyll Spring. It is also of great assistance in assessing applications for industrial discharge consents and renewals of existing consents within the catchment.

Information regarding the nature and extent of the Schwyll Spring catchment also assists the water company in managing the supply. Welsh Water has long been aware of the problem of turbidity at the spring and has constructed storage tanks to accommodate periods when the spring is taken out of supply due to high turbidity. Better knowledge regarding areas where rivers are influent and the likely volume of influent water can also help with water management if a pollution incident does occur. If the water company is aware that a pollution incident has taken place it can take precautions to ensure that any breakout at the spring is identified and the spring is taken out of supply if this represents a hazard to health. It also allows the company time to fill the on-site tanks if they are empty, and to source replacement water supplies if the pollution incident is significant and likely to take the spring out of supply for some time.

This work was completed by Enviros Aspinwall as part of a study for the National Rivers Authority of England and Wales (now part of the Environment Agency). The author is indebted to the Environment Agency for allowing it to be published. The figures have been adapted from an earlier paper (Hobbs 1993) from which some of the background text has also been distilled.

References

ALDOUS, P. J. 1986. *Water tracing experiments to determine the Schwyll Spring catchment area and the hydrogeology of the Carboniferous Limestone south of Bridgend, Glamorgan, Wales.* Water Research Centre Report **CO 1215–M/EV 8390**.

——1988. *Groundwater transport and pollution pathways in Carboniferous Limestone aquifers, The Cardiff Cowbridge block*: *Final Report*, Water Research Centre Report **CO 1820 M/I/EV 8663**.

DIXON, A. G., JONES, R. S. & MORGAN, P. 1986. *Schwyll potable water source, an investigation into the likely contribution made by water from the River Ogmore, Bridgend, Mid Glamorgan: Phase 1 the initial study*. Welsh Water Authority Internal Report **SW/18.86**.

FORD, D. & WILLIAMS, P. 1989. *Karst Geomorphology and Hydrology*. Unwin, London.

GLAMORGAN RIVER AUTHORITY. 1971. *First periodical survey of the water resource and demands.* Report produced as a requirement of the Water Resources Act 1963.

HOBBS, S. L. 1993. The Hydrogeology of the Schwyll Spring Catchment Area, South Wales. *Proceedings of the University of Bristol Spelaeological Society*, **19**, 313–335.

JONES, K. 1985. Schwyll Risings. *Journal ISCA Caving Club*, **4**, 36–41.

KNOX, G. 1933. Mid Glamorgan water supply. *Proceedings of the South Wales Institute England*, **69**, No. 3 251–282.

WELSH WATER AUTHORITY. Undated. *Schwyll Source – survey of likely catchment and recommendation or tracer investigation.* Welsh Water Authority internal report.

WILLIAMS, H. & BROWN, S. J. 1989. *Schwyll bacteriophage tracer studies.* Welsh Water Authority – South Western District Laboratory, Project **85/2**.

Shallow groundwater in drift and Lower Palaeozoic bedrock: the Afon Teifi valley in west Wales

N. S. ROBINS, P. SHAND & P. D. MERRIN

British Geological Survey, Maclean Building, Wallingford, Oxfordshire OX10 8BB, UK

Abstract: Detailed geological surveying in the Afon Teifi valley of west Wales has identified the distribution and geometry of a complex array of drift deposits over shales and greywackes of Silurian and Ordovician age. A well and borehole field inventory and groundwater sampling programme provides the basis for evaluating the groundwater flow patterns within these strata. Sustainable yields are mostly small with exceptional yields up to $345\,m^3\,d^{-1}$. Potential infiltration derived from an estimate of baseflow is $535\,mm\,a^{-1}$; much of this discharges locally and only $9\,Ml\,a^{-1}$ is estimated to flow longitudinally down the valley as groundwater throughflow. Total groundwater abstraction amounts to $760\,Ml\,a^{-1}$. Groundwater generally satisfies the EC maximum admissible concentration guideline levels and is Ca–HCO_3 to Ca–Cl type. Differences in bedrock groundwater chemistry between the western and the eastern parts of the study area are largely controlled by the drift type. By analogy, the groundwater potential demonstrated in the Afon Teifi valley is likely to be indicative of the potential in geologically similar valleys elsewhere in west and mid Wales.

The principal groundwater resources of Wales are contained in south Wales in Carboniferous rocks and in north Wales in Carboniferous and Permo-Triassic strata. Lower Palaeozoic rocks occur over large areas of central and west Wales, and many of these strata have traditionally been rejected as prospects for useable groundwater supplies (Bassett 1969). However, more recent studies are beginning to identify the importance of shallow groundwater in Devonian arenites and in Silurian and Ordovician shales and greywackes (Neal *et al.* 1997). For the most part, groundwater transport is limited to shallow cracks and joints in an upper weathered zone, and small springs and modest borehole supplies yield up to $345\,m^{-3}\,d^{-1}$.

Recent groundwater development work has shown that valley floor Quaternary deposits contain groundwater resources of strategic value. Springs, boreholes and dug wells draw groundwater from the more permeable granular drift deposits; an exceptional borehole in the Rheidol valley near Aberystwyth sustains a yield of $3460\,m^3\,d^{-1}$ to public supply and draws from gravels interconnected with the river and with bedrock. There are similar large abstractions in the Dyfi and Tywi catchments, although yields from these strata are more typically less than $170\,m^3\,d^{-1}$.

The hydraulic relationship between bedrock and the superficial aquifers is unclear. The complex nature of the superficial deposits includes argillaceous horizons which inhibit vertical connectivity, whereas groundwater in bedrock fractures may be locally confined. In the Silurian and Ordovician shales and greywackes of mid Wales, the majority of groundwater flow is essentially shallow and limited to an upper weathered layer wherever it has not been removed by glacial action (Edmunds *et al.* 1998). The same is likely to be the case elsewhere, i.e. three hydrogeological environments: the drift, relatively deep fractures in fresh bedrock and the shallow weathered zone of bedrock.

Drillers are keen to case off the drift deposits so that many boreholes draw on drift storage with abstraction through bedrock cracks and joints. The picture in the drift aquifers is further confused with local gaining and losing sections of river, depending on river stage and water table as well as river bed permeability, at any given point along the valley.

Similar geological terrain occurs in the Southern Uplands of Scotland and in counties Down and Armagh in Northern Ireland and the adjoining counties Monaghan and Louth in the Republic, where groundwater occurs principally in overlying Quaternary deposits rather than in the bedrock (Robins 1995). In the Wicklow and Wexford area of the Republic, boreholes in Ordovician strata, which are quite intensely folded, may yield up to $450\,m^3\,d^{-1}$.

From: ROBINS, N. S. & MISSTEAR, B. D. R. (eds) *Groundwater in the Celtic Regions: Studies in Hard Rock and Quaternary Hydrogeology*. Geological Society, London, Special Publications, **182**, 123–131. 1-86239-077-0/00/ \$15.00

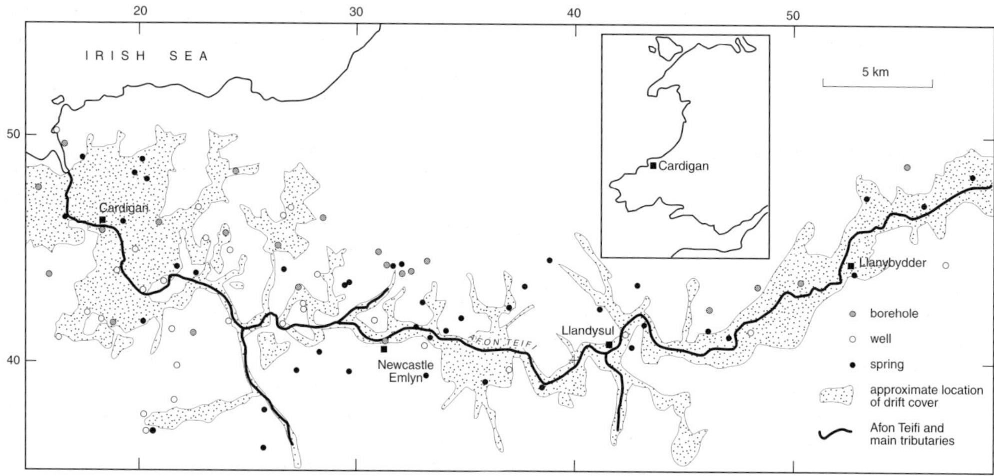

Fig. 1. Location of the Afon Teifi valley, outline geology, and borehole, well and spring locations at which measurements have been taken and a sub-set at which groundwater chemstry has been determined.

The best yields occur where the weathered bedrock is in hydraulic contact with overlying Quaternary material. In west Wales the drift deposits are also a valuable source of groundwater, but many of the boreholes draw directly from bedrock. It may be that the degree of weathering in Wales is greater than it is in the north, perhaps because of periglacial activity or reduced erosive action of the ice sheet margins.

The Afon Teifi valley in west Wales reaches the coast at Cardigan. It is one of a number of similar catchments on the western flanks of the Welsh Silurian and Ordovician hills, which lie between Snowdonia and Pembrokeshire (Fig. 1). These valleys typically contain a variety of Quaternary deposits ranging from coarse pebble breccias to silts and clays.

Conceptualization of groundwater flow in the Afon Teifi catchment illustrates a complex system which is strongly dependent on local characteristics. It also suggests that Afon Teifi and many other similar valleys in west Wales are capable of sustaining significant groundwater supplies to rural communities and farms alike.

Geological setting of the Afon Teifi valley

Geological mapping and associated geophysics in the Afon Teifi valley have defined the respective lithologies and their boundaries (Waters *et al.* 1997). The bedrock comprises a thick sequence of mudstones with subordinate siltstones and sandstones deposited during the Silurian and Ordovician age, that have been subsequently folded, faulted and uplifted during the Caledonian Orogeny. A shallow weathered zone of between 15 and 20 m thickness is present in many areas; this contains cracks and joints, many of which are dilated to some degree.

The valley floor contains a complex array of Quaternary glacial deposits (Fig. 1) that include till, glaciofluvial sand and gravel, and glaciolacustrine clays and silts. There are also a number of periglacial or post-glacial deposits such as head, peat, alluvium, alluvial fan deposits, river terrace deposits and tidal river deposits. The drift sequence locally exceeds 70 m in thickness within buried channels.

Identification of the depositional environment is a key to determining the geometry, distribution and hydrogeological character of the respective drift units. Two separate ice sheets occupied the valley, the Welsh Ice Sheet descending from the Cambrian Mountains and the Irish Sea Ice Sheet, sourced further to the north, but penetrating Cardigan Bay and much of its coastal tract. The latter contains shelly fragments and associated carbonate material.

The temporary blockage at the western end of the valley created Lake Teifi, with associated lacustrine deposits. Mounds and ridges were deposited whilst the Welsh Ice Sheet retreated, and these also created a number of blockages and associated ponding of meltwater. As the Afon Teifi began to cut through the new unconsolidated sediments, it abandoned its pre-glacial course in a number of places to flow over bedrock. These areas later developed to become the distinctive gorges that are present today at, for example, Cenarth Falls and Llanybydder.

With the establishment of the present day sea level the lower Teifi valley became drowned.

The estuary has subsequently silted up with salt marsh deposits and associated blown sand dune fields.

Groundwater occurrence and aquifer properties

Groundwater occurs in useable quantities in many parts of the shallow bedrock and in granular drift deposits. In order to delineate the more productive aquifer units, a well, spring and borehole inventory was carried out in the catchment. The data were placed over the geological framework in order to define discrete groundwater bodies. A total of 99 sources were recorded; at each source the production yield was estimated, the elevation of the standing water level measured, borehole or well depth was measured, the aquifer(s) identified, and the water use was recorded. The field study was constrained to the valley area with the 120 m elevation contour taken as an arbitrary limit of investigation.

Boreholes are typically 35–50 m deep, and yield on average only about 26 $m^3 d^{-1}$. The standing water level is generally shallow, particularly beneath the valley floor where it may be between 0.5 and 2.5 m below ground level. The water table is generally deeper on the valley sides but it is nowhere greater than 12 m below ground level in bedrock, although at one borehole in head gravel it is situated at 21 m below ground level reflecting the higher permeability of this material. Fifteen of the 24 boreholes were drilled into bedrock whereas the remainder penetrated only drift.

The wells and springs are more evenly distributed between drift and bedrock. Yields of wells are similar to the boreholes, and range up to 17 $m^3 d^{-1}$, whereas spring yields range from a spring described as a ‘damp patch’ to one source in head gravel and another two in bedrock with measured yields of 170 and 260 $m^3 d^{-1}$.

Three short duration single-borehole pumping tests were carried out in boreholes drawing on bedrock only. At Dan yr Allt (SN 187416), an impermeable barrier boundary or the dewatering of a fracture is evident from increased drawdown after 17 minutes elapsed time since pumping began. The Cooper–Jacob method of analysis for the earlier data suggests a transmissivity of 0.6 $m^2 d^{-1}$. The later data indicate a reduction in transmissivity to 0.3 $m^2 d^{-1}$. A borehole at Tandderi (SN 503435) shows a variable drawdown rate, with increased rate of drawdown after about 3 minutes, but increasing again after about 6 minutes. The data indicate transmissivity values of 1.1 $m^2 d^{-1}$ from the earlier data, and 0.6 $m^2 d^{-1}$ from the late data. The third test at Cyttirbach (SN 244483) indicates a transmissivity of about 0.5 $m^2 d^{-1}$.

Dwr Cymru (Welsh Water) operates one public supply borehole at Olwen, in the upper part of the catchment near Lampeter (SN 582496) which is 26.8 m deep and draws water from fluvioglacial deposits in the valley floor of the River Dulas, a tributary of the Teifi. The lithological log indicates that water is supplied from two sand and gravel aquifers, separated by a clay layer which confines the lower gravels. The borehole is licensed to abstract 395 $m^3 d^{-1}$, and typically takes close to the full licensed amount. A constant rate pumping test on the Olwen borehole indicated an aquifer transmissivity of 210 $m^2 d^{-1}$ from the early drawdown data. The effective porosity of the aquifer is assumed to be 0.1 and the saturated aquifer thickness 18 m. These values may not be representative of the whole aquifer due to its heterogenous nature.

Short duration pumping tests were also undertaken in two of three newly drilled exploratory/monitoring boreholes in the drift (Merrin 1999). These boreholes provide valuable insight to the range of hydraulic properties offered by the diverse spread of lithologies that were encountered. At Cenarth (SN 266417) a single gravel horizon between 23 and 24 m depth was screened in a borehole which was otherwise completed with plain casing. The succession penetrated above the gravel was almost entirely clay grade material. Pumping demonstrated almost no hydraulic response from the gravel. Nevertheless, the exploratory borehole at Lampeter (SN 577480) penetrated the same pair of gravel horizons as the Olwen public supply borehole and a transmissivity of 5000 $m^2 d^{-1}$ was indicated.

Water resources and flow patterns

Point source values of precipitation and runoff in the Teifi catchment at Glan Teifi (SN 244416) from 1959 to 1990, and at Llanfair (SN 433406) from 1971 to 1981 are reported in IH/BGS (1998):

	Precipitation ($mm\,a^{-1}$)	Runoff ($mm\,a^{-1}$)
Glan Teifi (1959–1990)	1349	999
Llanfair (1971–1981)	1446	988

The values for runoff include baseflow. The base flow index (BFI) is 0.54, and, therefore, some 535 mm of the average reported runoff derives from groundwater discharge. This represents a potential annual infiltration (renewable resource) of 535 Ml km^{-2} of aquifer.

Groundwater throughflow in the bedrock is modest by comparison. According to Darcy's law:

$$Q = Tyi$$

where T is the transmissivity, y is the width of the catchment and i is the hydraulic gradient. Using a transmissivity value of 0.6 m^2 d^{-1} (estimated from available short term pumping test analysis in bedrock boreholes), with an average catchment width of 17 km, and estimating the hydraulic gradient to be equal to the gradient of the river (2.5×10^{-3}), this gives:

$$Q = 0.6 \times 17\,000 \times 2.5 \times 10^{-3}$$
$$= 25\,\mathrm{m^3\,d^{-1}}\ \text{or}\ 9\,\mathrm{Ml\,a^{-1}}$$

A similar calculation for the granular drift deposits in the Lampeter area and the upper part of the study area can be made. Taking an overall aquifer width of 2.5 km, the same hydraulic gradient as for the bedrock and using the more conservative estimate of transmissivity from the Olwen borehole (210 m^2 d^{-1}) a throughflow estimate of 1310 m^3 d^{-1} or 480 Ml a^{-1} can be derived for the granular drift.

Figure 2 shows a schematic section of the catchment indicating the different flow components. Clearly the majority of the baseflow component of the river flow (535 Ml a^{-1}) derives from local recharge and lateral flow direct to the river, whereas longitudinal throughflow down the length of the (upper part) of the valley amounts to only about 490 Ml a^{-1}. Conversely, in some reaches of the river, groundwater may also locally gain from river flow although the overall balance is one of loss to the river.

Water resource potential and demand

The drift deposits in the west of the area are clay-rich and appear to have little groundwater potential. However, the sandier deposits below

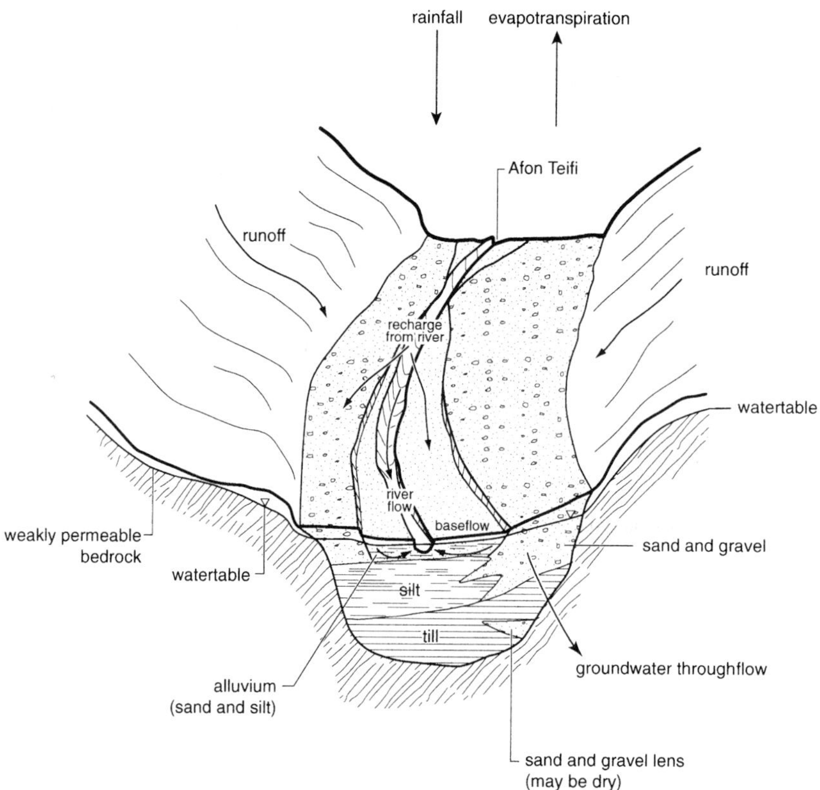

Fig. 2. Schematic section of the central part of the lower Afon Teifi valley showing the main flow components.

Table 1. *Groundwater consumption in the Afon Teifi catchment*

	Estimated daily consumption (m^3)	Estimated population	Total abstraction ($m^3\,d^{-1}$)
Domestic – single property	0.6	809	485.4
Domestic – <25 people	1.2	83	99.6
Farm (livestock)	1.5	68	102
Farm (dairy)	6.5	132	858
Commercial (hotel, youth hostel, abattoir, quarry, etc.)	8.0	20	160
Public supply – Olwen BH	377		377
Total		1112	2082 (760 $Ml\,a^{-1}$)

the clay may offer a significant resource potential. The Welsh Ice Sheet deposits in the east of the area, around Lampeter, offer better potential, and are already exploited for public supply at Olwen.

The bedrock aquifer supports most of the private supplies. Spring flows from the bedrock occur up to 260 $m^3\,d^{-1}$ giving the prospect of a supply of the same magnitude as the Olwen public source from a small group of springs or boreholes. The majority of the private groundwater sources are springs.

Most of the Teifi catchment is currently licence-exempt from restrictions imposed by Section 23 (1) of the Water Resources Act 1963. Licences are only required in the area contained within the flood limit of the river Teifi; apart from the Olwen public supply borehole, licensed abstractions in the catchment total some 34 $m^3\,d^{-1}$ from nine sources ranging from 2.3 to 8.2 $m^3\,d^{-1}$.

A best estimate of the quantity of groundwater abstracted can be made with the list of private groundwater sources maintained by the local authority Environmental Health departments (Table 1). Estimated abstraction rates are based on field evidence and on other studies in the UK (e.g. Wright 1995). Abstraction from the Olwen public supply borehole is based on the average abstraction rate from 1990 to 1995. There is insufficient evidence to estimate the proportions abstracted from the drift and bedrock aquifers. Mains water is available throughout most of the catchment, but

Table 2. *Range of waters analysed in the Afon Teifi catchment and EC maximum admissible concentration values*

	EC Maximum admissible concentration	Boreholes	Springs	Wells	Afon Teifi water	Detection limit
$mg\,l^{-1}$		$n = 19$	$n = 27$	$n = 6$	$n = 2$	
Ca	–	13–66	7–83	15–54	11	0.02
Mg	50	3–16	2–18	3–14	3	0.04
Na	150	10–52	5–35	6–18	9	0.02
K	12	0.5–17	0.4–34	0.6–9	1.6–1.7	0.3
Cl	–	14–92	11–63	12–43	13–16	0.2
SO_4	250	6–41	6–28	4–18	9–10	0.3
Al	0.5	0.003–0.19	0.001–0.31	0.003–0.032	0.026–0.044	0.0008
NO_3–N	50	<0.4–32	<0.4–19	0.9–41	<0.4	0.4
Fe	0.2	<0.006–0.48	<0.006–0.72	0.006–0.017	0.014–0.071	0.006
Mn	0.05	<0.001–0.217	<0.001–2.4	0.001–0.013	<0.001–0.016	0.001
$\mu g\,l^{-1}$						
Cu	–	0.5–676	<0.3–17	0.6–31	1.3–1.5	0.3
Zn	–	1–198	1–62	2–19	6–9	1
Ba	–	1–161	0.8–40	2–25	5.4–5.6	0.3
Ag	10	<0.06	<0.06	<0.06	<0.06	0.06
As	50	0.4–2	<0.3–2.9	0.4–6.5	0.7–0.8	0.3
Cd	5	<0.04–0.05	<0.04–0.17	<0.04–0.05	<0.04–0.05	0.04
Cr	50	<0.8	<0.8–1.9	<0.8	<0.8	0.8
Ni	50	<0.8–11	<0.8–6.7	<0.8–3	1	0.8
Pb	50	<0.4–3.6	<0.4–5	<0.4–3.5	<0.4	0.4
Sb	10	<0.1–0.4	<0.1–0.2	<0.1–1.6	<0.1	0.1

private supplies are preferred by many on the grounds of cost. The analysis shows that about $2\,Ml\,d^{-1}$ ($760\,Ml\,a^{-1}$) groundwater is being used in the catchment area. This is equivalent to the potential infiltration over an area within the catchment of $1.4\,km^2$.

There is, therefore, considerable potential for further groundwater development. The main limitation is abstraction capacity available from individual sources.

Groundwater chemistry

More than 50 groundwater samples were collected during the survey, 19 from boreholes, 27 from springs and six from wells in addition to two samples of the River Teifi at Cenarth. Temperature, specific electrical conductance (SEC), total alkalinity (as HCO_3) by titration, pH, redox potential (Eh) and dissolved oxygen (DO) were measured at site. A flowthrough cell was used where possible to measure pH, Eh and DO from boreholes and pumped wells to isolate samples from the atmosphere. In most cases the groundwaters were pumped via storage pressure tanks; although these were generally pumped until Eh and pH were stable before measurement, it is probable that DO and Eh are not representative of the groundwater at depth.

Samples for laboratory analysis were filtered through 0.45 μm membranes and collected in polyethylene bottles. The anions Cl, Br, NO_3–N, NH_4–N and F were analysed by automated colorimetry. Filtered and acidified samples (1% v/v HNO_3) were collected in polyethylene bottles for analysis of major cations, SO_4^{2-} and a wide range of trace elements by ICP–OES and ICP–MS (Table 2).

There is a range in pH from 5.2 to 7.6, however waters with pH less than 6 are confined to the catchment above easting 22 (Fig. 1). The waters are relatively low in total dissolved solids (TDS) and SEC varies from 117 to 662 $\mu S\,cm^{-1}$ with the highest values close to the coast. Eh is generally high but this may reflect sampling difficulties. This is the same for many of the DO measurements, but the majority of the flowing springs contained detectable DO.

In terms of major ion chemistry the groundwaters range from Ca–HCO_3 to Ca–Cl type (Fig. 3). The major cations show a tighter grouping than the anions on Piper diagrams, typical of upland waters. Mean concentrations of major elements for the boreholes are, in general, slightly higher than the springs and wells, but ranges are similar. Iron and Mn are low reflecting the presence of oxygen in the groundwaters.

The solute chemistry of Afon Teifi groundwaters varies along the length of the catchment. This is particularly pronounced for SEC and the major elements Ca, Mg, HCO_3 and Si, which show a wide range of concentrations in the western part of the catchment but less variation

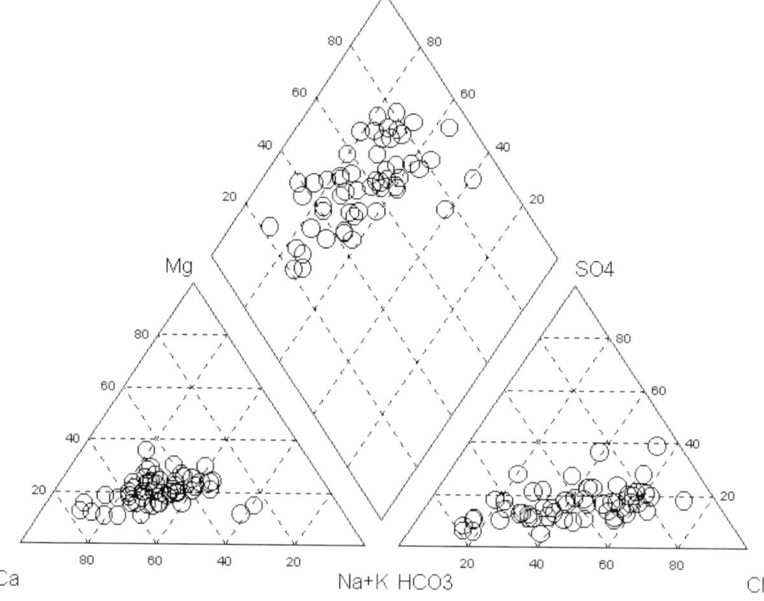

Fig. 3. Piper diagram of major ion groundwater chemistry.

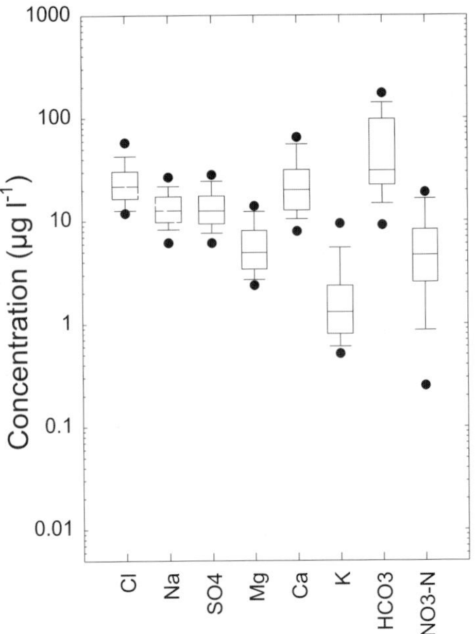

Fig. 4. Box plot summary of major element concentrations. Horizontal line is the median value, bars represent 25th and 75th percentile and dots 5th and 95th percentile. Diluted seawater curve is shown as line.

and a slightly lower baseline for some solutes in the east. There is also a tendency towards higher Na and Cl towards the coast. The western enrichment is also matched to a similar degree by many of the minor and trace elements e.g. Br, B and Sr, whilst others show no significant trend e.g. K, NO_3–N and Pb.

The influence of diffuse agricultural pollution of fertilisers or farm slurry on groundwaters can be seen at several sites through slightly enhanced concentrations of TOC, NO_3–N, K, Cl and K/Na ratio. At the other sources, however, agricultural pollution does not appear to be a significant problem to inorganic water quality.

Atmospheric inputs represent important contributions for the major elements Na and Cl but are generally insignificant for most other solutes including SO_4 and particularly for Ca and Mg. The dominant control on the inorganic chemistry is through water-rock interaction and the changes in groundwater chemistry along the Afon Teifi valley can be explained in terms of both the type and extent of reaction with minerals along the hydrological flowpaths.

Figure 4 shows the major element data plotted as a box plot along with a diluted seawater curve. The seawater curve has been normalized to the median Cl concentration of the groundwaters so that enrichments/depletions relative to

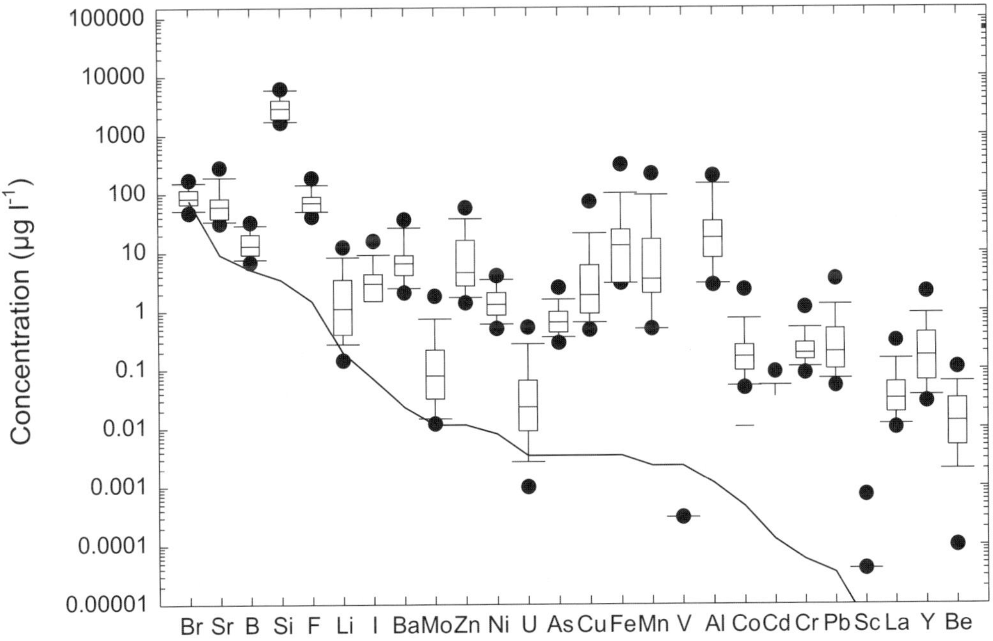

Fig. 5. Box plot of minor and trace element concentrations. Bars represent 25th and 75th percentile and dots 5th and 95th percentile. Diluted seawater curve is shown as line.

the rainfall (marine dominated) can be seen. It is clear that Na is dominated by marine inputs but that most other elements are enriched through inputs from the catchment. This is particularly the case for Ca, HCO_3 and NO_3 which are at least an order of magnitude higher than rainfall. The ranges and concentrations of minor and trace elements are shown in Fig. 5. Iron and Mn are generally lower than EC maximum admissible concentrations (MAC) reflecting the generally well oxygenated nature of the waters. Copper and Zn may be relatively high, probably reflecting local vein mineralization, but well below the MAC values. Other elements of potential concern, including the heavy metals (Cd, Pb, U) and F are present at low concentrations or below the limit of detection.

Hydrological factors play an important role in determining groundwater chemistry due to different flowpaths and changes taking place with increased residence time. In a typical catchment aquifer system, where a significant component of flow occurs along the valley from the higher to lower reaches, a gradual increase in baseline concentrations related to longer residence time would be expected. However, no such trend is apparent in the elements which typically show an increase with residence time (e.g. Li, Si, Na). This implies, in agreement with hydrogeological data, that flow in the catchment takes place through discrete fracture systems with no connected flow system along the length of the valley.

However, there are significant differences in the mean concentrations of some elements between the western and eastern parts of the study area. In the western part of the area, which is dominated by Irish Sea drift containing abundant carbonate material, concentrations of Ca, HCO_3 and Sr are much higher (Fig. 6) than in the east where Welsh derived drift is dominated by locally derived Lower Palaeozoic material with a relatively low carbonate content. Dissolution rates of carbonate minerals are high in comparison with silicate phases and dilute, relatively acidic recharge waters would rapidly increase concentrations of the dissolution products such as Ca, HCO_3, Mg or Sr. The correlation of high concentrations of these elements with the area containing Irish Sea drift implies that the type of drift may impart a distinct geochemical signature on inorganic groundwater quality.

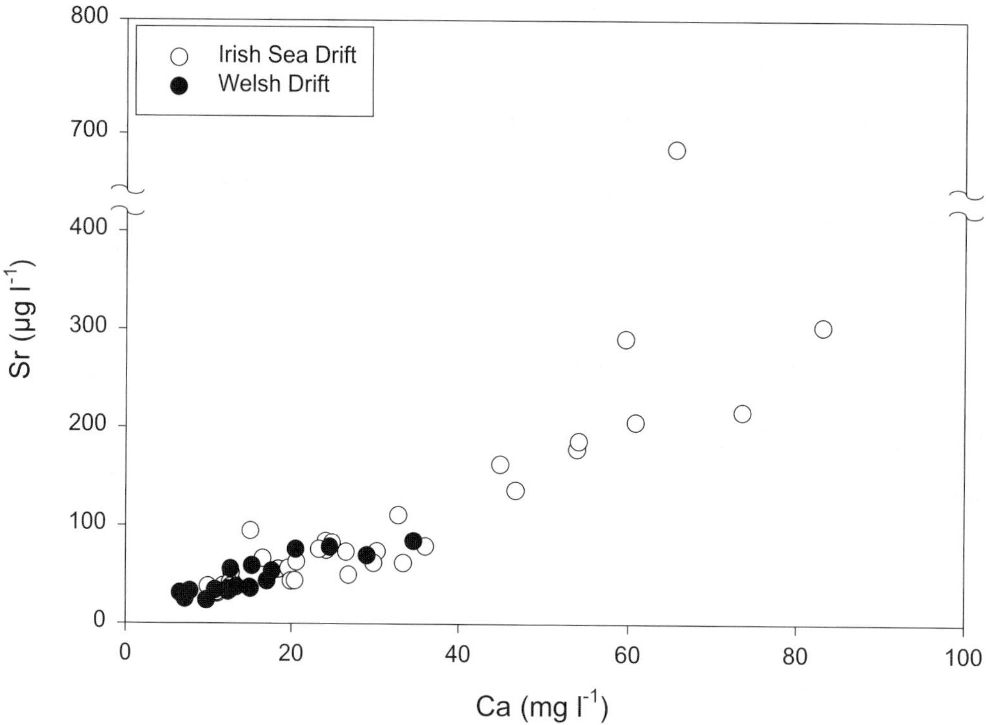

Fig. 6. Plot of Ca and Sr subdivided on the basis of where Welsh and Irish drift are present.

Conclusions

Although the calculations used to derive the conceptual groundwater flow model for the catchment are at best crude, they illustrate that a considerable volume of groundwater is present. The key features of the catchment are:

(a) Boreholes, wells and springs generally yield small volumes of groundwater from bedrock and drift due to low aquifer transmissivity. Exceptional yields of 170–345 $m^3 d^{-1}$ occur in granular drift and gravel deposits in which transmissivity may be two orders of magnitude higher than in bedrock.
(b) The water table is generally shallow beneath the valley floor, increasing beneath valley sides and is greatest within steeply sloping head deposits.
(c) Groundwater is contained within granular drift and in fractures in the shallow weathered saturated layer of bedrock.
(d) The Afon Teifi is both influent and effluent with regard to drift groundwater in different reaches depending on the differential head and stage of river and groundwater and river bed permeability.
(e) Groundwater flowpaths are essentially lateral to the river. Groundwater chemistry indicates that there is little longitudinal flow of groundwater down the valley; a Darcian calculation indicates longitudinal throughflow at only 490 $Ml\,a^{-1}$.
(f) Infiltration and discharge as baseflow to the river amounts to about 535 $Ml\,a^{-1}$ per km^2 of catchment.
(g) Only about 760 $Ml\,a^{-1}$ of groundwater is being abstracted annually.
(h) The groundwater is of the Ca–HCO_3 to Ca–Cl type.

Analogy with other geologically similar catchments in west and mid Wales suggests that there is considerable scope for groundwater development in these areas. The larger scale public water supply sources in granular drift deposits and smaller domestic and farm boreholes in bedrock are capable of satisfying a large component of the demand for this largely rural area with good quality water.

The geological mapping and interpretation was carried out by Jerry Davies, Dick Waters, Dave Wilson and Jo Prigmore from both BGS Aberystwyth and BGS Keyworth on behalf of the local county councils. Laura Coleby carried out much of the well and borehole inventory. Much encouragement and guidance was received from Wayne Davies at the Environment Agency Wales. The project was funded by the Environment Agency, and NERC. The paper is published by permission of the Director British Geological Survey (NERC).

References

BASSETT, D. A. 1969. The study of ground water, with particular reference to Wales and the Welsh Border. *Transactions of the Cardiff Naturalists' Society*, **94**, 62–87.

EDMUNDS, W. M., ROBINS, N. S. & SHAND, P. 1998. The saline waters of Llandrindod and Builth, Central Wales. *Journal of the Geological Society*, London, **155**, 627–637.

IH/BGS. 1998. Hydrological data UK, Hydrometric Register and Statistics 1991–1995. Institute of Hydrology/British Geological Survey, Wallingford.

MERRIN, P. D. 1999. Report on the drilling and installation of exploratory boreholes, Afon Teifi Valley, West Wales. Report British Geological Survey **WD/99/24**.

NEAL, C., ROBSON, A. J., SHAND, P., EDMUNDS, W. M., DIXON, J., BUCKLEY, D. K., HILL, S., HARROW, M., NEAL, M., WILKINSON, J. & REYNOLDS, B. 1997. The occurrence of groundwater in the Lower Palaeozoic rocks of upland Central Wales. *Hydrology and Earth System Sciences*, **1**, 3–18.

ROBINS, N. S. 1995. Groundwater in Scotland and Northern Ireland: similarities and differences. *Quarterly Journal of Engineering Geology*, **28**, 163–169.

WATERS, R. A., DAVIES, J. R., WILSON, D. & PRIGMORE, J. K. 1997. A geologcal background for planning and development in the Afon Teifi catchment. Report British Geological Survey **WA/97/35**.

WRIGHT, P. 1995. Water resources management in Scotland. *Journal of the Chartered Institution of Water & Environmental Management*, **9**, 153–163.

Public water supplies from alluvial and glacial deposits in northern Scotland

C. R. C. JONES & A. J. SINGLETON

Mott MacDonald, Demeter House, Station Road, Cambridge CB1 2RS, UK

Abstract: Surface runoff from upland areas in northern Scotland can be excessively coloured and turbid, making water treatment costs high. In contrast, groundwater has little colour or suspended solids, and has a more stable water quality, therefore treatment needs can be minimal. A number of public water supply schemes based on the development of groundwater in alluvial and glacial deposits are reviewed. These supplies range from major projects such as the 27 Ml/d Spey and 10 Ml/d Fort William schemes to the 0.16 Ml/d supply for Dalwhinnie. It has been demonstrated that yields of up to 2 Ml/d can be achieved from dug wells or boreholes even where the aquifer thickness is limited. Larger yields have been obtained from infiltration galleries and Ranney wells. Future challenges include improved siting techniques to locate thicker and more productive sediments and the avoidance of groundwater containing high iron and/or manganese concentrations.

A review of all of the water sources of the North of Scotland Water Authority (NoSWA) was undertaken in 1998 (Mott MacDonald 1999*a*) and a more detailed follow up study of the Tayside Region was started in February 2000. The original study included examination of the groundwater resources and involved a desk study, literature survey, data collection and review of case histories. The sites in alluvial and glacial deposits are shown in Fig. 1. The major groups of information that were reviewed were consultants' and contractors' reports, data held by NoSWA and the published records and reports of the British Geological Survey (BGS).

Superficial alluvial and fluvioglacial deposits occur locally throughout the region (Robins & Ball 1987) but are thickest and coarsest in the major valleys and thinner or finer grained in the headwater areas or valleys of minor streams. The fluvioglacial and alluvial aquifers have been investigated or developed at a number of sites. Ball (1997) discusses the geological potential at 18 of them. Site investigation of the valley aquifers can be used to improve yields and success rates, and can be based on a combination of geomorphology, trial pitting, test boring or surface geophysics. Investigation techniques range from simple trial pit and pump out tests as at Dalwhinnie, to construction of pilot wells and extended pumping trials as at Spey and Fort William. Geophysical methods have also been applied. However, electrical methods proved to be of limited use at Spey due to the lack of contrast between fresh water in the alluvium and the underlying bedrock. Seismic refraction was used more successfully at Spey. The recently developed electrokinetic survey (EKS) method was used at Fort William and at Spey as part of an unpublished MSc research project.

Development options include both shallow drilled boreholes and dug wells and, where close to rivers, infiltration galleries and Ranney wells. In terms of resource development the objective is not to draw heavily on storage within the aquifer but to intercept throughflow or induce infiltration from nearby streams and rivers. Therefore, in yield terms these groundwater schemes rarely add to exploitable catchment yield but are useful to smooth out short term fluctuations in runoff.

Groundwater quality varies greatly both between and within wellfields and, at present, there is no comprehensive predictive approach to site selection on the basis of water quality. Thus testing and sampling are always required during the investigation and construction phases. Studies in the mid-1980s for the Spey scheme (Watt *et al.* 1987) pointed to the significant role of residence time and temperature in colour removal by natural bacteria. A one-year pumping trial for the Fort William Scheme also confirmed the effect (Mott MacDonald 1989).

A potential benefit of groundwater development rather than direct river intakes is to achieve a more consistent and possibly improved water quality for some parameters, especially colour and turbidity. Other potential benefits are less

From: ROBINS, N. S. & MISSTEAR, B. D. R. (eds) *Groundwater in the Celtic Regions: Studies in Hard Rock and Quaternary Hydrogeology*. Geological Society, London, Special Publications, **182**, 133–139. 1-86239-077-0/00/ $15.00

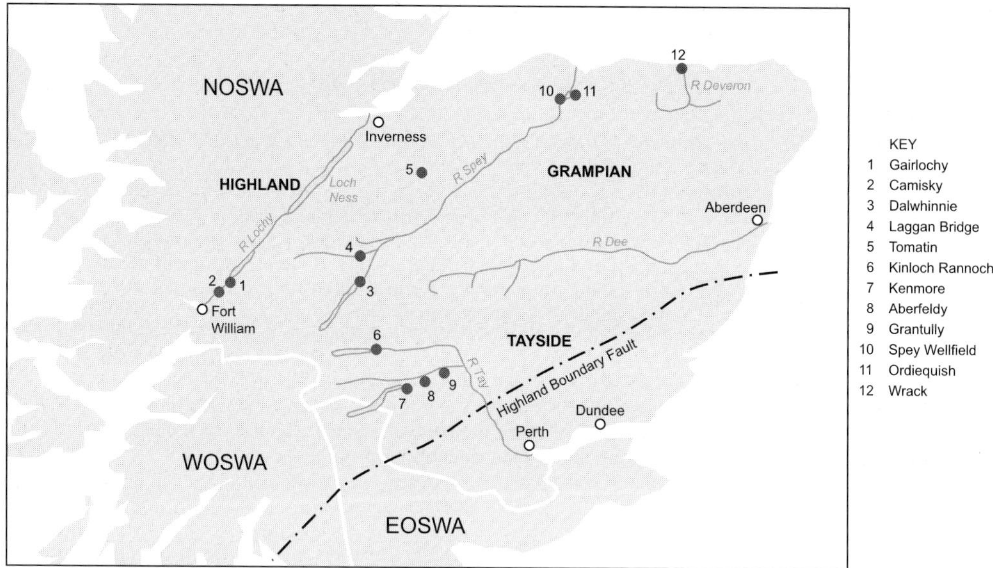

Fig. 1. Location of study areas.

sensitivity to stream bed movement, reduced impact on fishing, the financial advantage of being able to increase capacity incrementally to match rising demand, and the potential gain in reliable output at times of low flow. Disadvantages include the initial investment needed to overcome the uncertainty inherent in a groundwater scheme, costs of abortive investigations, the cost or time taken for an aquifer to clean up if subsurface pollution occurs, and the possibility of poor quality water, especially where high iron or manganese concentrations are found.

Details of selected groundwater development schemes

The Spey scheme at Fochabers

The Spey River was identified in the late 1970s as the most likely resource to satisfy the growth in demand in the Grampian Regional Council area. Studies in the 1980s indicated that the gravel aquifer in a 2.8 km reach upstream of Fochabers Bridge was the preferred source for a design yield of 27 Ml/d. Alternative options considered included direct intake(s), bed intakes and the local sandstone aquifers. Advantages of the Spey gravels included the notable reduction in turbidity and colour as river water was drawn through the aquifer materials and the likelihood that the bedrock aquifer would not be able to match demand (Watt *et al.* 1987).

The project was carried out between May 1992 and September 1994 under an £8M contract. The £0.6M drilling sub-contract covered eight deep and 36 shallow observation wells for initial and long term water level measurement, and 36 production wells drilled to rockhead in the underlying sandstone. The well design included 300 mm stainless wirewound screen drop-set within a 450 mm steel conductor casing. The wells were developed to a stable sand-free condition using high pressure jetting, airlifting and surge pumping. Well performance was tested using both step tests and 48 hour constant rate discharge pumping tests (CDT) with water quality measurements, CCTV and flow logging. The wellfield includes a problem area in the northern section where the gravel is generally less than 6 m deep, and shallow seismic refraction surveys were used to locate a deeper channel in the rockhead.

Construction and pumping test results for yield (Q), drawdown (s) and specific drawdown (s/Q) are summarized in Table 1. Pump testing was extended to around 15 days at two individual wells, and two groups of three wells. The results were used to assess the nominal (design) yields at which excessive drawdown would be avoided once all wells were operating for a period of 180 days and, as shown in Table 1, the resulting nominal specific drawdowns are around double those achieved in the CDTs.

Landscaping and attention to the environmental aspects, especially fishery interests, are an important aspect of the scheme and may have been instrumental in obtaining the Water Order (permit to abstract for public supplies) without objections.

Currently, up to around 21 Ml/d is being put into supply with treatment limited to chlorination. Once demand rises to the full 27 Ml/d, additional treatment will be provided on the discharge from 25–30% of the wells to reduce the iron and manganese content. The project has proved successful in meeting quality standards without excessive use of chemicals. This success continued despite the wellfield being partially flooded for a few days in September 1996 (Mott MacDonald 1997).

The Fort William water scheme

Investigations in 1984 concluded that abstraction of groundwater from sand and gravel deposits at Camisky, approximately 10 km north-east of Fort William, on the south-eastern floodplain of the River Lochy, would be the most cost-effective means of supplying 7–10 Ml/d of additional water demands (Johnstone & Rennie 1991). The wellfield area is approximately 800 m long and 170 m wide. The geology typically comprises a metre of superficial top soil and sand, underlain by a considerable thickness (in excess of 30 m) of variably graded sands and gravels. Sandy silt lenses are also evident across certain areas of the wellfield.

A £0.3M wellfield construction contract was completed in February 1999. Other components of the water supply scheme are programmed for completion by 2001. A total of 16 shallow observation boreholes and three deep observation boreholes were drilled using a one tonne cable-tool drill rig. The 10 production wells were constructed within a corridor set 50–75 m from the river bank, with at least 50 m between adjacent wells in order to prevent excessive interference effects during pumping. The production wells comprise a 525 mm surface conductor casing to a depth of 4 m and a 10–15 m length of 350 mm stainless steel well casing and screen below. The screened section within each well casing varied between 3 and 7 m in length. Drilling problems included abundant cobbles and boulders and spatially variable silty sand horizons liable to exert high skin frictions on the temporary drill string. This effect retarded drill progress at several locations and even led to abandonment and redrilling in two cases.

Testing was similar to that at Spey although there was only one extended constant rate test and one group test on three adjacent production wells. The final yield calculations suggested that sustainable operational discharges of between 0.5 and 2 Ml/d were attainable at all wells. The net wellfield yield at Camisky was estimated at around 13 Ml/d. Table 2 summarizes the construction and test results for the production wells.

Other borehole schemes

Data from a number of other shallow groundwater schemes in the NoSWA area are presented in Table 3. The schemes vary from exploration and pilot wells, including those in the Spey and Fort William areas, to the two newly commissioned production wells at Kenmore and Kinloch Rannoch. The Laggan Bridge and Dalwhinnie schemes are of special interest; the former since pumping trials failed to produce satisfactory water quality at any of the four drilled wells. The latter is an excellent example of a low cost scheme

Table 1. *Spey well data*

	Well depth (m below ground level)	Screen length (m below ground level)	Constant discharge test				delta s (m/log cycle of time)	Nominal yield Q (Ml/d)	Nominal s/Q ($m/m^3/d$)
			Yield Q (Ml/d)	Pumping water (m below flange)	Drawdown s (m)	s/Q ($m/m^3/d$)			
Average	13.16	8.17	1.44	6.40	4.36	0.0034	1.03	0.934	0.0066
Minimum	8.90	2.90	0.58	2.96	1.01	0.0006	0.13	0.631	0.0013
Maximum	16.90	12.10	2.11	11.15	9.04	0.0073	5.48	1.175	0.0138

Source: Mott MacDonald (1994).

Table 2. *Fort William well data*

	Well depth (m below ground level)	Screen length (m below ground level)	Constant discharge test		delta s (m/log cycle of time)	Nominal Yield Q (Ml/d)
			Yield Q (Ml/d)	Pumping water level (m below flange)		
Average	14.4	7.1	1.408	5.01	0.25	1.296
Minimum	12.2	5.0	0.648	3.23	0.08	0.432
Maximum	15.8	9.0	1.771	9.01	0.54	1.814

Source: Mott MacDonald (1999*b*).

Table 3. *Data for other wells*

Location	Well Nr	Year	Dist. from river (m)	Total depth (m)	SWL (m)	Casing diam. (mm)	Screen length (m)	Duration of test (hrs)	Disch. (m^3/d)	Draw-down (m)	Water quality
Spey Valley	W2	1983	56	14	0.22	150	11.24	25.3	173	4.55	OK
Spey Valley	W5	1983	17	12.75	1.55	150	11.28	48	1102	0.92	OK
Spey Valley	W8	1983		9.8	nil	nil	nil	nil	nil		
Spey Valley	W11	1983	145	13	1.17	150	11.24	13	1013	1.28	OK
Spey Valley	W12	1983	46	10	1.28	150	8.43	48.5	1140	2.62	OK
Spey Valley	W14	1983	17	17	1.71	150	11.24	54	691	2.54	1 nr Mn > MAC
Spey-Mosstodloch	W5	1982	>1 km	34	3.93	250/150	18	50.0	392	4.77	Mn, Fe >MAC
Spey-Red Burn	W3	1982	>1 km	23	3.89	250/150	9	50.0	398	3.7	Mn, Fe >MAC
Spey-Mannochburn	W2	1982	>1 km	24	2.9	250/150	12	75.0	1434	4.65	Mn, Fe >MAC
Camisky	Bh 5 TW	1984	30	13.3	2.28	200	8.4	74.2	1036	1.97	
Camisky	Bh 7 TW	1985	25	14.5	1.16	200	8.7	71.2	1728	1.4	
Camisky	Bh 9 TW	1987		15	2.3	200	9.05	8184	864	0.35	
Gairlochy	Bh 1	1991		20	3.5	100	14.3	3	389	0.02	
Gairlochy	Bh 2	1991		15	4.7	100	8.6	4	375	0.94	
Gairlochy	Bh 3	1991		7.85	4.2	100	2.85	3	13	0.6	
Gairlochy	Bh 4	1991		15	5.2	100	8.6	22	165	1.13	
Gairlochy	PW 1	1993		19.4		200	9.7	672	492	3.5	20 % Fe & Pb >MAC
Dalwhinnie	Well 1	1997		2.8		1200	1.4	None	0.16	<0.05	
Kinloch Rannoch	KR 1	1994	550	22.9	5.1	200 ?	7	2419	860	2.7	1 nr Fe >MAC
Kenmore	BG 1	1987		19	3.27	150	2.9	49.5	691	6.34	
Grantully	BG 3	1987		10.3		150		11	246	5.46	
Kenmore	PW 1B	1998		22		600	12.6	336	2358.72	2.65	
Laggan Bridge	LB 1	1998		10	1.84	205			156	4.05	Colour etc
Laggan Bridge	LB 2	1998		10	1.25	205			311	3.77	Colour etc
Laggan Bridge	LB 3	1998		10	1.7	205			86	4.26	Colour etc
Laggan Bridge	LB 4	1998		10	1.74	205			605	2.2	Colour etc
Tomatin	TO 1	1990 ?		20	1.3	205			1088	7.7	
Aberfeldy	AB 1	199 ?			Abandoned		0	ni			

SWL = static water level.
MAC = Maximum Admissable Concentration.
Source: Mott MacDonald (1999*a*).

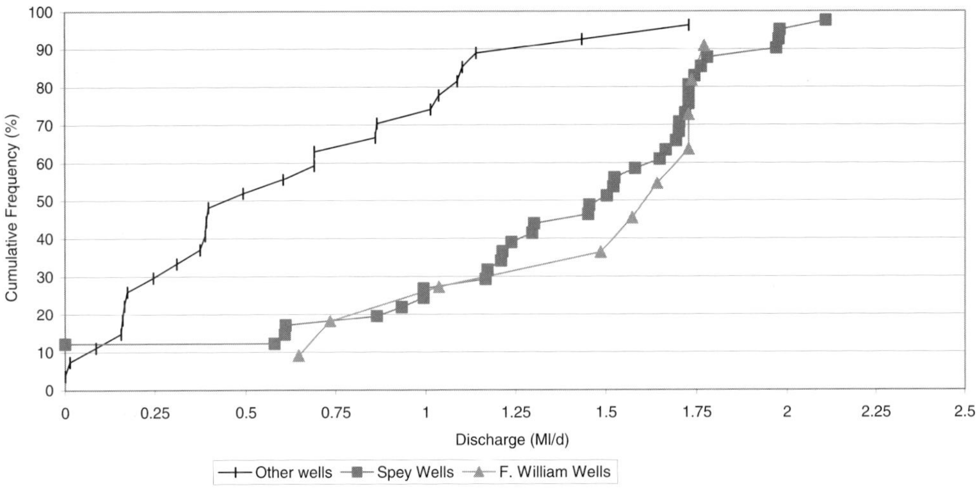

Fig. 2. Well test discharges.

suitable for rural supplies since both trial pits proved to have suitable water quality and discharge. Completion with a pump and concrete rings cost only £6000. In contrast, drilling at Aberfeldy failed to penetrate a suitable aquifer.

Comparison of borehole yields

Borehole yield data are presented in Fig. 2 as plots of cumulative frequency against test discharge. Inevitably, the data for other wells have a lower median (50%) discharge average (0.40 Ml/d) than either the Spey (1.5 Ml/d) or Fort William (1.6 Ml/d) scheme since the first data include all the narrow diameter test wells and sites with limited potential. However, there is still a three to fourfold variation in test discharge within the limited areas selected for the two major schemes. This demonstrates the significance of variations in the thickness of superficial deposits, depth to the base of the effective aquifer (as confirmed by flow logging) and aquifer permeability.

Infiltration galleries

Two of NoSWA's public water supply galleries were studied. Inflow rates of 2.85 Ml/d and 3.39 Ml/d are quoted for Ordiequish (on the opposite bank of the River Spey to the Spey wellfield) and the Wrack (on the River Deveron), respectively. The latter gallery was constructed as a Ranney well (Anon 1961). Since the overall gallery lengths are 316 m and 130 m the average inflow rates are 9 m^3/d per m length and 27 m^3/d per m length, respectively. These results compare with another gallery in river gravel in Cardiff that the authors have studied previously where the infiltration rate is 8.2 m^3/d per m length.

The use of 0.75 m or larger diameter perforated concrete pipes laid in a trench is common for major galleries. The cost of this type of construction rises dramatically with depth if this is beyond the reach of an excavator, and with increased depth below the watertable.

For low discharge private schemes, the gallery section may comprise a few lengths of 75 mm or 100 mm agricultural field drain feeding to a collector chamber. If the land gradient permits, gravity flow may be possible from the collector chamber so that a pump is not required. This arrangement would normally be classed as a spring chamber. If ground conditions are poor (for example, if clays and peats are dominant rather than sands or gravels) the field drain trenches can be filled with an imported sand as a surround to the pipe. This may increase the yield and degree of filtration achieved. The smaller diameter pipes are less accessible for routine inspection for incrustation than the large diameter pipes used in the bigger schemes.

In contrast to a multiple borehole scheme, there is less collector pipework and valving and so pumps, power supply and control equipment can all be simpler at an infiltration gallery. There

Table 4. *Water quality results*

Scheme	Colour	Turbidity	Iron	Manganese
River Spey	40% F*	90% F	0% F	70% F
Spey Scheme (PW1 to PW36) (post-commissioning)	0% F	0%† F	0% F	0%‡ F
Spey W12	10–15% F	2% F	0% F	50% F
Ordiequish Gallery	5% F	1% F	2% F	2% F
River Deveron	80% F	40% F	80% F	40% F
Wrack Gallery	<1% F	<1% F	20% F	90% F
River Lochy	100% F			
Camisky Bh9 TW	0% F	0% F	0% F	0% F

Note: 2% F means 2 % failures of MAC for data examined.
* field data.
† one or two individual wells failed in floods.
‡ two wells are consistently high at up to 5 times MAC but are diluted when in use.
Source: Mott MacDonald (1999*a*).

is, however, little opportunity for a gallery to exploit local variations in water quality.

Water quality

Groundwater from properly sited boreholes and wells in shallow alluvial aquifers is usually better in several respects than in the adjacent rivers or streams, for both filtered and unfiltered, acidified samples. Data from operation of the Spey scheme from 1995 to 1997 and Spey test well W12, Ordiequish (1984/1985) and the Wrack from the Galleries and the one year extended test at the Fort William wellfield, are shown in conjunction with corresponding data from the adjacent rivers in Table 4.

The results for the Spey scheme are for final water into supply. Thus the variance between wells is averaged out. It was also decided not to operate the worst well (PW 26), which has a very high manganese concentration, until demand rises and a treatment process is installed.

The improvement in water quality at the Spey wellfield takes place despite a short residence period in the aquifer as determined by groundwater modelling studies (Chen *et al.* 1997) and tracer experiments.

Water quality at a few other borehole/well schemes is as follows:

Tomatin	In use, no data reviewed
Laggan Bridge	Four test bores, high metals in most, high colour in some
Gairlochy	Generally good
Kenmore	Quality understood to be good
Dalwhinnie	Excellent
Kinloch Rannoch	Two sample results exceeded the MAC during construction but no problem in 14 day test in August 1994. The production well is now in use.

Although no data on biological quality, particularly coliforms or oocysts, have been examined for this study, it is clear that groundwater is usually of a higher quality than nearby surface water. However, it may be worse in some respects, i.e. metal concentrations. Whereas chlorination alone is not always guaranteed to meet potability requirements, the use of groundwater may reduce the treatment requirements from a relatively complex or expensive task for say, colour, to that for a more straightforward parameter such as iron. Time series data for groundwater sources (not presented herein) show much less pronounced 'spikes' in raw water quality than surface sources and this can assist treatment. Outage due to colour or turbidity 'spikes' is thus reduced significantly.

Discussion and conclusions

Potential well yields of 0.5 Ml/d to 2 Ml/d are a reasonable planning guide and are feasible in many of the alluvial and fluvio-glacial aquifers in northern Scotland. However, there are a few yield failures where the aquifer is effectively absent, too thin or too shallow. A rough estimate of the 'success' rate (i.e. demand met) is around 80 % and this figure should be used for planning purposes in the absence of more specific site investigation data.

Production well design and costs mainly reflect borehole geometry and access. Factors include depth to pumping water level (affects choice of pump type), discharge (affects pump and rising main diameter) and delivery head (affects motor power in submersible pumps). Depth to the main aquifer is a critical factor in the design since well chambers may be suitable for depths within the reach of an excavator (say 5–6 m maximum). Thermoplastics are readily available and are highly suitable for screen and casing diameters up to around 300 mm and depths down to about 50 m. The more expensive stainless steel well screens may be required for larger diameter wells, for deeper boreholes, for wells where the screen length/aquifer thickness is short and where provision has to be made for energetic workover/rehabilitation operations.

The cost of site investigation can be balanced against the relatively low cost of construction and testing a production borehole particularly if the exploratory borehole is used for observation purposes. The operating costs of raising water to the surface are usually low since drawdown is limited by the depth of the boreholes and the need to maintain some minimum length of saturated wellscreen.

Successful schemes which could be used as a model for future developments range from Dalwhinnie, with a yield of around 0.16 Ml/d and construction cost of around £6000, to the 20 Ml/d (rising to 27 Ml/d once a treatment plant is installed to handle the lower quality production wells) Spey Scheme.

It is also possible that infiltration galleries operating under gravity could be cost-effective at reducing treatment needs at small isolated sites with poor access and no immediate source of power. Although no such schemes are reported to be in use for public water supply in the NoSWA area, there are some in use for private supplies.

The data reviewed show that there is a high and unpredictable variance in individual borehole discharges, colour removal and iron and manganese concentrations both within and between different wellfields. The variance is probably a function of geology or geomorphology and so there appears to be some potential to improve well siting techniques and reduce the present reliance on test wells.

The authors gratefully acknowledge the permission of North of Scotland Water Authority to publish this paper. The authors would like to record their thanks to the referees for their comments and acknowledge the cooperation of Mr Derek Ball at the British Geological Survey, especially for making reports and records available. The authors would also like to acknowledge the various unpublished reports consulted during this work, including reports by Grampian Soil Surveys, Ritchies and Wimpey Environmental.

References

ANON. 1961. *Ranney collector for the Lower Deveron Water Board.* Water and Water Engineering, January 1961.

BALL, D. 1997. *The North West Highlands: the potential of groundwater at selected sites for rural water supply.* British Geological Survey Technical Report **WD/97/42**.

CHEN, M., SOULSBY, C. & WILLETTS, B. 1997. Modelling river–aquifer interactions at the Spey abstraction scheme, Scotland: implications for aquifer protection. *Quarterly Journal of Engineering Geology*, **30**(2), 123–136.

JOHNSTONE, J. M. C. & RENNIE, G. M. 1991. Development of a new water supply for Fort William. *Journal of the Institution of Water & Environmental Management*, **5**, 503–512.

MOTT MACDONALD. 1989. *Fort William water supply, investigations into the proposed River Lochy scheme, main report.* Mott MacDonald report for Highland Regional Council, Department of Water and Sewerage.

——1994. *Spey abstraction scheme – record of wellfield construction.* Mott MacDonald report for Grampian Regional Council.

——1997. *Spey abstraction scheme – record on wellfield performance.* Mott MacDonald report for NoSWA.

——1999*a*. *Water resources study – final report.* Mott MacDonald report for NoSWA.

——1999*b*. *Fort William area water supply – report of wellfield construction.* Mott MacDonald report for NoSWA.

ROBINS, N. S. & BALL, D. F. 1987. *Groundwater in the superficial deposits of Scotland – a review.* Applied Research and Development Report Number **ARD**.

WATT, G. D., MELLANBY, J. F., VAN WONDEREN, J. J. & BURLEY, M. 1987. Groundwater investigations in the lower Spey valley, near Fochabers. *Journal of the Institution of Water & Environmental Management*, **1**, 89–103.

Groundwater resources in the Quaternary deposits and Lower Palaeozoic bedrock of the Rheidol catchment, west Wales

K. HISCOCK[1] & A. PACI[1,2]

[1] *School of Environmental Sciences, University of East Anglia, Norwich NR4 7TJ, UK (e-mail: k.hiscock@uea.ac.uk)*

[2] *Present address: Environment Agency, Bromholme Lane, Brampton, Huntingdon PE18 8NE, UK*

Abstract: An assessment of the groundwater potential of the River Rheidol catchment in west Wales is presented, a region where groundwater is an important source of water to Aberystwyth. An alternative approach to a conventional well inventory was undertaken based on the manipulation of data within a geographic information system (GIS). The GIS application revealed that, in the lower reaches of the catchment, 25% of the total recorded abstractions fall on the boundaries between different geological units and, of these, two-thirds relate to the contact between Quaternary deposits and bedrock. These abstractions are typically correlated with spring sources and represent small domestic supplies from shallow, brick-lined pits dug into the valley sides. Simple catchment water balance calculations showed that groundwater abstractions are minor in comparison to total combined surface and groundwater runoff.

Compared with the weathered and fractured bedrock, the floodplain gravel aquifer is of greater significance and is further enhanced by an unquantified amount of induced recharge from the River Rheidol. However, the resource contained in the Quaternary gravels is vulnerable to mainly agricultural contaminants derived either from direct leaching through the thin unsaturated zone, especially on the floodplain, or from induced recharge from the River Rheidol.

Water supply in the upland regions of Britain and Ireland is usually considered to be dominated by surface water resources, with groundwater of only minor, local importance. As a consequence, the groundwater potential of these remoter areas has received less attention than the major sedimentary aquifers of England, which are heavily relied upon for public water supply. However, previous work on the groundwater resources of Scotland and Northern Ireland (Robins 1990; Robins 1996) has shown that groundwater is of major significance in the agricultural, fish farming, chemical, manufacturing and food industries.

An important consideration in terms of groundwater supplies in the upland areas of Britain and Ireland is the role of Quaternary deposits. Alluvial channel fills, glaciofluvial sand and gravel deposits and raised beaches are often relatively accessible for exploitation. The main constraints on groundwater development are the limited thickness of the Quaternary cover and poor storage capacities which induce large drawdown on pumping. Yields of up to 8.5×10^2 m^3 per day have been obtained from these superficial deposits in Scotland (Robins 1987).

This paper is a study of the groundwater resources in the bedrock catchment of the River Rheidol in west Wales. The Rheidol catchment (see Fig. 1 for location) is an important source of water for public supply in the Aberystwyth area, Ceredigion. In recent years, the experience of a drier climate and an increase in water charges, combined with a favourable geological situation, has led to a dramatic rise in the number of private groundwater abstractions.

Previous detailed geological mapping and hydrogeological investigation work in west Wales has been undertaken by Robins *et al.* (this volume) in the River Teifi catchment, south of the present study area. The resource potential was determined by conducting a borehole, well and spring inventory accompanied by hydrochemical sampling and analysis with a sample size representing 70% of groundwater abstractions in the catchment. Robins *et al.* (this volume) concluded that the shales and greywackes was the most productive aquifer unit, with almost all of the recently constructed boreholes developed in the bedrock.

In contrast to the physiographically similar Teifi catchment, the Rheidol catchment is

From: ROBINS, N. S. & MISSTEAR, B. D. R. (eds) *Groundwater in the Celtic Regions: Studies in Hard Rock and Quaternary Hydrogeology*. Geological Society, London, Special Publications, **182**, 141–155. 1-86239-077-0/00/ $15.00

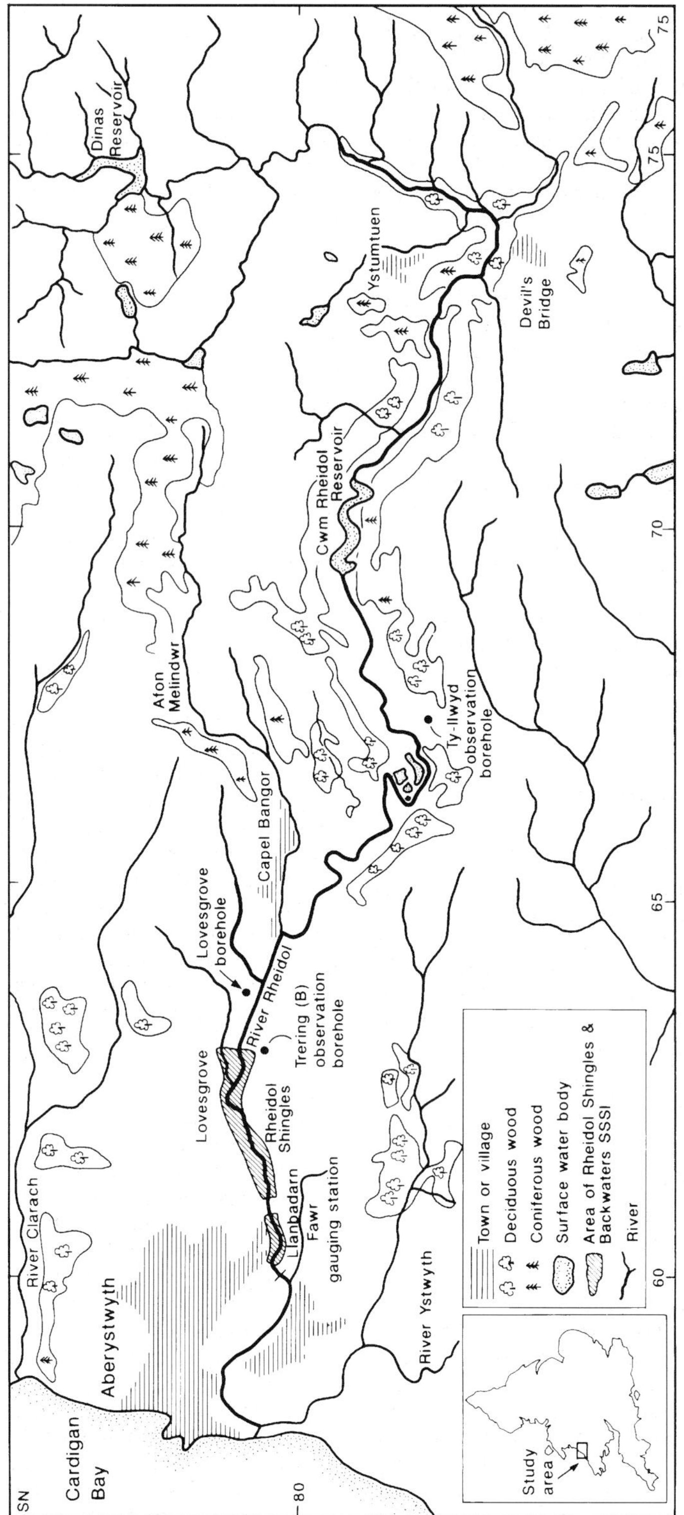

Fig. 1. Location map of the River Rheidol catchment, west Wales.

unusual in that there are no surface water supplies, due to the contamination of surface water by acid mine drainage from abandoned mines in the catchment. The Cefnllan Reservoir, south of Aberystwyth, combines surface water from outside of the catchment with groundwater abstracted from clean gravel on the Rheidol floodplain. This gravel aquifer produces almost half of the water supply in the Aberystwyth district.

To provide an assessment of the groundwater potential of the Rheidol catchment, an alternative approach to a conventional borehole and well inventory was undertaken based on the manipulation of available data within a geographic information system (GIS). The GIS software ARC/INFO (Environmental Systems Research Institute, Inc., Redlands, CA, USA) was employed as a tool to analyse the spatial relationship between the mapped geology and points of groundwater abstractions in order to interpret the occurrence and usage of groundwater in the Rheidol catchment. A field excursion to carry out 'ground truthing' of the GIS results and to collect data for catchment water balance calculations was also undertaken to provide an overall environmental appraisal of the groundwater resources.

Catchment characteristics

Topography, land use and climate

The Rheidol catchment covers an area of 182 km^2 with the topography rising from sea-level at Aberystwyth to *c.* 300 m above sea-level at Devil's Bridge (Fig. 1) and *c.* 500 m at the source of the river in the Cambrian Mountains. The predominant land use is upland farming, mainly sheep grazing and a few isolated dairy farms. About 30% of the catchment area is forested (National Rivers Authority 1996). Soils in the district are generally shallow, acidic and peaty. The area experiences a temperate climate with a mean rainfall of 1790 mm.

Hydrology

The River Rheidol has its source in Llyn Llygad Rheidol (SN 792 877), a small corrie lake on the western flanks of Plynlimon (Jones & Pugh 1935). In the vicinity of Devil's Bridge, the river turns westwards, as a result of glacial capture, and flows into Cardigan Bay at Aberystwyth. The River Rheidol is a rapid flowing, meandering river with a sand/cobble base. The channel width is 30 m in its lower course. The floodplain width varies from 800 m in the lower reaches to 30 m in the upper reaches. The river flow is modified by releases of water from the Cwm Rheidol Hydro-Electric Power Station, served by the Dinas and Nant-y-moch reservoirs, and causes the accumulation and erosion of sediment along the channel profile. The mean annual discharge of the River Rheidol at Llanbadarn Fawr gauging station (SN 601 803) is 8.91 m^3 s^{-1} (Institute of Hydrology 1988).

Geology

Table 1 provides a stratigraphical sequence for the Rheidol catchment. A general map of the solid geology is shown in Fig. 2 while Fig. 4 shows the distribution of Quaternary deposits. The area east of Aberystwyth is composed predominantly of Silurian rocks, with inliers of Ordovician age found in the upper reaches of the Rheidol near Plynlimon. The rocks comprise thinly-bedded Lower Palaeozoic sedimentary mudstones, siltstones and sandstones deposited when the Welsh Basin was situated in the Iapetus Ocean (Woodcock 1984). During a transgressional phase of the Silurian, sub-marine landslips formed turbidity currents that spread over the sea bed from east to west depositing sequences of turbidites, which are recognized by distinctive graptolite biozones. The rhythmic sequences of turbidite sediments fine upwards as the energy of the density currents diminished (Bouma 1962). At the end of the Silurian, the direction of the turbidity currents changed to south–north and the associated sediments (Aberystwyth Grits Formation) are much coarser.

During the Caledonian Orogeny the Iapetus Ocean closed and the sediments that had accumulated in the marine basin underwent compaction, de-watering, buckling and folding, with cleavage developed in a NNW–SSW direction, sub-parallel to the folding (Cave & Haines 1986). Structurally, as a result of the tectonic activity, the whole region is a tightly folded series of synforms trending north-north-east. A series of mineralized tectonic fractures trending east–west/WSW also developed and contain ore-bearing galena and sphalerite, which have been extensively mined for lead and zinc in the Ystumtuen region.

Quaternary deposits overlie the bedrock in most areas of the valley bottom. During the Pleistocene glaciations, ice masses originating as ice caps on Plynlimon flowed radially, fault-guided, along the major valleys cutting through the shales and sandstones. Due to the higher metamorphic grade of the bedrock in the upper

Table 1. *Stratigraphical sequence for the Rheidol catchment (after Cave & Haines 1986)*

Age	Series	Formation	Lithology	Thickness (m)
Quaternary	Flandrian and Devensian		Peat, Head/Scree, Alluvium, Morainic Drift, Glacial Till, River Terrace Deposits	<30
Silurian	Llandovery	Aberystwyth Grits	Medium grey mudstone interbedded with pale grey medium to coarse grained sandstone	240
		Borth Mudstones	Medium to dark grey mudstone thickly bedded. Thin beds of siltstone and fine grained sandstone	350
		Devil's Bridge	Grey thinly bedded mudstone, sandstone and siltstone; base shows thicker sandstone units	10–600
		Cwmsymlog	Mainly grey mudstone, thin colour-banded (green and maroon). Siltstone and sandstone sometimes found	0–140
		Derwenlas	Medium and pale grey mudstone, thinly bedded. Thin siltstone beds in places and sandstone	20–80
		Cwmere	Dark grey pyritous mudstone thinly bedded. Thin siltstone beds	70–160

reaches of the valley, there was less erosion, resulting in a sparse distribution of glacial deposits. As the glaciers advanced and retreated in the lower reaches, great thicknesses of glacial and glaciofluvial drifts were deposited in the U-shaped valleys that formed. In the Rheidol valley, glacial till and morainic drift of Devensian age, deposited in contact with the ice, comprise heterogeneous mixes of clay, silt, sand and gravel. The till deposits (boulder clay) are stiff blue/grey clays and silt with clasts of poorly sorted mudstone debris ranging from grain to boulder size and sub-rounded to sub-angular in shape. These deposits form a minor valley fill and are found in kame terraces along the floodplain; they are identified from the river terraces by not possessing a continuous thalweg. Glaciofluvial outwash deposits occur and form clean gravels deposited on the floodplain by glacial meltwaters.

The river terrace deposits have been described in detail by Macklin & Lewin (1986). In composition they are similar to the morainic drifts and are found on the flanks of the valley. The basal sediments are poorly sorted, sub-angular sands, silt and clay, grading upwards into well-sorted, sub-rounded sands and gravels. The deposits are very variable in thickness and extent. Most of the floodplain is filled with alluvium. During the glacial period the valley was overdeepened and thick sequences of clay and silt overlying sands and gravels were deposited. It is neither clear what the thickness of this alluvium is, nor the distribution and geometry of the underlying glacial deposits.

On the steep valley sides, highly porous head deposits, 3–4 m thick, have developed in hollows. The angular rock debris, set within a silt/clay matrix, has undergone solifluction. In limited areas, slumped regoliths (scree) found overlying the bedrock also comprise angular rock debris of local origin.

Hydrogeology

The Survey of Existing Water Management and Use by the Welsh Water Authority (1979) described the Lower Palaeozoic successions as impervious and of little value as aquifers, but regarded the glacial and river gravels in the floodplains of several Welsh rivers to be important sources of water. Table 2 provides a summary of the hydrogeological characteristics of the geological formations present in the Rheidol catchment.

In general, the aquifer potential of the Silurian rocks is very limited. Across the Llandovery Series, the turbidite sediments become coarser and the proportion of sand increases. Between the Cwmere and Aberystwyth Grits Formations the metamorphic grade decreases and the increase in the thickness of weathered zones and the presence of thickly bedded sandstone increase the potential for groundwater abstraction.

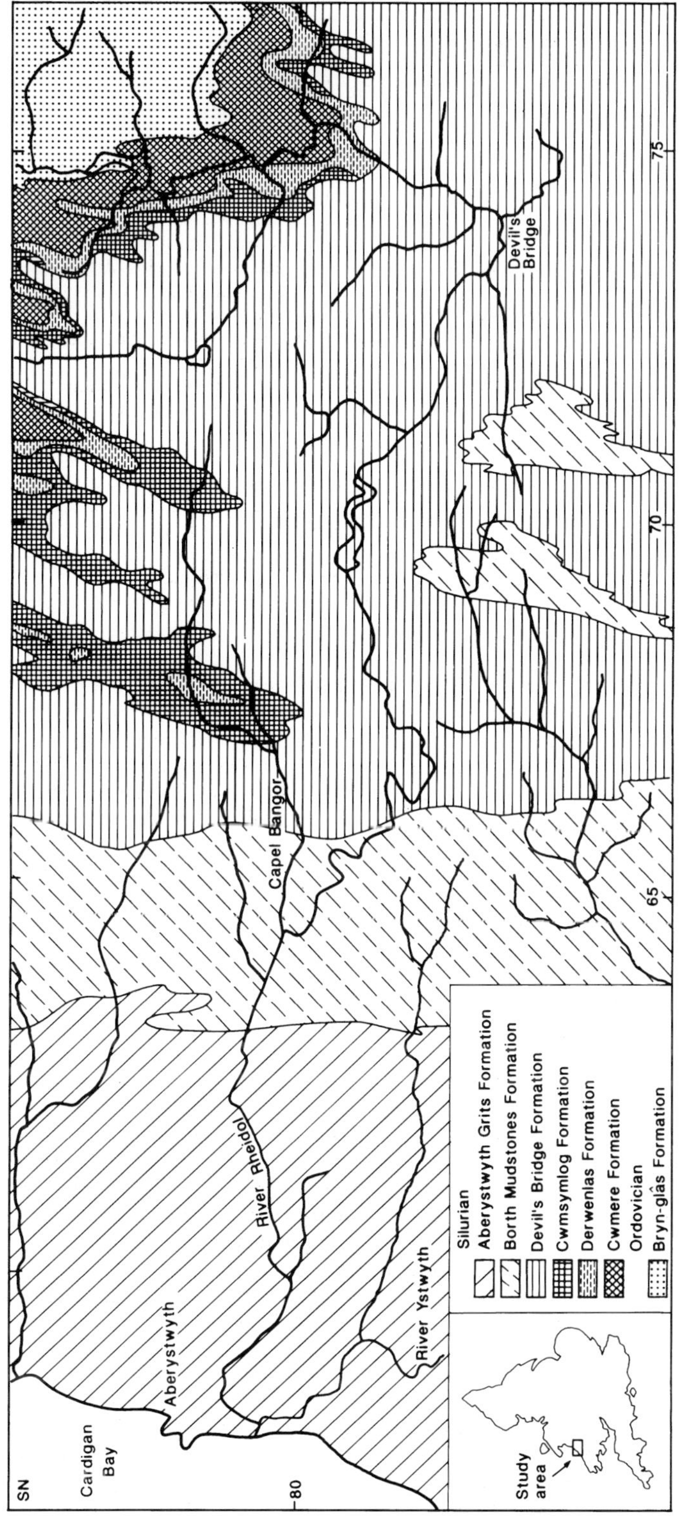

Fig. 2. Simplified solid geology map of the River Rheidol catchment, west Wales, based on the 1 : 50 000 Solid Edition Sheet 163, Aberystwyth and the 1 : 50 000 Solid with Drift Edition Sheet 178, Llanilar. Reproduced by permission of the Director, British Geological Survey. NERC copyright reserved.

Table 2. *Hydrogeological characteristics of the geological formations present in the Rheidol catchment*

Age	Series	Lithological sub-division	Hydrogeological characteristics	Hydrogeological classification	Aquifer classification*
Quaternary	Flandrian and Devensian	Alluvium Till	Negligible permeability due to high clay fraction.	Aquitard	Non-aquifer
		Head/scree	Variable porosity. Supports locally important abstractions.	Aquifer	Minor aquifer
		River Terrace Morainic Drifts	Variable porosity. Supports locally important abstractions.	Aquifer	Minor aquifer
Silurian	Llandovery	Aberystwyth Grits Formation Borth Mudstones Formation Devil's Bridge Formation	Negligible permeability. Fracture flow supports minor abstractions.	Solid bedrock/ impermeable base	Non-aquifer
		Cwmsymlog Formation Derwenlas Formation Cwmere Formation	Negligible permeability, no groundwater in exploitable quantities	Solid bedrock/ impermeable base	Non-aquifer

* After the National Rivers Authority (1992).

Boreholes and wells for private supply are first recorded in the Devil's Bridge Formation but yields are variable, dependent on the degree of secondary porosity. The frost-shattered Aberystwyth Grits Formation found underlying the surficial deposits and in contact with the Borth Formation is an important unit in connection with development of the bedrock aquifer. Overall, the Silurian sequence in the Rheidol catchment can be considered a non-aquifer with minor thin productive horizons. The solid bedrock acts as a very low permeability base to the Quaternary aquifers.

In the bedrock, the degree to which the fractures are interconnected determines the effective porosity. Evidence from the speed with which abandoned mines flood with water indicates that the joints in the weathered zone are open and well connected. Values of porosity derived from laboratory experiments on rock samples from the Llandovery Series are of the order of 2 to 4% on average.

In general, the near-surface weathered zones, which can be up to 10 to 20 m thick, provide enough water to support small domestic supplies from shallow, brick-lined pits dug into the valley sides that tap into natural spring lines issuing from either the contact between head deposits and bedrock or the contact between the Aberystwyth Grits and Borth Mudstones Formations (P. Armstrong pers. comm.). Groundwater is typically abstracted on demand and is often gravity fed.

The Quaternary geology presents a complex hydrogeological situation; the extreme heterogeneity of the deposits inhibits the definition of continuous aquifer units. In general terms, the glacial till, a stiff clay unit, forms a discontinuous confining or leaky layer approximately 1–5 m thick, which results in the development of perched water table conditions. Like the glacial till, the valley alluvium is considered a non-aquifer due to the presence of silt, clay and peat layers. The most significant and extensive aquifer sediments are the gravel deposits. Clasts that have been sorted and deposited by aqueous processes, such as the meltwater or outwash deposits, are very clean, containing little clay or silt content, and form the most productive aquifer material. At Lovesgrove (SN 638 808), transmissivities are in the order of 4000–6000 m^2 per day, specific yields are about 5% and the sustainable yield is around 50 $l s^{-1}$ (Hyder Consulting Ltd. 1998). The thickness of the gravel deposits varies from 15–30 m. Elsewhere, morainic drift, deposited in contact with former ice sheets, has lower transmissivities due to the greater percentage of clay and silt fractions present.

The extent of the individual glacial deposits is difficult to establish and they are therefore classed as a single hydrogeological unit. The primary constraints on the porosity of these unconsolidated deposits are the degree of sorting and the proportion of clay and silt material. The borehole logs given in Fig. 3 demonstrate the existence of silt and clay bands that act to reduce the yield of the sand and gravel aquifer.

The gravel to boulder size clasts that comprise the river terraces along the Afon Melindwr and River Rheidol are also significant in terms of water supply and aquifer potential (thickness between 14–30 m, transmissivities of the order of 500–1000 m^2 per day, storage coefficient about 2% and yield about 35–45 $l s^{-1}$) and can be considered to extend the gravel aquifer despite the geological and depositional differences. The head (scree) deposits which form discrete aquifer units in hollows on the valley sides are very limited in extent and only locally important for groundwater supplies.

Environmental constraints on groundwater development

The continually changing profile of the River Rheidol has created a number of environmentally sensitive backwaters within the floodplain, the most important of which is the Rheidol Shingles and Backwaters (SN 617 807), a site of special scientific interest (SSSI) and the largest unmodified lowland water body in North Ceredigion. Unless groundwater abstractions are located in close proximity to such ecologically-important sites, then the size of current groundwater abstractions and the small drawdowns demonstrated for the gravel aquifer do not pose a significant problem.

A further, largely untested concern, is the reduction in baseflow to the River Rheidol as a result of groundwater abstractions located on the floodplain and the effect this may have on river water chemistry. The alkaline nature of the groundwater runoff can potentially dilute and buffer the acidic stream water chemistry. The presence of bicarbonate, even in low concentrations in crystalline bedrock, can have an influence on stream chemistry (Reynolds *et al.* 1986). The buffering capability of the Lower Palaeozoic slates and shales in Wales has been demonstrated by Neal *et al.* (1997) to provide groundwater with sufficient bases to buffer acidic water. In the Rheidol catchment, the peaty soils, coniferous forest plantations and acid mine drainage contribute to the acidity of surface runoff, with most concern centred on the impact of acid mine drainage on surface water quality.

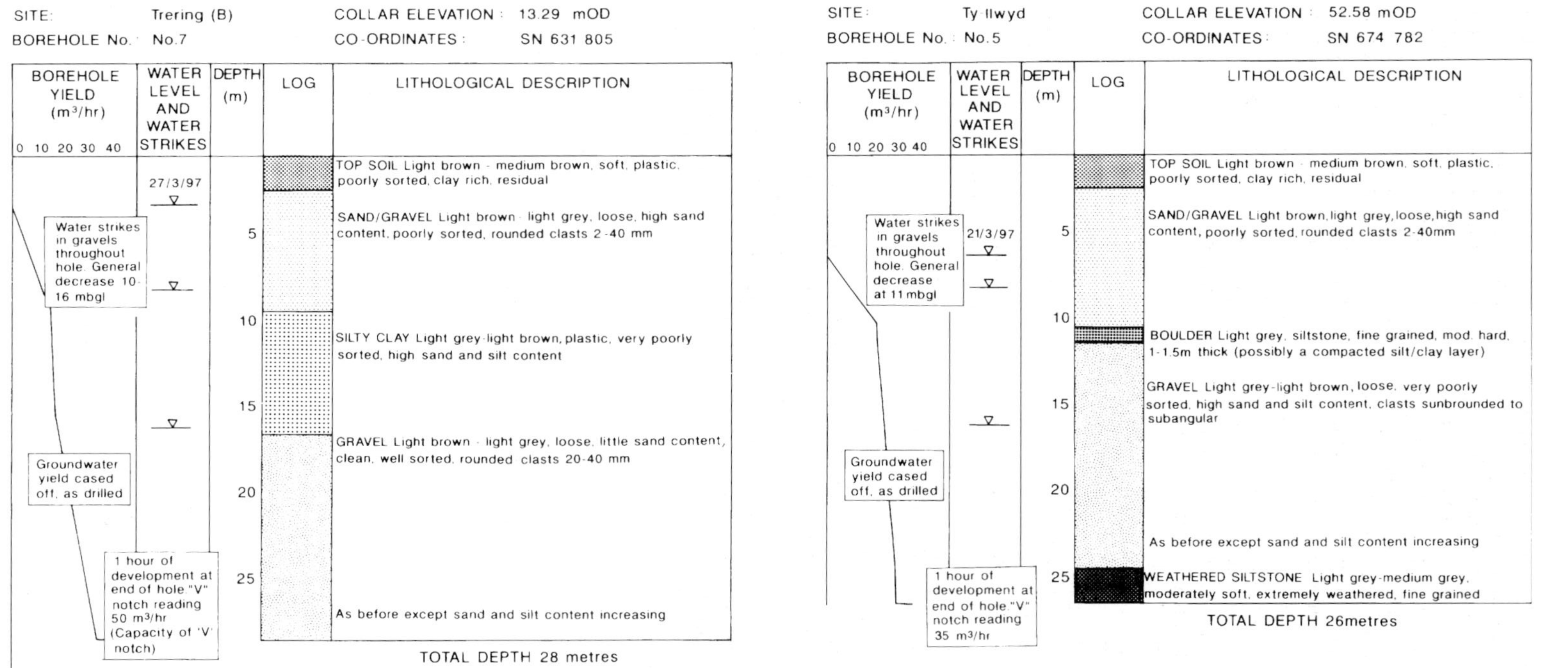

Fig. 3. Representative lithological logs and hydrogeological details for two Environment Agency observation boreholes in the Quaternary aquifer of the Rheidol catchment. See Fig. 1 for borehole locations.

The mineralized veins of the Ystumtuen region supported a prosperous mining industry between about 1750–1920. Ochreous acid mine drainage resulting from the flushing of ferrous sulphate and sulphuric acid (the products of oxidation of pyrite and marcasite during dry-working of the mines) and issuing from the abandoned mine adits and spoil heaps previously killed all aquatic life in the river. The acidified water dissolved heavy metals such as lead, zinc, cadmium and aluminium which entered the river system, the effect of which has been detected 16 km downstream of the mined area with a pH as low as 2.6 (Fuge *et al.* 1991). Installation of a limestone filter upstream of Cwm Rheidol to precipitate out metals has ameliorated the situation and trout and salmon have returned to the river. However, despite attempts to channel all surface drainage waters into the limestone filter, it is apparent that appreciable volumes of drainage reach the river directly via springs and seepages. One such spring occurring in the river bank below the Ystumtuen spoil area typically has a pH of 3.9 and a zinc content of 19 $mg\,l^{-1}$ (Fuge *et al.* 1991).

The quality of groundwater in the Quaternary deposits is generally regarded as good although the shallow thickness of the unsaturated zone means that the groundwater is vulnerable to contamination, especially from point sources such as agricultural spills of fertilizer, pesticides, oil and fuel. Analysis of drawdown data from pumping tests conducted at Lovesgrove provides evidence for induced recharge from the River Rheidol. High concentrations of metals have been found bound to the river sediment, the legacy of mining activity in the catchment, although groundwater sampling by Hyder Consulting Ltd. (1998) showed levels of heavy metals close to zero (see Table 3). A greater water quality concern for those wells and boreholes likely to be influenced by river water infiltration is associated with the microbiological quality of the river water and groundwater as a result of sewage disposal and sheep grazing activities on the floodplain deposits.

Methodology for assessment of groundwater potential

To provide a relatively rapid assessment of the groundwater potential of the Rheidol catchment, a GIS approach was adopted which, in general terms, permits the capture, storage and manipulation, analysis and display of spatially referenced information. An important feature of a GIS is the ability to generate new information by the integration of existing diverse datasets sharing a compatible spatial referencing system (Goodchild 1993).

In this study of the Rheidol catchment, the following information was combined within ARC/INFO: polygon coverage delineating the alluvial floodplain, river terrace, glacial and head/scree deposits and bedrock; and point coverage of

Table 3. *Water quality parameter values for surface water and groundwater measured at the Lovesgrove abstraction site, August 1998. Data based on Hyder Consulting Ltd. (1998)*

Parameter	Lovesgrove test production borehole	River Rheidol, Lovesgrove	Water quality standard (CEC 1980, 1998)
pH	6.0	7.0	6.5–9.5
E.C.† ($\mu S\,cm^{-1}$)	99	73	2500
NO_3 (mg/l)	1	1	50
Hardness (mg/l as Ca)	30	19	min. 60
Ca (mg/l)	8	5	100*
Mg (mg/l)	3	2	50
Cu (μg/l)	<10	<10	2
Zn (μg/l)	n.d.	n.d.	0.1*
Cd (μg/l)	<0.5	<0.5	5
Al (μg/l)	<20	n.d.	200
Pb (μg/l)	n.d.	n.d.	10
Cr (μg/l)	<1	<1	50
Mn (μg/l)	19	n.d.	50
Fe (μg/l)	123	n.d.	200
Ni (μg/l)	n.d.	n.d.	20

Notes: * Guide level.
† Electrical conductivity.
n.d. Not detected/determined.

groundwater abstraction points, with the addition of attribute data on abstraction rates, source type and whether abstractions are licensed. Overlay of the geological and abstraction data using the GIS software, together with catchment water balance calculations, formed the basis for the interpretation of the groundwater potential. To verify the results of the GIS overlay operation and water balance calculation, a field excursion was undertaken in summer 1998 to trace the geological succession and assess the hydrogeological characteristics of the main deposits in the area. Selected private wells were visited to obtain information, unavailable from Ceredigion County Council, on well construction, depth and usage. Following the field visit, and for the purpose of this study, the area between Aberystwyth and Devil's Bridge was considered to have the greatest groundwater potential because it contained the greater majority of groundwater abstractions.

Geological information was digitized from the 1:50000 Sheets 163 (Aberystwyth) and 178 (Llanilar). Licensed abstraction data were obtained from the Environment Agency. Private abstractions are recorded by Ceredigion County Council for water quality testing purposes, with abstraction rates calculated on the assumption that five persons consume water at a rate of 1 m^3 per day (Department of the Environment 1991). Further classification and estimation of water abstraction depends on water usage, for example food production and dairy farming, and abstraction rates were applied as appropriate. Additional, published hydrological data (Department of the Environment 1978, 1982, 1983; Institute of Hydrology 1988) were compiled in order to complete the water balance calculation.

Results and Discussion

Groundwater occurrence and usage

Table 4 shows the information collected during the field excursion and demonstrates that spring, well and borehole supplies are obtained from both surface peat and gravel deposits, and the bedrock. It is evident that, in this predominantly rural district, the bedrock aquifer is an important source for those individuals not on mains water supply. Table 5 shows the quantities of groundwater abstracted for various categories of supply and it is clear that abstraction for public water supply is two orders of magnitude more important than for private water supplies.

Figure 4 depicts the map produced by the GIS overlay operation. On examination, it is evident that the relationship between geology and hydrogeology is variable, although in general terms the groundwater abstraction locations can be classed into one of three categories: (a) boreholes and wells in the Quaternary deposits (gravel, moraine and head); (b) boreholes and wells at the contact between the Quaternary deposits and bedrock; and (c) valley-side springs at the contact between different bedrock formations; and trapped interflow in peaty areas, collected on the steep valley sides in the higher reaches of the catchment.

Table 4. *Representative details of selected groundwater abstraction sites visited in the Rheidol catchment, 1998*

Location	National Grid Reference	Aquifer unit	Source type	Depth (m)	Construction
Woodlands Caravan Park, Devil's Bridge	SN 746 773	Bedrock	Borehole	198	Unknown
Tycam	SN 683 795	Head?	Spring	Unknown	Enclosed high on hill, gravity fed to reservoir behind farm
Tycam	SN 683 795	Bedrock	Borehole	61	Unknown
The Briars	SN 662 789	Gravels	Well	7	Dolomite filter installed to increase hardness
Min Rheidol	SN 674 790	Bedrock	Borehole	122	Unknown
nr Pond Yr Oerfa	SN 732 801	Peat	Spring	Unknown	Piped to concrete chamber, gravity fed to house
nr Pond Yr Oerfa	SN 732 801	Bedrock	Well	3.7	Dug by mechanical shovel, lined with concrete rings, pumped to house

Table 5. *Categories of supply and quantities of groundwater abstractions in the Rheidol catchment, 1998*

Category of supply	Quantity abstracted ($m^3 a^{-1}$)
Public supply	1300156
Private supply	28823
Licensed	1308685
Unlicensed	20294
Total	1328979

Inspection of Fig. 4 and further interrogation of the GIS permits the calculation of polygon areas for each geological unit and the identification of the number of abstractions associated with each geological unit and contact (Table 6). Of the total number of abstractions, 25% fall on the boundaries between different geological units and, of these, two-thirds relate to the contact between Quaternary deposits and bedrock. For licensed abstractions, and using the GIS application INFO, the locations of the spring source type are found to be correlated with geological contacts. It should also be noticed that the 29 abstractions that appear in bedrock are all in close proximity to the Quaternary deposits. Within the accuracy of the map digitization, a number of these abstractions could also be positioned on a geological boundary or represent interflow within the peat developed on the valley sides. In the higher reaches of the Rheidol catchment, between Capel Bangor and Devil's Bridge, there is an apparent series of private abstractions all approximately equidistant from the river, again indicative of a spring line discharge feature.

Table 6 lists 12 abstractions as being located on alluvium. Unfortunately, information concern ing well or borehole length and deposit thickness is incomplete. These 12 abstractions presumably penetrate the morainic or river terrace deposits below the alluvium. Of the licensed abstractions, boreholes located on the alluvium correlate with the gravel deposits.

Compared with the licensed abstractions, it is more difficult to interpret the type of groundwater abstraction from the unlicensed private supplies. It is inferred that abstractions on geological boundaries are spring-fed. From field observations, wells located on the contacts between the Quaternary deposits and bedrock tap into the permeable, weathered upper layers of bedrock, while those on the contact between the different Quaternary deposits depend on the presence of clay and silt layers which result in the emergence of seepages or springs. Abstractions on the glacial deposits are most likely to be drawing water from gravely morainic drift.

Catchment water balance calculations

Figure 5 presents available, published data of mean annual rainfall in the Rheidol catchment,

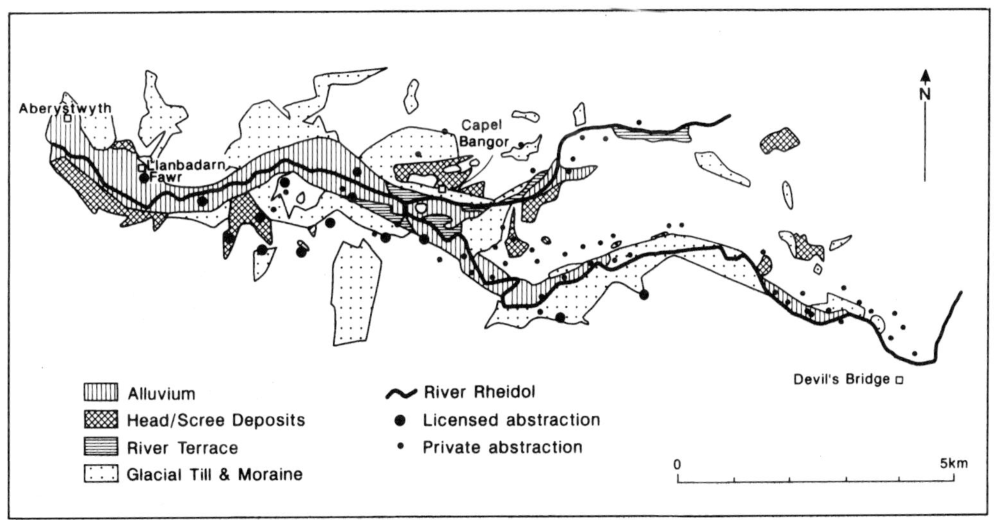

Fig. 4. Map of the distribution of Quaternary deposits in the Rheidol catchment together with the locations of private and licensed groundwater abstractions. Geological information is based on the 1:50000 Drift Edition Sheet 163, Aberystwyth and the 1:50000 Solid with Drift Edition Sheet 178, Llanilar. Reproduced by permission of the Director, British Geological Survey. NERC copyright reserved.

Table 6. *Numbers of groundwater abstractions and associated surface areas of different geological categories in the Rheidol catchment*

Geological category	Number of abstractions	Surface area of category (km^2)
Alluvium	12	6.5
Glacial	5	10.2
River Terrace	0	9.1×10^{-1}
Head/Scree	3	2.3
Bedrock	29	–
Glacial/Bedrock	7	–
River Terrace/Glacial	1	–
Alluvium/Bedrock	2	–
Alluvium/Glacial	3	–
Head/Bedrock	1	–
Alluvium/River Terrace	2	–

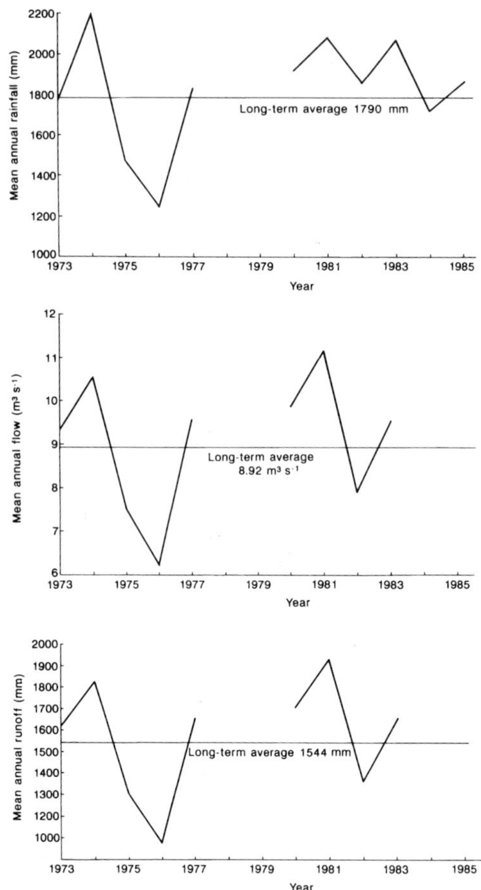

Fig. 5. Mean annual rainfall, mean annual flow at Llanbadarn Fawr gauging station and mean annual runoff for the Rheidol catchment. The annual long-term averages are for the period 1965–1985.

mean annual discharge of the River Rheidol and mean annual runoff from the catchment, together with the long-term averages for the period 1965–85. Although hydrological records used in this study are available from 1965–85, unfortunate gaps in these temporal datasets occur as a result of discarding unreliable data.

Adopting a simple water balance approach that equates hydrological inputs and outputs at the total catchment scale, and assuming negligible changes in storage over the chosen time period, Table 7 presents a water balance for the Rheidol catchment. Two calculations are provided: the first calculates the evapotranspiration loss based on the difference between rainfall input and available data relating to outputs; and the second includes an estimate based on the difference between rainfall input and the effective precipitation value ($1037\,mm\,a^{-1}$) used by the Environment Agency. The discrepancy between the two evapotranspiration figures ($514\,mm\,a^{-1}$) is mostly accounted for by error in the measurement of discharge at the Llanbadarn Fawr gauging station, in particular the error associated with the significant regulation effect imposed by the Cwm Rheidol Hydro-Electric Power Station, releases from which act to increase the frequency of medium discharges at the expense of low flows (Jones 1997). At all times, there is a minimum discharge of $1.835\,m^3\,s^{-1}$ ($318\,mm\,a^{-1}$) below the power station into the River Rheidol, rising during times of electricity generation.

Taking into account the effect of the power station releases, the second water balance calculation given in Table 5 shows that there is a large groundwater potential in the catchment of $363\,mm\,a^{-1}$. In comparison, current total groundwater abstractions ($7.3\,mm\,a^{-1}$) are only 2% of this potential resource.

Table 7. *Long-term average annual water balance for the Rheidol catchment for the period 1965–85. The catchment area is 182 km²*

Component	Depth of water (mm)	
	Calculation 1	*Calculation 2*
Input		
Rainfall	1790	1790
Outputs		
Evapotranspiration	239*	753†
Surface runoff	757	667
Groundwater runoff	787‡	363§
Groundwater abstraction‖	7.3	7.3

Notes:
* Calculated by difference between rainfall input and outputs.
† Calculated by difference between rainfall input and the effective precipitation value (1037 mm a^{-1}) used by the Environment Agency.
‡ Based on hydrograph separation using a baseflow index of 0.51 (Institute of Hydrology, 1980).
§ Groundwater runoff corrected for the effect of reservoir water releases (318 mm a^{-1}) imposed by the Cwm Rheidol Hydro-Electric Power Station.
‖ Total abstraction figure for 1998 taken from Table 5.

The above water balance calculation assessed the whole catchment area. In reality, the most significant area for groundwater abstraction is restricted to the river valley below Devil's Bridge. In the upper reaches of the valley there are no recorded licensed abstractions, and although there is a high density of wells and boreholes in the bedrock, these are mostly small, private abstractions. Current groundwater abstraction for public water supply is concentrated on the lower floodplain between Llanbadarn Fawr and Capel Bangor, a distance of 5.5 km.

Given the importance of the gravels and river terrace deposits in sustaining public water supply, it is appropriate to estimate the amount of recharge to the gravel aquifer. Employing an approach applied to shallow aquifers in Scotland (Robins 1990), groundwater recharge can be estimated by calculating the flow through an aquifer using Darcy's Law ($Q = -AiK$, where Q is total groundwater flow, A is the cross-sectional area of flow (floodplain width × saturated depth), i is the hydraulic gradient (negative in the direction of flow) and K is the hydraulic conductivity). For the Rheidol gravel aquifer, representative values are $A = (800 \times 25 = 2 \times 10^4\ m^2)$, $K = 200$ m per day, and $i = 3 \times 10^{-3}$ (Acer Consultants Ltd. 1996) which yield a value of $Q = 12 \times 10^3\ m^3$ per day. The total areal extent of the gravel aquifer $= 5500 \times 800 = 4.4 \times 10^6\ m^2$ and so the estimated groundwater replenishment (recharge) is $Q/A = 2.73 \times 10^{-3}$ m per day or 995 mm a^{-1}, which is in very good agreement with a value of 993 mm a^{-1} used by the Environment Agency and calculated using infiltration data. Again, in comparison with the total groundwater abstraction (7.3 mm a^{-1}), it appears that the available groundwater resource in the gravel aquifer is sufficient to support further development. This resource is further enhanced by the unquantified amount of induced recharge from the River Rheidol.

Summary and conclusions

To summarize the hydrogeological conditions encountered in the lower reaches of the Rheidol catchment, Fig. 6 is a conceptual cross-section illustrating the principal findings of this study. The sequence of Silurian mudstones and sandstones can be considered a non-aquifer but with minor thin productive horizons resulting from the presence and weathering of interconnected fractures. In general, the near-surface weathered zone is up to 10–20 m thick, with groundwater issuing from either contact between head deposits and bedrock or between the Aberystwyth Grits and Borth Mudstones Formations.

The Quaternary geology presents a complex hydrogeological situation and includes poorly permeable glacial till, morainic drift and valley alluvium together with significant and extensive gravel deposits that form an important minor aquifer. Clean gravels that originated as meltwater or outwash deposits contain little silt or clay and form the most productive aquifer

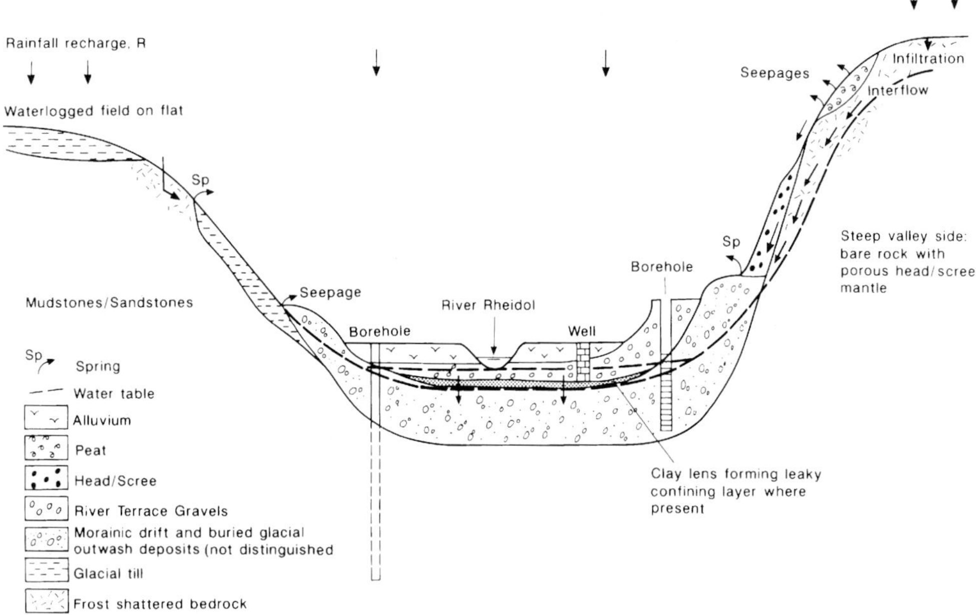

Fig. 6. Conceptual hydrogeological cross-section of the Quaternary and bedrock aquifers of the Rheidol catchment.

material. Under favourable conditions, transmissivities of the order of 4000–6000 m^2 per day and specific yields of about 5% are possible from gravel deposits 15–30 m thick.

A general restriction on the development of the Quaternary deposits is their degree of heterogeneity; particularly in terms of extent and depth, and the degree of sorting and the proportion of clay and silt material present. The quality of groundwater in the Quaternary deposits is vulnerable to surface, mainly agricultural, contaminants derived either from direct leaching through the thin unsaturated zone, especially on the floodplain, or by induced recharge from the River Rheidol. Of concern for those wells and boreholes likely to be influenced by influent conditions is the microbiological quality of the river water linked to sewage disposal and sheep grazing activities in the catchment.

In examining the groundwater potential of the Rheidol catchment, the GIS application revealed that of the total number of recorded abstractions, 25% fall on the boundaries between different geological units and, of these, two-thirds relate to the contact between Quaternary deposits and bedrock. In this situation, it was found that the abstractions are typically correlated with the spring source type and represent small domestic supplies from shallow, brick-lined pits dug into the valley sides, with groundwater abstracted on demand and often gravity fed.

The catchment water balance calculations showed that present groundwater abstractions are only a small percentage of the total available resource in the Rheidol catchment. Given its local significance in the coastal district of West Wales, the groundwater resource of the Rheidol catchment should continue to be managed in terms of implementing regulations for protecting the quantity and quality of this valuable resource. In particular, the development of deep boreholes in the weathered bedrock appears less sustainable than further development of the productive gravel aquifer.

The authors are grateful to the British Geological Survey (Aberystwyth and Wallingford), Environment Agency (Cardiff), Ceredigion County Council (Environmental Health Department, Aberystwyth) and Hyder Consulting Limited (Penarth) for their considerable assistance with fieldwork, data provision and useful discussions in completing this work. Andrew Lovett and Gilla Sunnenberg at the University of East Anglia are thanked for their assistance with the GIS application. This paper is based on the MSc Hydrogeology project completed by Anna Paci and the views expressed in this paper are those of the authors.

References

Acer Consultants Ltd. 1996. *Lovesgrove Borehole no. 2 Rheidol Valley, Ceredigion. Data review and hydraulic predictions*. Report to Dwr Cymru Welsh Water.

Bouma, A. H. 1962. *Sedimentology of some flysch deposits: a graphic approach to facies interpretation*. Elsevier, Amsterdam.

Cave, R. & Haines, B. A. 1986. *Geology of the country between Aberystwyth and Machynlleth*. Memoir of the British Geological Survey **Sheet 163** (England and Wales).

Council of European Communities. 1980. Directive relating to the quality of water intended for human consumption (80/778/EEC). *Official Journal of the European Communities*, **L229**.

——1998. Directive relating to the quality of water intended for human consumption (98/83/EC). *Official Journal of the European Communities*, **L330**.

Department of the Environment. 1978. *Surface water: United Kingdom 1971–73*. HMSO, London.

——1982. *Surface water: United Kingdom 1974–76*. HMSO, London.

——1983. *Surface water: United Kingdom 1977–80*. HMSO, London.

——1991. Private water supplies: a guide to the laws controlling private water supplies. Department of the Environment and the Welsh office.

Fuge, R., Laidlaw, I. M. S., Perkins, W. T. & Rogers, K. P. 1991. The influence of acid mine and spoil drainage on water quality in the mid-Wales area. *Environmental Geochemistry and Health*, **13**, 70–75.

Goodchild, M. F. 1993. The state of GIS for environmental problem solving. *In*: Goodchild, M. F., Parks, B. O. & Steyaert, L. T. (eds) *Environmental modelling with GIS*. Oxford University Press, New York.

Hyder Consulting Ltd. 1998. *Lovesgrove Borehole no. 2 Rheidol Valley, Ceredigion*. Final Report and Recommendations to Dwr Cymru Welsh Water.

Institute of Hydrology. 1980. *Low flow studies report*. Natural Environment Research Council, Wallingford.

——1988. *Hydrological data UK: Hydrometric register and statistics 1981–1985*. Natural Environment Research Council, Wallingford.

Jones, J. A. A. 1997. *Global hydrology*. Addison Wesley Longman Ltd., Harlow.

Jones, O. T. & Pugh, W. J. 1935. The geology of the districts around Machynlleth and Aberystwyth. *Proceedings of the Geological Association*, **46**, 247–300.

Macklin, M. G. & Lewin, J. 1986. Terraced fills of Pleistocene and Holocene age in the Rheidol valley Wales. *Journal of Quaternary Science*, **1**, 21–34.

National Rivers Authority. 1992. *Policy and practice for the protection of groundwater*. National Rivers Authority, Bristol.

——1996. *Rheidol, Ystwyth and Clarach Catchment Plan*. National Rivers Authority, Welsh Region.

Neal, C., Hill, T., Alexander, S., Reynolds, B., Hill, S., Dixon, A. J., Harrow, M., Neal, M. & Smith, C. J. 1997. Stream water quality in acid sensitive UK upland areas: and example of potential water quality remediation based on groundwater manipulation. *Earth System Sciences*, **1**, 185–196.

Reynolds, B., Neal, C., Hornung, M. & Stevens, P. A. 1986. Baseflow buffering of streamwater acidity in five mid-Wales catchments. *Journal of Hydrology*, **87**, 167–185.

Robins, N. S. 1987. Development of groundwater resources in Scotland. *Proceedings of the Institution of Civil Engineers*, Part 2, **83**, 747–754.

——1990. *Hydrogeology of Scotland*. HMSO, London.

——1996. *Hydrogeology of Northern Ireland*. HMSO, London.

——, Shand, P. & Merrin, P. D. 2000. Shallow groundwater in drift and Lower Palaeozoic bedrock: the Afon Teifi valley in west Wales. *This volume*.

Welsh Water Authority. 1979. *Report of survey of existing water management and use*.

Woodcock, N. H. 1984. Early Palaeozoic sedimentation and tectonics in Wales. *Proceedings of the Geological Association*, **95**, 323–335.

Conceptual modelling of data-scarce aquifers in Scotland: the sandstone aquifers of Fife and Dumfries

I. GAUS & B. É. Ó DOCHARTAIGH

British Geological Survey, Maclean Building, Wallingford, Oxon OX10 8BB, UK (e-mail: iga@bgs.ac.uk)

Abstract: Hydrogeological conceptual modelling is an essential tool in aquifer management, particularly in areas where groundwater abstraction is increasing. Conceptualization requires extensive (and expensive) amounts of data, for which resources are often lacking. This paper describes conceptual models of the Permo-Triassic sandstones in the Dumfries basin and the Upper Devonian sandstones in Fife. Data scarcity means that each aquifer requires an individually tailored approach to formulate the conceptual model and calculate the water balance. This approach can differ considerably from the standard techniques used when more data are available. The implications of data scarcity on the confidence in results are discussed and the conceptual models for both aquifers are compared.

Current legislative emphasis focuses on integrated surface and groundwater river basin management. The proposed EC Water Framework Directive prescribes the need to understand the processes relating to the water budget of each catchment as a component of a river basin. Approximate runoff and the baseflow component of runoff have been calculated for significant water courses in Scotland (Marsh 1996), but lack of data means that little specific evaluation has yet been made of the individual groundwater flow regimes that occur within those same catchments. Large tracts of the Southern Uplands and the Highlands, where groundwater plays a small, although often significant, part in the water budget, are data-scarce. For some of these areas there are no contemporary geological maps, and for most there is no useful drift geology mapping. The same is largely true of Scotland's most important groundwater resource catchments, including the Eden valley of Fife and the sandstone basin at Dumfries.

The conceptual model approach is the best technique to increase the understanding of catchment-wide groundwater flow systems, and to identify areas where data urgently need to be collected in order to satisfy the new environmental legislation. This paper highlights this concept for two data-scarce aquifer systems in Scotland, the Permo-Triassic sandstones of the Dumfries basin and the Upper Devonian sandstones of the Eden valley and Loch Leven basin in Fife (Fig. 1). The aquifers are currently intensively exploited for public, industrial and domestic water supplies, and groundwater quality is an issue in both catchments.

Developing conceptual models

Groundwater forms only one part of the total hydrological system. Any investigation of groundwater on a catchment basis must include the surface water components so as to provide a complete understanding of the system. This process normally begins with the development of a catchment conceptual model. This takes the form of a description of the current understanding of the hydraulic regime of the catchment, including the geological, topographical,

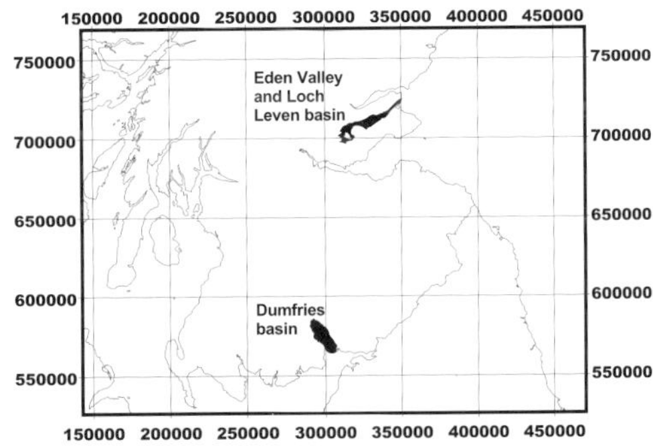

Fig. 1. Location of the Dumfries and Fife aquifer outcrops.

From: ROBINS, N. S. & MISSTEAR, B. D. R. (eds) *Groundwater in the Celtic Regions: Studies in Hard Rock and Quaternary Hydrogeology*. Geological Society, London, Special Publications, **182**, 157–168. 1-86239-077-0/00/ $15.00

surface water and groundwater components. The data for the conceptual model may come from a variety of sources, such as field mapping (e.g. topographical, geological and geomorphological), field observations and monitoring (e.g. pump tests, groundwater level and hydrogeochemical measurements and river flow gauging), and laboratory experiments (e.g. porosity and permeability core sample tests).

One of the most data-intensive parts of developing a conceptual model is the production of a catchment water balance. A number of authors discuss the formulation of the catchment water balance (e.g. Rushton & Ward 1979; Alley 1984; Ponce & Shetty 1995; Rushton 1998). All the approaches rely on one or more equations to separate precipitation into its various components. A simple water balance equation for an undisturbed catchment for a specified time period is:

$$P = E + R + \Delta S$$

where P, is precipitation (in mm); E, is actual evapotranspiration (mm); R, is runoff (in mm over the catchment); and ΔS, is the increase in stored water in the catchment (Rushton & Ward 1979). This equation does not take into account abstractions whereby water is lost to the catchment.

Water balance computations may be carried out on daily, weekly or monthly time scales (Alley 1984). In general, the shorter the time scale the more accurate the calculations: e.g. recharge estimations done on a monthly basis have been shown to underestimate recharge estimated on a daily basis. In practice, compromises between accuracy and data availability mean that many water balances are performed on a monthly or annual basis. Annual, steady state, lumped parameter water balances, such as the two presented in this paper, provide only an approximation of the hydrogeological regime in a catchment. In particular, they do not take into account areal variations in precipitation, evapotranspiration, recharge and discharge, all of which may vary significantly even within a small catchment. The effect of an annually averaged approach also masks the importance of time-dependent effects, particularly on precipitation, evapotranspiration and stream flow. Although, in principle, recharge and discharge can be determined by flow net analysis, in practice the large uncertainties attached to aquifer property values which have been derived mean that a wide range of feasible recharges and discharges values can be obtained.

Estimating the volume of recharge to an aquifer is one of the most difficult parts of hydrogeological modelling. It is generally not possible to make direct measurements of the quantity of water entering an aquifer system (Simmers 1998). If flow measurements are available it may be possible to make reliable estimates of indirect recharge, due to losses from rivers or other water bodies (Rushton 1998). However, in most cases the major inflow to an aquifer is direct recharge from precipitation.

Soil moisture balance techniques are widely used to estimate recharge, using measured and/or calculated values of precipitation, evapotranspiration and soil moisture deficit. This approach is useful for aquifers with a shallow water table and either a permeable cover or no cover at all. The methods can be modified to take account of low permeability cover deposits by applying reduction factors to limit the volume of recharge infiltrating through the cover.

Case studies

Dumfries Basin sandstone aquifer

The Dumfries basin (Fig. 2) is an outlier of Permian sandstones and breccias measuring approximately 25 km long by 10 km wide, with the long axis oriented to the NW. The topography in the basin is generally subdued, apart from a prominent ridge trending south-west. Surrounding areas, underlain by Silurian greywackes, siltstones and mudstones (and some granites in the west), have markedly steeper topography. The extreme south-eastern margin of the basin abuts Carboniferous rocks consisting predominantly of sandstone and limestone. To the south, the basin opens to the Solway Firth, which is underlain by Permo-Triassic strata. Gravity surveying and modelling indicate that the thickness of the sediments in the centre of the basin exceeds 1000 m. In the eastern part of the basin, the Permian deposits consist of aeolian sandstones which lie on rocks of Lower Palaeozoic age. Towards the west, coarser sediments, comprising breccias and water-lain sandstones were deposited along the mountain front along the western margin of the basin. Contemporaneous faulting and subsidence along the western boundary have allowed accumulation of breccias in a wedge, which thins towards the east.

Quaternary drift cover is both spatially and vertically heterogeneous. Much of the drift comprises glaciofluvial deposits. The river flood plain is covered with alluvium, which contains silt, sand and some gravel. Higher ground is covered by till. Bedrock outcrops only in small

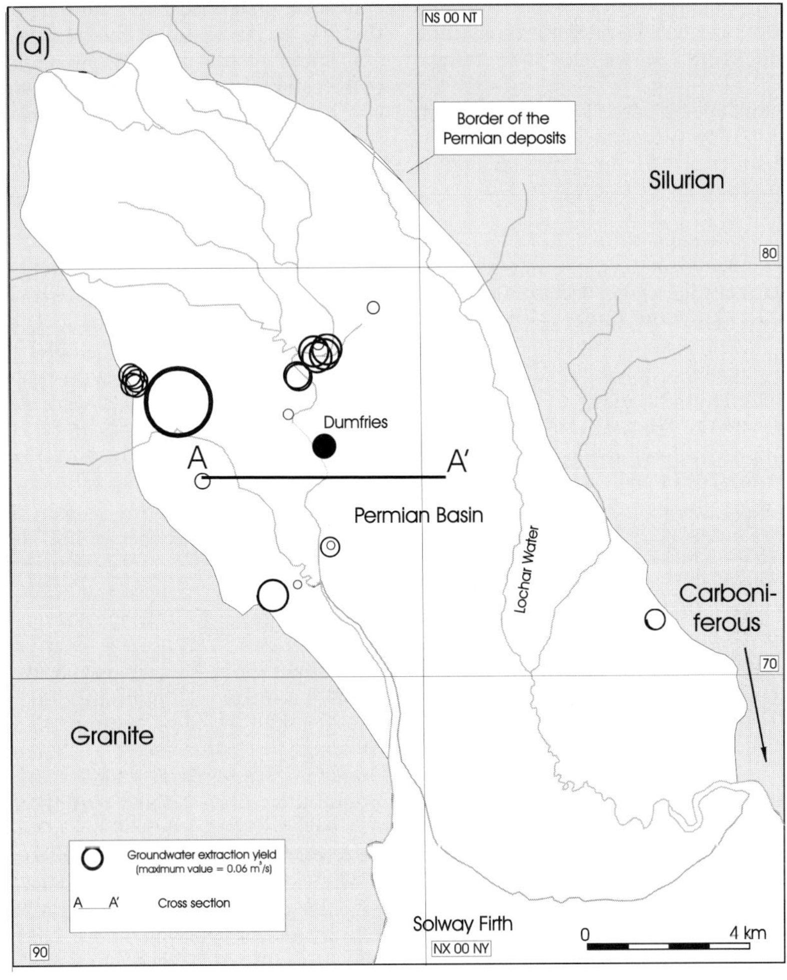

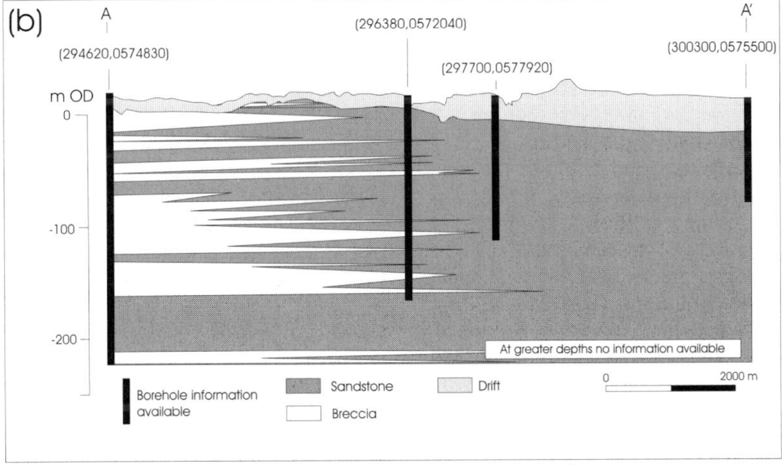

Fig. 2. The Dumfries aquifer: (**a**) Major surface water bodies and abstractions; (**b**) cross-section (W–E) through the Dumfries basin.

areas in the higher parts of the basin. Towards the coast, there are tidal river deposits consisting of muds and sands, as well as river terrace deposits.

The Permian deposits act as a single aquifer in which fracture flow dominates. According to Robins & Buckley (1988), intergranular flow is of minor importance.

The dominance of fracture flow has two main consequences for aquifer behaviour:

(1) *Groundwater flow is limited to preferred flow paths in the aquifer*. Geophysical logging at several locations in the SW of the basin indicates distinct horizons in which the majority of groundwater flow occurs. These correlate with sandstone partings and junctions between sandstone and breccia. Data are scarce in the eastern part of the basin, where the Permian deposits mainly comprise sandstones, and the pattern of flow may be different.

(2) *The permeability of the aquifer is highly variable and depends on the location and orientation of the fractures*. Because flow largely occurs in fractures, the transmissivity value obtained by test pumping is influenced by the type and number of fractures intersected by a borehole. Based on pumping test results, calculated transmissivities vary from $270\,m^2/d$ to $4000\,m^2/d$ over a few kilometres, but variations over even shorter distances are likely. These tests also demonstrate the anisotropic behaviour of the aquifer with large drawdown differences in different directions over a scale of less than 2 km. Pumping tests were mainly carried out in the SW of the basin and indicated higher transmissivities along a NW–SE trend.

The high variability of the drift cover, both in thickness (0–30 m) and lithology, makes it extremely difficult to estimate its influence on the aquifer system. Thick drift infill in the river valleys can control the interactions between groundwater in the Permian aquifer and Quaternary deposits, and with surface waters.

In general, there must be significant hydraulic contact between the rivers and the Permian aquifer. Flow measurements indicate that the rivers gain water from the aquifer, especially in lower lying areas. In addition, groundwater level measurements indicate that groundwater generally flows towards the rivers. Therefore, the rivers may control discharge from the aquifer. However, the drift cover may act locally as a barrier between a river and the groundwater system, as suggested by a pumping test close to the river in the centre of the basin.

Combining the information on topography, piezometry and nature of the drift deposits from existing maps, an estimate of the main recharge areas in the basin was made (Gaus 1999). Recharge was assumed to be greatest where there is a thick unsaturated zone and a permeable drift cover, areas which generally coincide with the higher ground of the basin.

Piezometric measurements were insufficient to produce a reliable piezometric map but some major conclusions can be drawn. In the western side of the basin, groundwater flow takes place from the recharge areas towards the River Nith. In the eastern part of the basin, groundwater flow is towards the sea, parallel to the Lochar Water, although information in this area is scarce. In areas of lower elevation groundwater levels are within the drift, and in higher areas the water table may be within the Permian deposits, with a thick unsaturated zone. It is only in this upper part of the basin that rivers might recharge the drift (or the Permian bedrock). Zones of high groundwater velocity exist along the River Nith, particularly along the sandstone ridge in the centre of the basin. The groundwater flow pattern is significantly influenced by groundwater abstraction NW of Dumfries. This area acts as a local sink, with groundwater flow directed to the centre of the area and reduced hydraulic gradients towards the River Nith. The close proximity of boreholes in this zone results in interference. Preferred flow along fractures and bedding planes extends drawdowns in a SE direction, along the flow lines, towards the Nith. Artesian conditions occur in several boreholes, but it is unclear if the drift is the confining layer or if the water is confined in the fractures or highly permeable zones of the Permian aquifer (Jenkins 1992).

The limited amount of available data makes large extrapolations necessary in order to construct a water balance. This was especially the case for the calculation of the effective rainfall and the outflow from the aquifer. An overview of the available data is listed in Table 1, and the calculations are summarized in Table 2.

To gain an insight into the amount of recharge an attempt was made to estimate the amount of runoff based on the flow rates in the Lochar Water (Marsh 1996). The flow in the Lochar Water is a measure of the runoff (flow and baseflow) within the capture area. Dividing the runoff by the area of the Lochar Water catchment leads to an amount per km^2 and it was further assumed that this amount is an estimate for the recharge in the Lochar Water

Table 1. *Availability of data for the water balance for the Dumfries aquifer*

Parameter	Data available	Length of record
Rainfall	Yes	1961–1999
Evapotranspiration	Yes	1961–1999
Riverflow leaving catchment	No	
Baseflow	No	
Surface water abstraction	N/a	
Groundwater abstraction	Yes	Current data
Groundwater levels	Yes	Scattered measurements, three regularly measured piezometers

Table 2. *Water balance for the Dumfries aquifer (numbers rounded to the nearest 1 000 000 m^3 a^{-1})*

Inflow		Outflow	
	Amount (m^3 a^{-1})		Amount (m^3 a^{-1})
Recharge from rainfall	100 000 000	Groundwater abstraction	11 000 000
Influent from rivers	0	Discharge towards river/sea	88 000 000
Error on the water balance (IN-OUT): 1%			

catchment. Extrapolating this calculation to the whole of the Dumfries basin, recharge was estimated at 436 mm a^{-1}, which is 32.5% of the mean annual rainfall of 1343 mm a^{-1} (based on three observation points within the Dumfries basin (Marsh 1996)). Taking into account the area of the basin this equals a total inflow due to rainfall of 100 000 000 m^3 a^{-1}.

Groundwater inflow from the Silurian deposits surrounding the Permian basin can be neglected due to the relatively low permeability of these strata, while surface runoff contributions from these strata are assumed to be incorporated in the calculation in the previous paragraph. The boundaries of the aquifer are assumed to be impervious. Inflow can therefore only occur from the rivers which enter the basin and lose water to the aquifer. River gauging indicates only negligible gains to the aquifer from surface water in the upper parts of the basin. Any such gains were therefore not taken into account during water balance calculations.

Most groundwater discharge occurs to the lower reaches of the rivers, close to the sea, or directly to the sea itself. Since no regular river gauging takes place in the lower part of the basin, discharges to the sea and into the river system are difficult to separate. A (steady state) estimate of the discharge is based on piezometric data (averaged in time and space) and Darcy's law. It is assumed that discharge from the aquifer to the river only takes place in the Nith between the town of Dumfries and the sea, and in the Lochar Water close to the coast. A constant hydraulic conductivity in the aquifer of 5.5 m/d (obtained from pumping test data) and a uniform aquifer thickness of 250 m are assumed, giving an outflow of 55 000 000 m^3 a^{-1}. The assumed aquifer thickness is a rough estimate and it is likely that upward flow occurs in the discharge area making the flow regime more complex.

Intense groundwater withdrawal in the Dumfries basin has only occurred since the late 1970s. Since then many boreholes for public supply as well as private use have been commissioned. The estimated current total abstraction from 27 boreholes in the Dumfries area is 348 l s^{-1}, or approximately 11 000 000 m^3 a^{-1}.

Leakage from water mains is expected to be of minor importance due to the low population density in the area, but main sewers are known to drain groundwater in places where the water table is shallow.

The inflow and outflow imbalance in the basin, based on the assumptions stated above, is 34%. This suggests that some of the assumptions are not accurate. That only discharge towards the lower part of the Nith is taken into account almost certainly underestimates the inflow, but it is difficult to estimate the discharge in the upper part because of inadequate data. Another uncertainty is the assumed aquifer thickness over which the aquifer discharges, to which the balance is very sensitive. Increasing the initial assumed aquifer thickness of 250 m to 400 m reduces the error on the water balance from 34% to 1% (Table 2). Although a uniform aquifer thickness was assumed for both the River Nith and the sea, it is likely that the real aquifer thickness to the sea is larger. In this calculation the amount of water discharging to the sea relative to the River Nith is likely to be underestimated.

In conclusion, the discharge figure for the aquifer is a fitted value to 'close' the water balance. The reliability of this value is determined by the reliability of the estimated inflow in the water balance. The water balance calculations suggest that some 10 % of aquifer recharge is abstracted.

Upper Devonian sandstone aquifer of Fife

The Upper Devonian sandstone outcrop in Fife forms a belt of low-lying land stretching some 40 km from Loch Leven north-east along the Eden valley to the North Sea coast (Figs 3 and 4). The sandstones dip south-eastwards beneath Lower Carboniferous strata, and their base is marked by an unconformable contact with a largely impermeable Lower Devonian lava succession to the north (Figs 4 and 5). The outcrop width varies from less than 2 km to over 10 km, the narrowest section lying where the Loch Leven basin meets the Eden valley.

A new stratigraphy for the Upper Devonian and Lower Carboniferous in Fife is described in Ó Dochartaigh *et al.* (1999) and illustrated in Table 3. The Upper Devonian aquifer comprises mainly fine- to coarse-grained, feldspathic, weakly to well cemented sandstones, with subordinate siltstones, mudstones and conglomerates. Most of the succession is fluvial in origin, but the Knox Pulpit Formation is thought to be aeolian (Browne *et al.* 1987). The structure of the aquifer outcrop appears to be largely controlled by extensional faulting. Much of the aquifer is fault-bounded, in particular by the SW to NE trending Fernie and Dura Den faults. A number of smaller faults strike more or less perpendicularly across the Eden valley, particularly in the Falkland–Freuchie area (Fig. 4).

Quaternary cover comprises a heterogeneous sequence of glacial till and glaciofluvial and alluvial sands and gravels. Much of the bedrock is overlain by a few metres of till, consisting of pebble to boulder sized particles in a silty-clay matrix. It is generally poorly permeable, but the presence of fractures provides conduits for groundwater flow. Glaciofluvial sand and gravel are often present above till, but locally rest directly on bedrock. In the Loch Leven basin thick deposits of glaciolacustrine clay rest either on till or sand and gravel (Foster *et al.* 1976; Aitken & Ross 1982).

The Knox Pulpit Formation is the most productive aquifer unit of the Upper Devonian in Fife, although each of the aquifer units plays

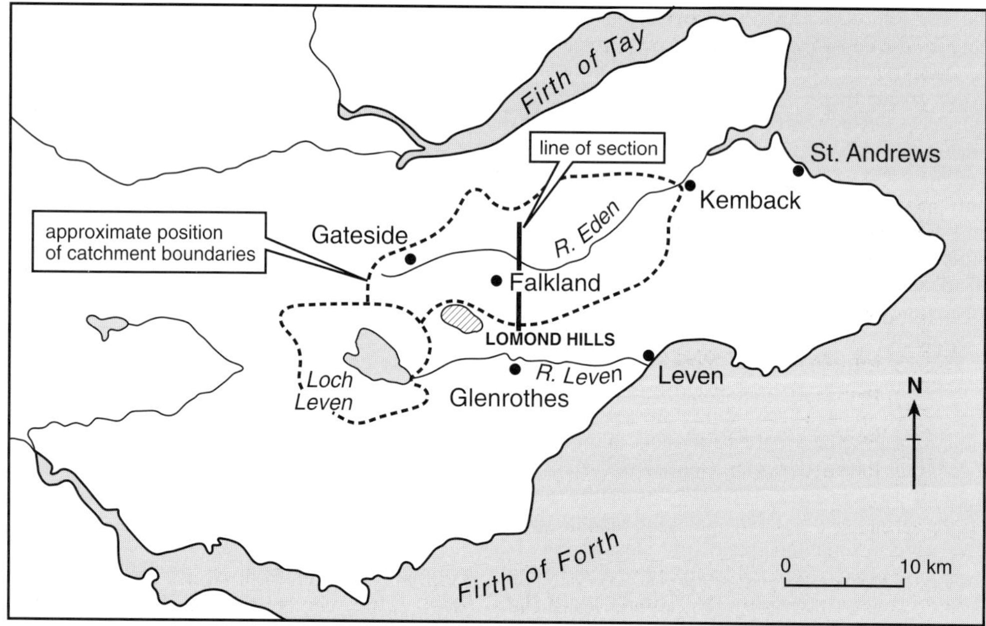

Fig. 3. Location map and general hydrology of the Fife aquifer area, showing line of cross-section in Fig. 5.

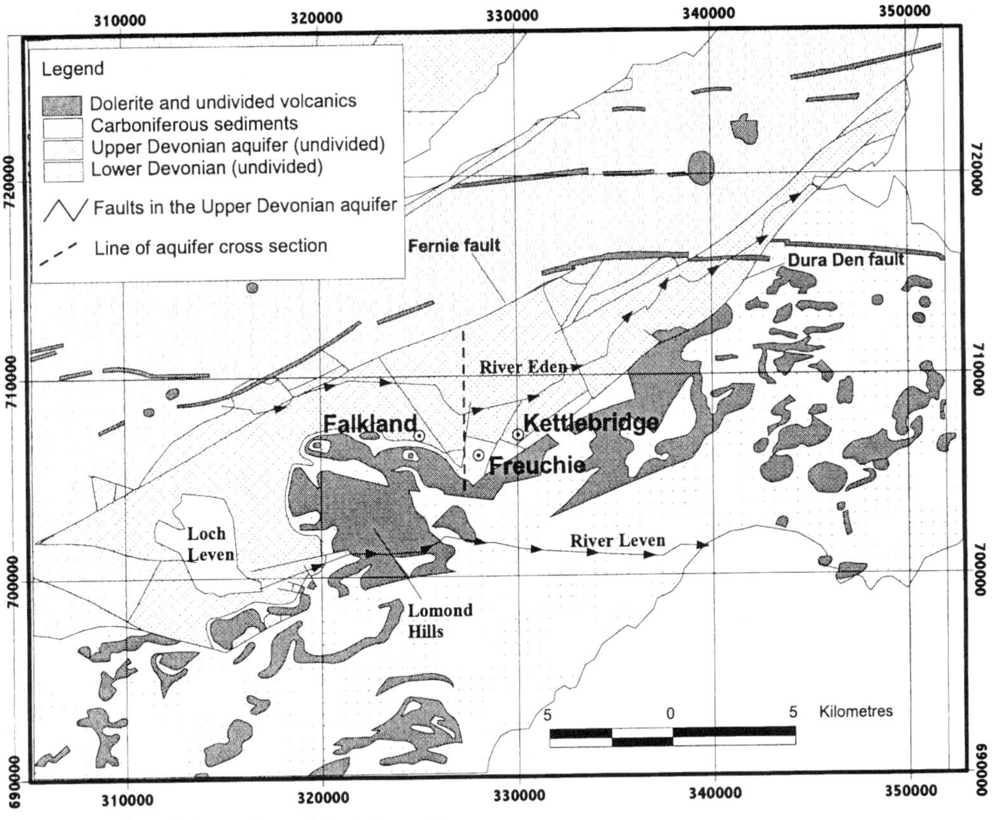

Fig. 4. Overview of the geology of the Fife aquifer area.

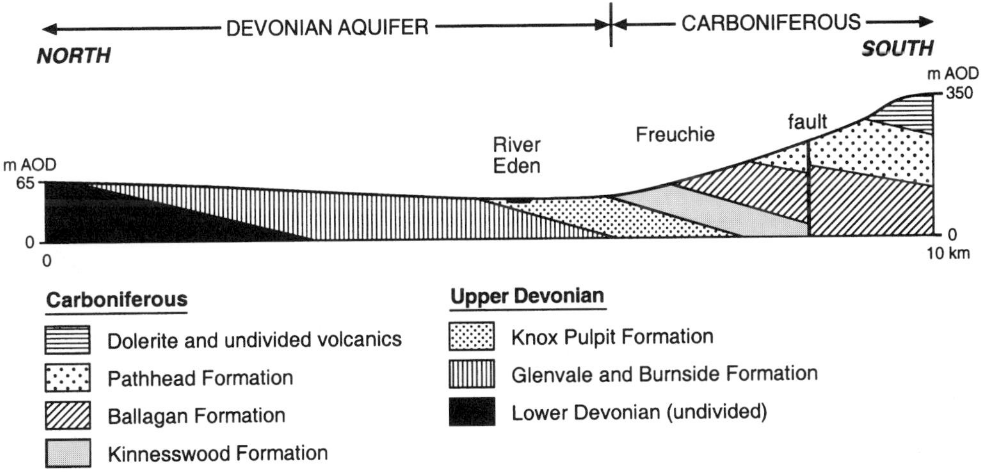

Fig. 5. Schematic cross-section through the Fife aquifer at Freuchie.

a role in the overall hydrogeological regime. For the purposes of this study the boundaries of the Eden valley aquifer system are marked by the Eden surface water catchment as monitored at Kemback, an area of 307 km^2. The Loch Leven basin sub-catchment is not defined explicitly by any previous studies. It forms the upper part of the catchment of the River Leven as

Table 3. *New stratigraphy for the Upper Devonian and Lower Carboniferous of Fife*

Age	Group	Unit	Approximate maximum thickness (m)	Lithology of aquifer units
Carboniferous	Strathclyde Group	Pathhead Formation	40 to 130	(Not part of Upper Devonian aquifer)
	Inverclyde Group	Ballagan Formation	160	(Not part of Upper Devonian aquifer)
		Kinnesswood Formation	20 to >130	Weak to well cemented sandstones with mudstones and nodules or thin beds of concretionary dolomite or calcitic carbonate
Upper Devonian	Stratheden Group	Knox Pulpit Formation	130 to 180	Very fine- to coarse-grained weakly cemented sandstones
		Glenvale Formation	350	Fine- to coarse-grained weakly to well cemented sandstones with occasional silty claystone and siltstone beds
		Burnside Formation	160	Fine- to coarse-grained moderately or well cemented feldspathic sandstones with occasional conglomerate beds and rare siltstone and mudstone beds

After Ó Dochartaigh *et al.* (1999).
Note: The Kinnesswood Formation, formerly placed at the top of the Upper Devonian, has been reclassified as the lowermost unit of the Carboniferous. However, it is considered as part of the main Fife aquifer.

monitored at Leven. The base of the sub-catchment is taken as the Loch Leven sluices, which drain to the River Leven, and the total area is approximately 190 km^2. The approximate positions of the catchment boundaries are shown in Fig. 3. The aquifer outcrop makes up 40% of the Eden valley catchment and just over 50% of the Leven sub-catchment.

The water table is generally shallow, particularly in the valley bottom where the unsaturated zone is normally less than 5 m thick. Borehole water levels commonly react within hours to rainfall events. Annual fluctuations of groundwater levels on the valley floor are small, in the range 1–4 m, increasing at higher elevations on the slopes of the surrounding hills.

Laboratory tests on core samples from the Knox Pulpit Formation at outcrop give an average porosity of over 20%. In the other formations porosity varies from 10 to 20%. The highest porosity is seen in the near-surface weathered zone, where cementation is weakest. Porosity appears to decrease with depth: samples from the Knox Pulpit Formation at over 400 m below ground level (mbgl) at Glenrothes give a mean porosity of 15%.

Horizontal intergranular permeability in the aquifer is on average twice as high as vertical. Core sample tests give a horizontal permeability range of 10^{-5} to 4 m/d, with a geometric mean of 0.43 m/d. Vertical permeability ranges from 10^{-5} to 3 m/d, with a mean of 0.19 m/d. Pumping tests generally give transmissivity values for the Knox Pulpit and Kinnesswood Formations of around 200 m^2/d, and as high as 800 m^2/d at Freuchie, where the aquifer is highly faulted. Tests in the Glenvale and Burnside Formations suggest that transmissivity in these units is generally less than 50 m^2/d, although higher values may exist locally due to faulting or weathering. The highest overall permeability in each aquifer unit tends to be in the uppermost 15 m of the saturated zone, where weathering has enhanced secondary permeability. Flow logging of boreholes also shows significant inflows at depths of up to 80 mbgl, indicating that fracture flow in the upper part of the aquifer is an important component of permeability. In total, secondary permeability is estimated to account for up to 70% of total transmissivity. Transmissivity declines with depth: in the Kettlebridge borehole only 10% of the total yield derives from the interval between 100 m and the base at 123 mbgl.

Baseflow estimates for the River Eden are only available for Kemback, at the mouth of the catchment. This gauging station lies below the confluence of the Eden with a number of streams running off the Carboniferous hills to the south. The baseflow estimate is, therefore, likely to include a proportion of baseflow to tributary streams from the Carboniferous sandstones on the southern flanks of the catchment, as well as from the Upper Devonian aquifer. Flows into and out of Loch Leven are unknown. The data available for the water balance are listed in Table 4 and an overview of the balance calculations is given in Fig. 6.

The main input of water to the catchment is rainfall, and the main output is evapotranspiration. Average rainfall in the Eden valley is between 730 and 795 mm a^{-1}. In the Loch Leven basin, much of which is at a higher altitude, average rainfall ranges from 910 to 940 mm a^{-1}. Evapotranspiration in the Eden valley is estimated at 470 mm a^{-1}, and in the Loch Leven basin at about 440 mm a^{-1}.

As well as direct rainfall recharge to the Upper Devonian aquifer, indirect recharge through the beds of losing streams descending from the surrounding hills is thought to occur. In the absence of flow gauging, and of information on stream-bed characteristics, accurate estimation of the volume of indirect recharge is impossible. The water balance does not therefore consider indirect recharge. However, it is likely that it is a minor component in comparison with direct recharge. The total direct natural recharge was estimated by integrating recharge estimates over the surface areas of the two sub-catchments. Average values obtained were 140 mm a^{-1} for the Eden valley and 210 mm a^{-1} for the Leven sub-catchment. The contrast between estimated recharge for the two sub-catchments is in part a reflection of the difference in elevation, leading to divergent rainfall and evapotranspiration, and partly a function of drift cover variation across the area.

There are unlikely to be significant cross-boundary flows into the aquifer from other geological units. The Lower Devonian volcanic rocks to the north and west are fine- to medium-grained with no intergranular permeability, and although occasional fractures may transmit groundwater, the overall volume will invariably be small.

Shallow groundwater flow within the Eden valley appears to be dominantly cross-valley, discharging to the River Eden. Baseflow is estimated to make up over 60% of long term flow in the river at Kemback. The overall hydraulic gradient and aquifer properties suggest that while there is a component of groundwater flow down the valley, this is minor in comparison, probably representing only 1% of the total flow in the river. Although the regional stratigraphical dip of the aquifer sandstones is beneath the Carboniferous strata to the SE, the decline in their porosity and permeability away from their surface outcrop suggests that very little groundwater flow out of the Eden valley occurs in this direction (Browne *et al.* 1987).

Average annual flow in the River Eden at Gateside, near the head of the Eden valley, is much smaller than that at the mouth of the catchment: only some 8% of the annual flow at Kemback. Flow in the Eden at Kemback is less variable seasonally than that in the upper catchment. Average winter flow at Gateside is also 8% of that at Kemback, but average summer flow at Gateside is less than 5% of

Table 4. *Availability of data for the water balance for the Fife aquifer*

Parameter	Description	Data available	Length of record
Rainfall	Average of Leuchars, Pitlair, West Hall and Gateside; Pitlessie; Loch Leven Sluices; MORECS; Institute of Hydrology	Yes	1931–60 1941–70 1961–90 1967–94
Evapotranspiration	Average of Leuchars, Pitlair, West Hall and Gateside; MORECS	Yes	1931–60 1961–90
Riverflow leaving catchment	Eden at Kemback	Yes	1967–94
	Leven at Loch Leven sluices	No	
Baseflow	Eden at Kemback	Yes	1967–85
Surface water abstraction		N/A	
Groundwater abstraction		Yes	Current data
Groundwater levels		Yes	1979–1982 1997–1999

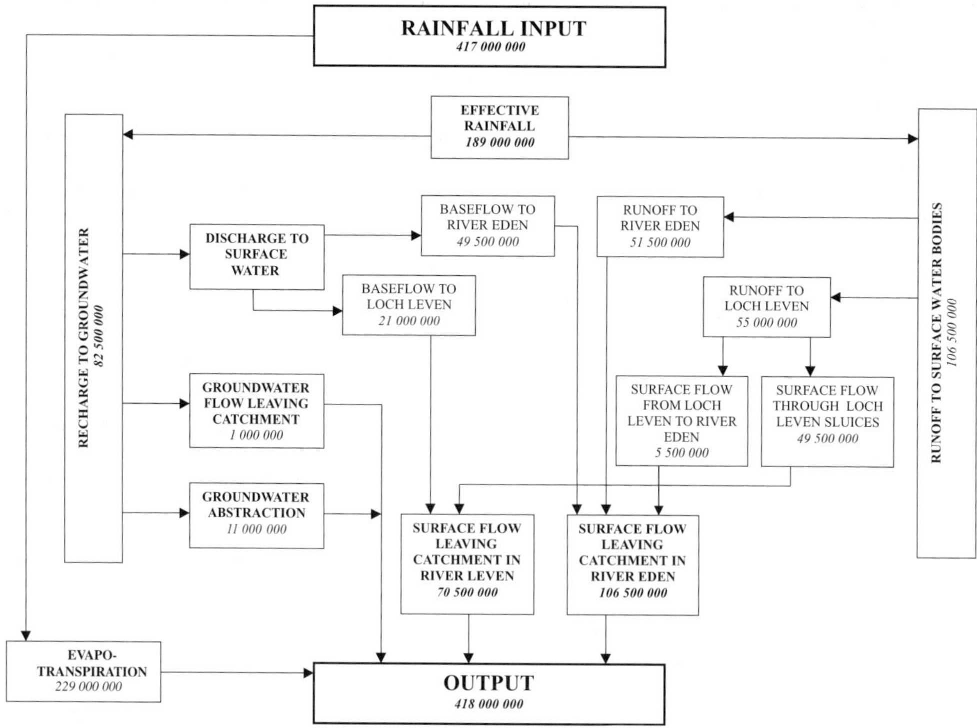

Fig. 6. Flow chart illustrating the water balance for the Fife aquifer. All figures are in $m^3 a^{-1}$.

that at the mouth of the catchment. The average summer flow to winter flow ratio is only 25% at Gateside, but 40% at Kemback, implying that during the summer there is a significant baseflow input to the river as it flows through the catchment.

Although the topography and available groundwater level observations for the Loch Leven basin imply that part of the shallow groundwater regime discharges to the loch, they also suggest that a significant proportion flows to the Eden valley. Water levels in the loch are regulated by means of an engineered channel leading to the River Leven, and most of the water which enters the loch is likely to leave the basin this way. However, flow through the Loch Leven sluices is not monitored. The lack of data for the Loch Leven sub-catchment precludes a confident assessment of the surface water–groundwater relationship.

It is assumed that effluent discharges to streams in the catchments are largely imported, as the majority of public water supply is derived from distant reservoirs. The total volume of effluent discharge is insignificant in comparison to the total flux of water through the system, and is not taken into account in the current water balance.

Abstraction from shallow wells and springs for domestic and farm water supplies has been common in Fife for centuries. More systematic hydrogeological exploration began in the early 1970s, with the construction of deeper, higher yielding boreholes for public supply and industrial use. During the 1990s there was a significant increase in the number of farm irrigation boreholes, particularly in the productive Falkland–Kettlebridge area, as well as westwards along the northern slopes of the Lomond Hills to Gateside. The water balance suggests that groundwater abstraction currently makes up only 13% of total groundwater recharge to the River Eden and Loch Leven catchments, although locally abstraction may be more significant. All groundwater abstraction is from the Devonian aquifer; however, the aquifer comprises only 40–50% of the catchment area. Abstraction may therefore account for some 20–25% of recharge to the Devonian aquifer.

In general, the water balance implies that there is a fairly rapid shallow groundwater flow system in the Eden valley, with most recharge, at least in the Eden valley, discharged as baseflow. There appears to be little longitudinal groundwater flow down the valley towards the coast, and any flow is likely to be at depth, away from the influence of the near-surface regime. Because the porosity and permeability of the sandstones are much lower at depth than near the surface, deep groundwater flow is likely to be slow.

The relative lack of data for the Leven basin, in particular on surface flows, makes it more difficult to characterize the groundwater regime in this part of the catchment. A proportion of recharge to the aquifer in the basin will discharge to Loch Leven, while the remainder flows to the Eden valley. However, the poor understanding of the relationship between the loch and groundwater precludes an accurate depiction of the true groundwater flow pattern.

Discussion and conclusions

Despite the differences between the two aquifers, the process of devising a conceptual model for each of the aquifer systems is broadly similar. The main differences were dictated by data availability, which is the principal control on the conceptualization process. Data gaps mean that varying numbers of assumptions must be made as part of the water balance.

Reliability of the conceptual models

In Fife, stream flow measurements and baseflow calculations for the Eden valley catchment are available which allow a more complete, and hopefully more accurate, picture of the hydrological regime to be developed for this part of the aquifer. However, the relative scarcity of data for the Loch Leven basin means that a number of assumptions must still be made as part of the water balance calculations. These include estimates of the inflow to, and outflow from, Loch Leven, and the groundwater flow from the Loch Leven basin to the Eden valley. The combination of observed and inferred data makes it difficult to estimate the total error within the aquifer water balance.

In Dumfries, only groundwater data are available, so that a number of simplifications and assumptions have had to be made as part of the water balance, including surface water inflow to the catchment. The presence of an aquifer boundary with the sea, and the need to estimate groundwater outflow to the sea, renders the situation even more complex. The generalizations mean that there is a large degree of uncertainty over the results of the water balance.

The scarcity of data affects a number of different aspects of an aquifer investigation, including characterization of the groundwater flow regime and definition of the coastal/salt water boundary. In particular, a major weakness of the water budget calculations for both Fife and Dumfries is the estimation of recharge. Improving the estimate requires a large amount of data and involves both complex calculations and the introduction of new parameters. One option to reduce uncertainty is to estimate recharge using different methods and to compare the results, as was done for Fife.

Comparison of the conceptual models

In both aquifers, primary permeability is low and fracture flow dominates. Preferential flow is, therefore, an important issue and makes both aquifers vulnerable to agricultural and industrial contamination.

The boundaries of both aquifers are fairly well defined, and groundwater inflow across these boundaries is expected to be of minor importance. Natural outflow is towards the sea (Dumfries), an inland lake (Fife) or to the river system (both aquifers), and the groundwater flow pattern is largely governed by these surface water bodies. In both cases, the main river draining the catchment is shown to be an important control on the groundwater system. Based on the water balance calculations, it was found that baseflow to the main river is a major outflow for both aquifers. Quantifying this control more accurately is not possible since direct measurements of the river–aquifer interaction are lacking.

Recharge in each of the aquifers, which is one of the main controls on the hydrological system, seems to differ. For Dumfries, located in the maritime SW of Scotland, a value of $436\,mm\,a^{-1}$ has been derived; in Fife, on the drier east coast, recharge is estimated to be between 140 and $210\,mm\,a^{-1}$. These large differences in recharge have important consequences for the water balance calculations and the flow regime. Larger volumes of recharge promote a quick flushing of the unsaturated zone and higher groundwater velocities. The higher transmissivities observed in the Dumfries basin are also likely to contribute to more rapid overall groundwater flow. It would therefore be expected that the mean age of the groundwater in the Dumfries basin is less than that in the Fife aquifer, although there are insufficient data to check this hypothesis.

Even with numerous data gaps, constructing a catchment water balance is still a valid and useful exercise. The resulting conceptual model of the catchment provides an insight, however general, into the local hydrological and hydrogeological regime, and this is an important first step towards managing the water resources. The process of conceptualization of a groundwater system also highlights what data gaps exist and where, and so points the way for further data collection. This is an essential precursor for meeting the requirements of the proposed EC Water Framework Directive. However, it is important to be aware of the limitations of conceptual models of data-poor aquifers. Any attempt to characterize an aquifer further by using a distributed numerical model must be backed up by reliable and sufficient hydrogeological data.

This paper is published with the permission of the Director of the British Geological Survey. The authors are grateful to Derek Ball, the reviewers and the editors for valuable comments. SEPA and NERC funded the research which formed the basis for this paper.

References

AITKEN, A. M. & ROSS, D. L. 1982. *The sand and gravel resources of the country around Glenrothes, Fife Region: Description of 1:25 000 sheet NO 20 and parts of NO 21, 30 and 31*. British Geological Survey. Mineral Assessment Report **101**.

ALLEY, W. M. 1984. On the treatment of evapotranspiration, soil moisture accounting and aquifer recharge in monthly water balance models. *Water Resources Research,* **20**(8), 1137–1149.

BROWNE, M. A. E., ROBINS, N. S., EVANS, R. B., MONRO, S. K. & ROBSON, P. G. 1987. *The Upper Devonian and Carboniferous sandstones of the Midland Valley of Scotland.* Investigation of the Geothermal Potential of the UK, British Geological Survey.

FOSTER, S. S. D., STIRLING, W. G. N. & PATERSON, I. B. 1976. *Groundwater storage in Fife and Kinross, its potential as a regional resource.* Report Institute of Geological Sciences, No. **76/9**.

GAUS, I. 1999. *The Dumfries basin sandstone aquifer: a review of current knowledge and the scope for future needs in understanding.* Report British Geological Survey No. **WD/99/49**.

JENKINS, K. 1992. *A study of groundwater resources of the Dumfries basin of Scotland, with special attention to a possible source of chemical contamination.* Unpublished MSc Thesis, University of Birmingham.

MARSH, T. J. (ed.). 1996. *Hydrological Data United Kingdom:1995 Yearbook*. Institute of Hydrology-British Geological Survey.

Ó DOCHARTAIGH, B. É., BALL, D. F., BROWNE, M. A. E., SHAND, P., MACDONALD, A. M., ROBINS, N. S. & MCNEILL, G. W. 1999. *The Upper Devonian Sandstone Aquifer of Fife.* British Geological Survey Report No. **WD/99/39**.

PONCE, V. M. & SHETTY, A. V. 1995. A conceptual model of catchment water balance: 1. Formulation and calibration. *Journal of Hydrology*, **173,** 27–40.

ROBINS, N. S. & BUCKLEY, D. K. 1988. Characteristics of the Permian and Triassic aquifers of south–west Scotland. *Quarterly Journal of Engineering Geology*, **21**, 329–335.

RUSHTON, K. R. 1998. Groundwater at risk – a reflection. *In*: ARNELL, H. & GRIFFIN, J. (eds) *Hydrology in a Changing Environment. Poster papers and index*. British Hydrological Society International Conference, Wiley.

—— & WARD, C. 1979. The estimation of groundwater recharge. *Journal of Hydrology*, **41,** 345–361.

SIMMERS, I. 1998. Groundwater recharge: an overview of estimation 'problems' and recent developments. *In*: ROBINS, N. S. (ed.) *Groundwater Pollution, Aquifer Recharge and Vulnerability*. Geological Society, London, Special Publications, **130**, 107–115.

QSC graphs: an aid to classification of data-poor aquifers in Ireland

G. R. WRIGHT

Geological Survey of Ireland, Beggars Bush, Haddington Road, Dublin 4, Ireland (e-mail: geoffwright@gsi.ie)

Abstract: The Geological Survey of Ireland's aquifer classification system recognizes three main categories: Regionally Important, Locally Important and Poor Aquifers. This system is increasingly used to assist local authorities and state agencies to make decisions on planning applications and integrated pollution control licences, by prioritizing areas according to the value of their underlying groundwater resources. Most aquifers in Ireland are unconfined fractured hard rock aquifers, often of limited extent, which can exhibit a wide range of properties. Pumping test data are scarce, patchy and often of uncertain quality, and reliable transmissivity or permeability values are unavailable for many aquifers. Under these circumstances, the classification of a given geological formation in a given region can be difficult. The 'QSC Graph' compares the specific capacity (SC) for a borehole, determined by a pumping test, with the abstraction rate during the test (Q), and indicates a 'borehole productivity index', in five classes (I, II, III, IV and V from highest to lowest). From the relative frequency of productivity classes for a given aquifer, the appropriate aquifer category can be inferred. However, other types of information for the aquifer should also be considered. The current QSC data set comprises about 1100 boreholes, and for individual formations up to 150 boreholes. The minimum data set required for an aquifer, depending on the diversity or compactness of the data, is between 20 and 50. Examples are given of the application of the approach to a number of Irish aquifers.

The Geological Survey of Ireland (GSI) has created a system of aquifer classification for use in groundwater protection schemes (Daly & Warren 1998; Department of Environment and Local Government *et al.* 1999). Three basic aquifer categories are recognized – Regionally Important Aquifers (R), Locally Important Aquifers (L) and Poor Aquifers (P), further subdivided as follows:

- Regionally Important Sand/Gravel Aquifers (Rg)
- Regionally Important Fractured Bedrock Aquifers (Rf)
- Regionally Important Karstified Aquifers (Rk)
- Locally Important Sand/Gravel Aquifers (Lg)
- Locally Important Bedrock Aquifers which are generally moderately productive (Lm)
- Locally Important Bedrock Aquifers which are moderately productive only in local zones (Ll)
- Poor Bedrock Aquifers which are unproductive except in local zones (Pl)
- Poor Bedrock Aquifers which are generally unproductive (Pu)

This classification acknowledges that wells in almost any type of rock in Ireland can yield sufficient water to supply at least a single household, and therefore no bedrock type is termed a 'non-aquifer'. The Republic of Ireland lacks any thick clay or shale lithologies (cf. the Oxford Clay of England) which are sufficiently unproductive to be unequivocally 'non-water-bearing'. The GSI classification also avoids the use of such terms as 'aquitard', 'aquiclude' or 'aquifuge', which are largely unknown to the wide range of people who are involved in water supply and protection in Ireland.

Because hydrogeological data are often scarce and patchy, various criteria are used in aquifer classification, including lithology, karstification, structural setting, occurrence/size of springs, baseflow estimation, and drainage density. However, where data are available on borehole yields and specific capacities, these are normally the main evidence whereby an aquifer is classified. While transmissivity (T) estimates from pumping tests would be desirable, there are currently too few of these for any given aquifer, and they are particularly scarce for the poorer aquifers. Even where pumping test data are available, they often fail to yield unambiguous T values, and often only a range can be suggested.

GSI's initial efforts to use borehole yield data for aquifer classification were based on the relative occurrence of 'excellent', 'good', 'moderate'

From: ROBINS, N. S. & MISSTEAR, B. D. R. (eds) *Groundwater in the Celtic Regions: Studies in Hard Rock and Quaternary Hydrogeology*. Geological Society, London, Special Publications, **182**, 169–177. 1-86239-077-0/00/ \$15.00

and 'poor' yields (>400, <400 > 100, <100 > 40, and <40 m^3/d). However, it was known that the reported yield of a well could be quite different from its maximum sustainable yield. Specific capacities were also taken into account, where available, but there was no simple or consistent way to do this. There was a need for a means of integrating borehole yield data with specific capacity data in a consistent manner (Wright 1997).

Methodology

All available specific capacity data for Irish boreholes (currently almost 1100) have been compiled. The data come from four main sources: (1) local authority public supply boreholes, mostly with relatively high yields (>100 m^3/d), and mainly from counties where specific studies have been carried out; (2) boreholes drilled by the Irish Land Commission, mainly for single farms or small group schemes between 1950 and 1980, typically 30–60 m deep, with generally low test rates (<100 m^3/d); (3) consultants' well records, normally for deeper boreholes (60–120 m) with higher yields (often over 400 m^3/d); and (4) GSI records, including boreholes drilled for GSI projects.

Two important features of Irish aquifers need to be borne in mind when reviewing the data:

(a) Most Irish aquifers are unconfined.
(b) Except for Quaternary gravel/sand deposits, Irish aquifers depend almost entirely on fracture permeability, and well losses are very significant. Since fracture frequency and openness tend to decrease with depth, permeability also tends to decrease with depth, and transmissivity and specific capacity can vary substantially according to the water table level. Thus a given borehole can show very different specific capacities at different times of year, due to water table changes, and may also show sharp decreases in specific capacity at higher drawdowns.

From the data set, graphs were prepared for the commonest geological formations, plotting well 'yield' (Q) against specific capacity (SC), hence the graphs are termed 'QSC Graphs'. The QSC graphs allow the available yield and specific capacity data to be viewed simultaneously and in the context of similar data from other aquifers.

Results

The QSC graph for all boreholes (Fig. 1) covers a range of several orders of magnitude for both parameters, showing an obvious general trend within a broad 'envelope' of data points. The upper boundary of the envelope is very 'fuzzy', but the lower boundary is sharper and can be seen as controlled by two main factors, one artificial and one natural: (a) borehole depths limit the available drawdown, and (b) permeability tends to decrease with depth.

In general, a data point close to the lower boundary indicates a deep borehole which is being

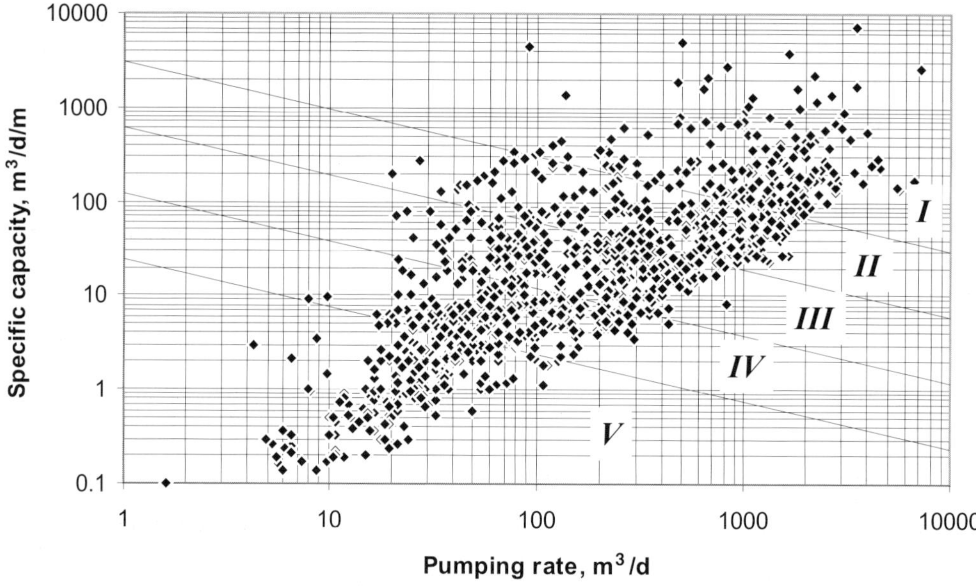

Fig. 1. QSC graph for full data set (*c.* 1100 boreholes), showing productivity classes I II, III, IV & V.

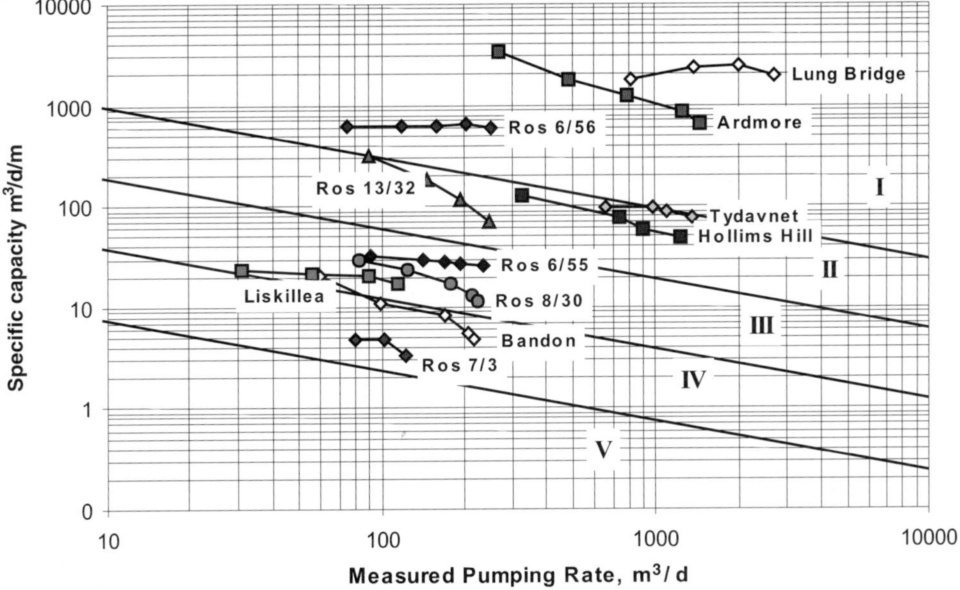

Fig. 2. QSC data for selected step tests in Ireland.

pumped near its maximum sustainable rate (i.e. most available drawdown has been used). Conversely, a data point in the upper part of the envelope indicates a borehole which could be pumped at a much higher rate than indicated, or a shallow borehole with little available drawdown.

The QSC axes can also be used to plot data from multiple tests (e.g. step tests) from the same borehole, in which case the points normally plot along a line which falls from left to right, i.e. as the pumping rate increases the specific capacity decreases (Fig. 2). The gradient will normally be quite gentle until a critical point is reached, after which it may become much steeper as well losses increase. This critical point may indicate the maximum sustainable yield of the borehole, due to increasing well losses and/or decreasing permeability down the hole.

If the general slope of such curves from individual boreholes in a given aquifer can be estimated, it is possible to extrapolate from a single data point on a QSC graph in order to predict the approximate maximum yield of a given borehole. However, it should be borne in mind that step tests may give different curves at different times of year.

By plotting QSC data from a number of boreholes in a given geological formation or aquifer unit, a classification of the aquifer can be attempted. In some cases, the graph indicates that the formation is too variable for a single classification, or that no single classification applies in all regions. In such cases the data must be examined to see if a regional or stratigraphical–lithological sub-division is valid (or both), i.e. is the wide spread of the data due to lithological variations within the formation or due to regional (perhaps structural) variations, or both?

In considering the data, too much attention must not be paid to any single data point, because the data quality may be suspect. The validity of any conclusions depends on the totality of the data, and therefore depends on having enough data points, so that occasional extreme values can be discounted. A number of factors may detract from the data quality:

- data collection by untrained personnel, low accuracy of measurements;
- pumping tests of variable length, sometimes too short: in some cases it is possible to extrapolate drawdown to, say, one week, in order to derive a realistic specific capacity value;
- in many cases, pumping rates were not constant;
- often, drawdown may not have stabilized;
- pumping tests may have been carried out at a time of relatively high water table, thus giving an unduly high SC; this may be particularly important in the poorer aquifers;
- the geological formation/lithology may not have been identified/inferred correctly;
- borehole depths are very variable (graphs plotting depth against yield show virtually no correlation, but in an individual borehole the depth will often be significant);

- borehole construction is often unknown, is very variable and may adversely affect the yield.

Comparing well yields and specific capacities highlights the difficulties of using well yield categories alone, because each category includes wells with a very wide range of SC and, by inference, a wide range of transmissivities.

Borehole productivity index

To simplify comparisons between aquifers, and to supplement the four existing well yield categories, the QSC graphs were used to create a new index ('Productivity'), with five classes: I (highest), II, III, IV and V (lowest), with boundaries as shown on Fig. 1. The boundary lines have a gradient of 1:10 (corresponding to the approximate gradient given by step tests in unconfined Irish aquifers) and are half an order of magnitude apart on the y-axis. The boundaries were set so that each class contained approximately equal numbers of data points: in practice, this worked out as follows: (I) 19%, (II) 21%, (III) 22%, (IV) 19% and (V) 19%. A five-class system seems to offer a suitable balance between simplicity and discrimination.

Fig. 3. Map of Ireland, showing counties.

For any given formation or aquifer unit, the numbers of wells in each productivity class are plotted as bar charts, to provide a productivity 'profile' which should be attributable to a particular aquifer category. Boreholes in a Regionally Important Aquifer should plot largely within classes I & II, and those in a Poor Aquifer should plot predominantly within classes IV & V. Locally Important Aquifers will plot largely in classes II to IV. However, boreholes in limestone aquifers in particular may plot across all classes because of the extreme variability produced by karstification.

Figure 3 shows a map of Ireland, with counties identified for reference. The relative frequency of occurrence of productivity classes are shown by means of bar charts for each aquifer (Figs 4–8). To illustrate the usefulness of these charts, the data are discussed below, beginning with aquifers already known to be either very good or very poor.

Regionally Important Aquifers

Quaternary Deposits (Sands & Gravels) (Fig. 4a). Well-sorted sand/gravel deposits, if sufficiently thick and saturated, are good aquifers. There are 98 data points from Quaternary deposits, mainly known sand/gravel aquifers. As expected, both Q and SC are generally high, and many data points indicate higher ultimate well yields. Lower values are probably from very thin aquifers, glacial tills, or poorly constructed wells. The great majority of data points fall into productivity classes I or II. These aquifers, if sufficiently extensive, are categorized as Regionally Important (Rg), or if too small, as Locally Important (Lg).

Wexford Formation (Limestone) (Fig. 4b). This is a small dolomitized limestone formation in County Wexford in SE Ireland, which has been quite intensively developed. Only 17 data points exist, but the data are of good quality. This is the most compact data set, with a relatively small logarithmic range of both Q (600–3000 m^3/d) and SC (15–350 $m^3/d/m$), and the aquifer is classed as a Regionally Important Aquifer (Rk). Most wells plot near the lower boundary of the QSC envelope, indicating that they have been tested near their maximum yield.

Campile Formation (Ordovician volcanics) (Fig. 4c). This is a highly fractured aquifer in the SE corner of Ireland (Counties Wexford and Waterford), providing 39 data points. The data set is fairly compact, and even the few lower yielding wells have rather high SCs, indicating much higher ultimate yields.

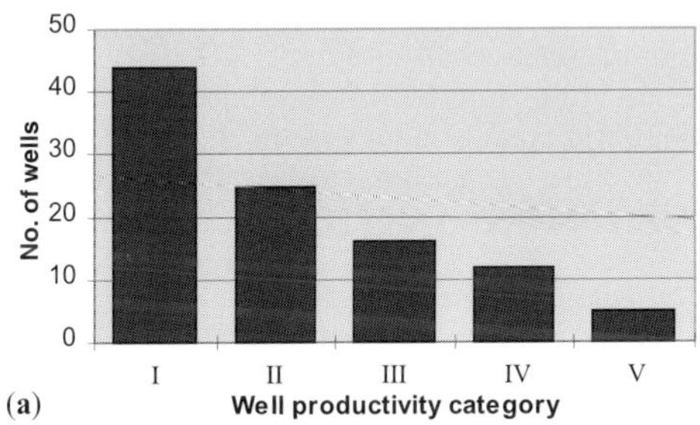

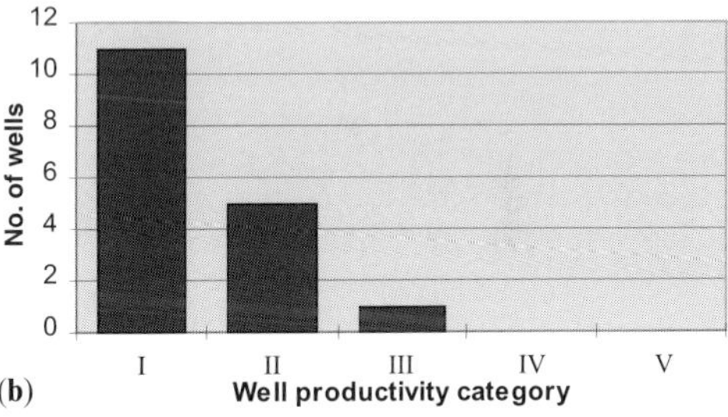

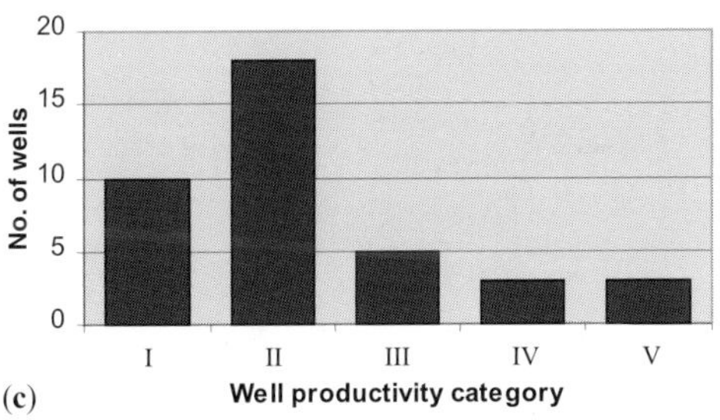

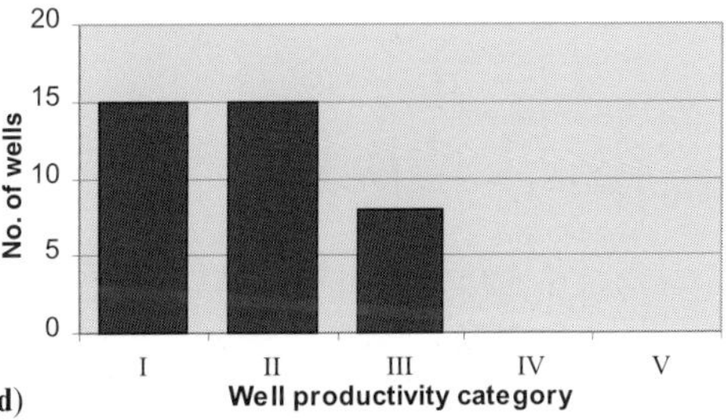

Fig. 4. Productivity bar charts for Regionally Important Aquifers (**a**) Quaternary deposits, (**b**) Wexford Limestone Formation, (**c**) Campile Formation, (**d**) Kiltorcan Sandstone Formation.

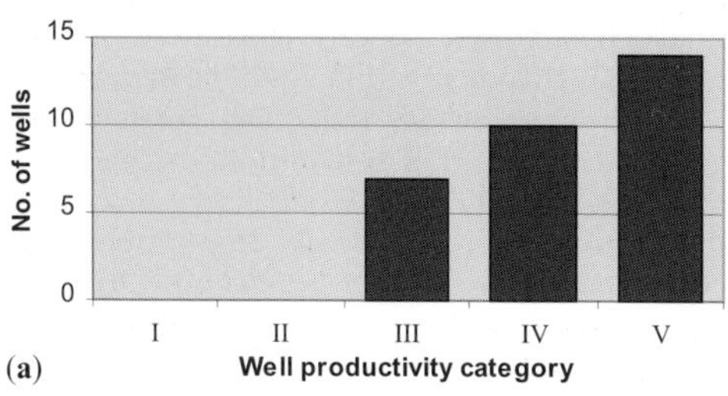

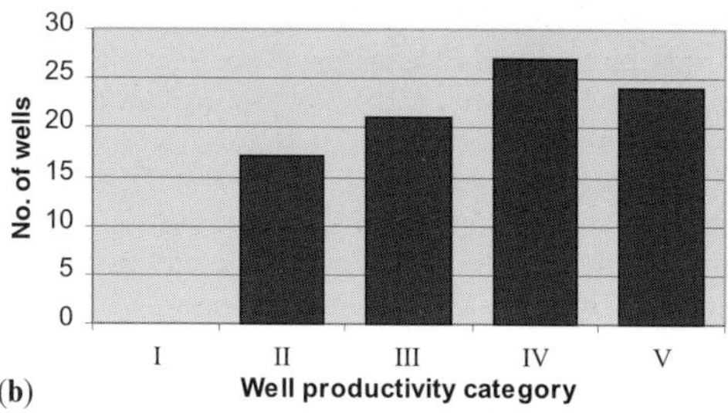

Fig. 5. Productivity bar charts for Poor Aquifers (**a**) Granites and metamorphic rocks, (**b**) Lower Palaeozoic rocks (excluding Campile Formation).

Kiltorcan Formation (Sandstone) (Fig. 4d). This Upper Devonian–Lower Carboniferous sandstone formation provides 37 data points, almost all in the upper half of the envelope, and mostly in the upper third. This aquifer is often confined. Step test results from confined aquifers show that the SC values decline only slightly with increases in discharge, so the use of the 'standard' QSC graphs tends to underestimate their productivity index. However, most values still fall into productivity classes I or II. Ideally, a different set of productivity classes should be created for confined aquifers.

Poor Aquifers

Granites and Metamorphic rocks (Fig. 5a). These lithologies, which would be expected to be poor aquifers, provide 31 data points, clustered in the lower half of the total range, and mostly in the lower third. Some SCs are higher than might be expected (these probably derive from pumping tests at high water tables). Median yield is about 33 m^3/d. These data can be taken as characteristic of a Poor Aquifer.

Lower Paleozoic rocks (excluding Campile Formation) (Fig. 5b). These fine-grained sedimentary rocks, mainly slates and greywackes, provide 89 data points. The spread of data is surprising: Q ranges up to 1000 m^3/d and more, but most values are under 100 m^3/d, and the median is about 55 m^3/d. Median SC is about 7 m^3/d/m. The data need further examination for regional or lithological factors, but a general classification of Poor Aquifer (Pu or Pl) is supported, with some Locally Important (Ll).

Locally Important Aquifers

Cork Group (Fig. 6a). These Lower Carboniferous rocks, mainly fine sandstones, siltstones and mudstones, are found only in County Cork, and are quite intensely fractured. Twenty five data points exist, generally in the middle third of the envelope, suggesting a Locally Important (Ll) classification.

Ballysteen Formation (Limestone) (Fig. 6b). This Lower Carboniferous argillaceous bioclastic limestone formation provides 75 data points over a wide range, but with few yields over 400 m^3/d.

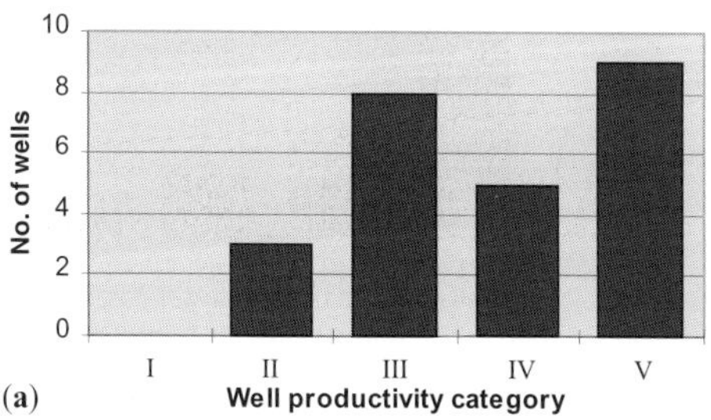

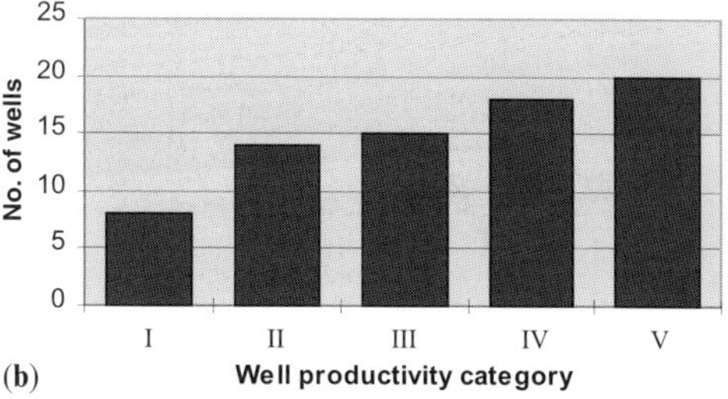

Fig. 6. Productivity bar charts for Locally Important Aquifers (**a**) Cork Group, (**b**) Ballysteen Limestone.

Median Q is about 60 m^3/d, and median SC about 8 m^3/d/m. The data need further study to distinguish regional (structural) and lithological (clean/muddy/dolomitized) variations. Higher yields (e.g. in counties Kilkenny and Tipperary) are probably from cleaner oolitic and/or dolomitized members. In general, a classification of Locally Important (Ll) is supported.

Variable aquifers

Devonian 'Old Red Sandstone' (excluding Kiltorcan Sandstone) (Fig. 7a). This provides 61 data points spread widely through the range, except at the very top and bottom. Median Q is 70–80 m^3/d and median SC is about 13 m^3/d/m. A general classification of Locally Important (Ll) is supported, but there is evidence that the coarser formations (conglomerates/coarse sandstones) are more productive, whereas finer-grained formations (siltstones, etc.) may be Poor Aquifers (Pl).

Calp Formation (Limestone) (Figs 7b–d). The Calp is a Lower Carboniferous argillaceous limestone sequence occurring widely across central Ireland from Dublin in the east to Galway in the west. The 122 data points cover a very wide range. Median Q is about 100 m^3/d, and median SC about 12 m^3/d/m. The data show substantial regional variations. In County Galway, in the west of Ireland, values are almost invariably low (Poor) (Fig. 7b). In County Meath in the east, Q and SC are fairly high, justifying a Locally Important (Lm) rating (Fig. 7d). Values in the intervening counties (Offaly, Westmeath, Dublin, and Kildare) are intermediate, justifying a Locally Important (Ll) rating (Fig. 7c).

Waulsortian Formation (Limestone) (Figs 8a–b). The Waulsortian Formation is a clean, fine-grained, massive Lower Carboniferous limestone which was deposited as large mud-mounds. Ninety data points exist, covering a very wide range, but mostly in the upper half of the envelope. A marked regional contrast is evident. In the south (Counties Cork, Kilkenny, Limerick, Tipperary and Waterford), where it was intensively fractured by the Variscan orogeny, and is extensively karstified and often dolomitized, a classification of Regionally Important (Rk) is amply justified (Fig. 8a). Elsewhere it is no better than Locally Important (Ll) (Fig. 8b). Further investigation may extend the Regionally Important area.

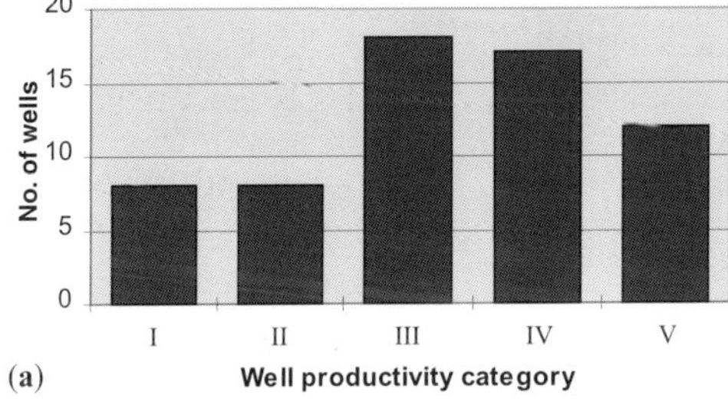

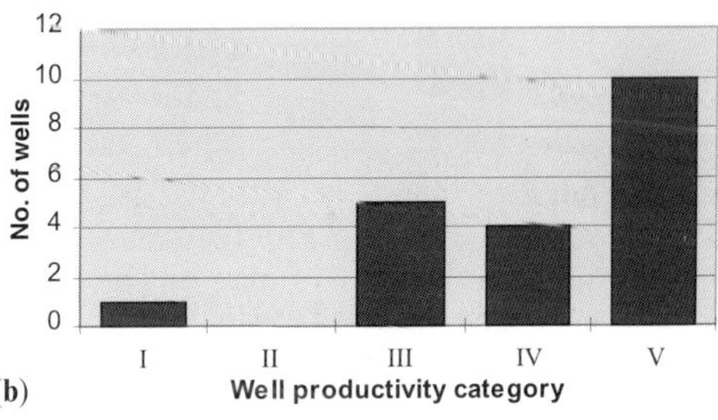

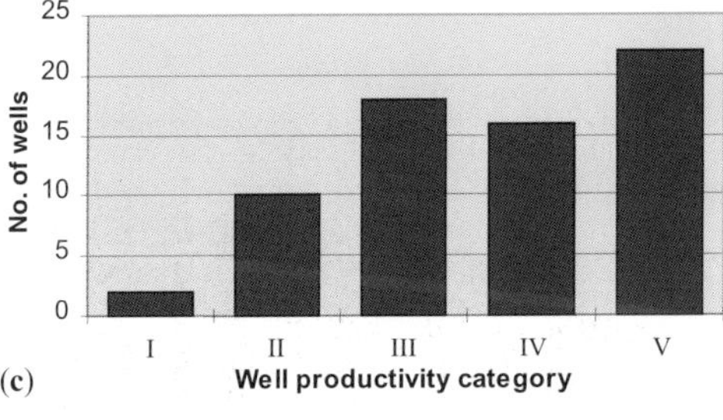

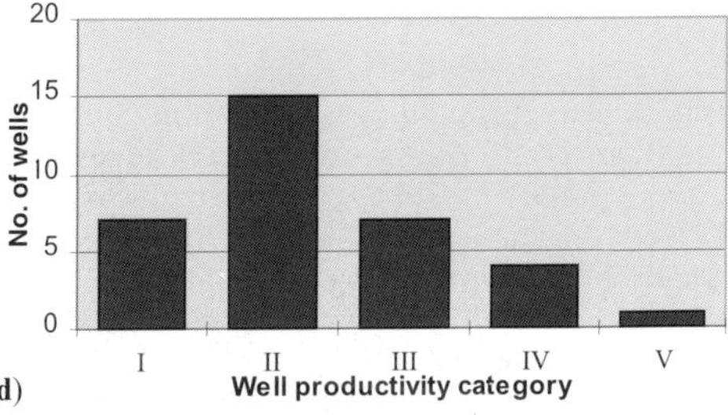

Fig. 7. Productivity bar charts for Variable Aquifers (sandstones and muddy limestones) (**a**) Old Red Sandstone (excluding Kiltorcan Formation), (**b**) Calp Limestone (Galway), (**c**) Calp Limestone (Midlands–Dublin), (**d**) Calp Limestone (Meath–North Dublin).

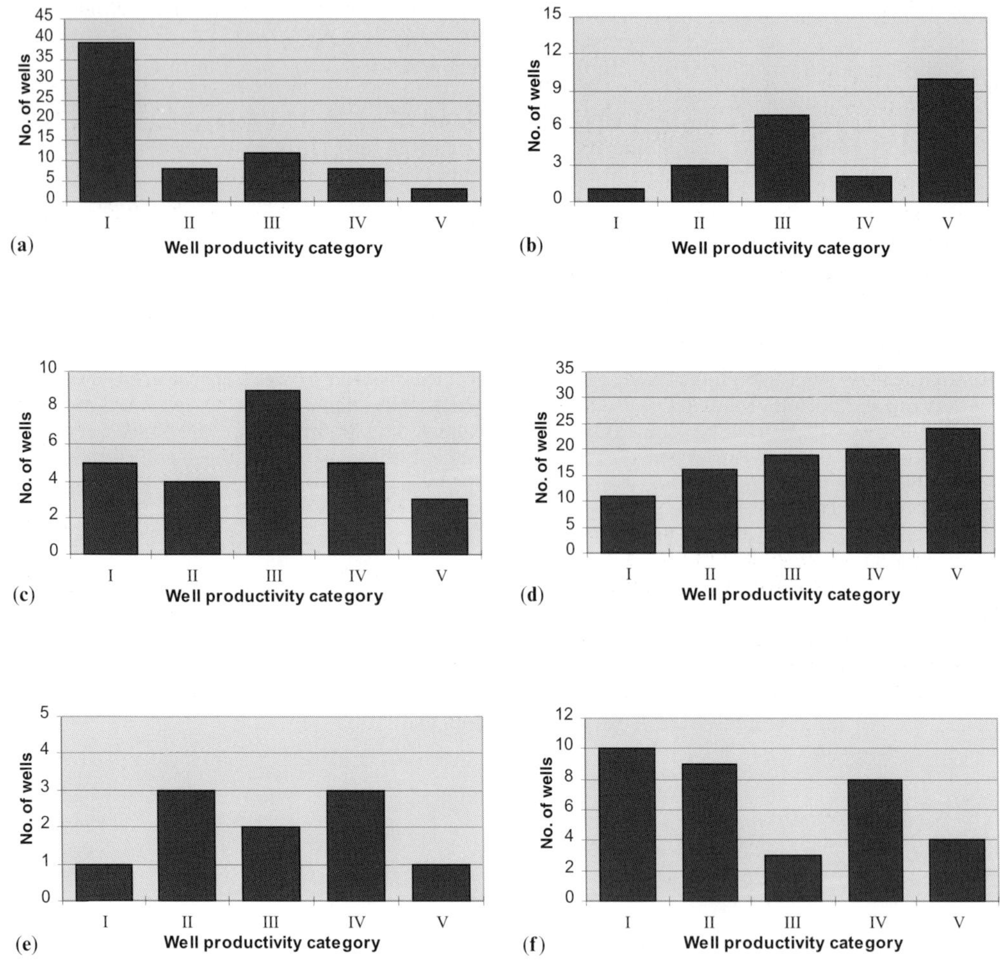

Fig. 8. Productivity bar charts for Variable Aquifers (clean limestones) (**a**) Waulsortian Limestone (South), (**b**) Waulsortian Limestone (North), (**c**) Ballyadams Limestone, (**d**) Burren Limestone (Galway–Mayo), (**e**) Burren Limestone (Laois–Offaly), (**f**) Burren Limestone (north-central region).

Ballyadams Formation (Limestone) (Fig. 8c). This Lower Carboniferous formation comprises clean, generally coarse-grained, bedded limestones, often karstified and/or dolomitized. Twenty six data points cover quite a wide range. The occurrence is limited to the southern and SE midlands (Counties Carlow, Laois, Kilkenny, and South Tipperary), so further regional analysis is difficult. Generally it is regarded as Regionally Important (Rk), but the QSC data throw some doubt on this, and in places Locally Important (Ll) may be more appropriate.

Burren Limestone Formation (and equivalents) (Figs 8d–f). These are also clean, generally coarse-grained, bedded Lower Carboniferous limestones. One hundred and forty one data points cover almost the whole range of Q and SC. Counties Galway and Mayo (Fig. 8d) show lower values than the Laois–Offaly region (Fig. 8e) and the 'north-central' region (counties Cavan, Leitrim, Longford, Meath and Roscommon, Fig. 8f). Classification is currently Regionally Important (Rk), but to some extent this derives from the high aquifer throughput rather than the QSC data.

Conclusions

QSC graphs and bar charts offer a useful semi-quantitative method of using limited pumping test data to evaluate the appropriate aquifer categories in Ireland. It is envisaged that their principal use will be to support aquifer categorization which may come under challenge where important planning decisions are involved.

A similar approach may be applicable in other countries where groundwater mainly occurs in unconfined fractured formations and where good hydrogeological data are scarce. However, the boundaries between productivity classes used in Ireland are essentially arbitrary and would probably require adjustment in other environments.

The size of the data set and its distribution are important. A total data set of at least 500 is probably needed for an area of the size and complexity of Ireland, and (depending on variability) a data set of at least 20 to 40 data points is probably needed for a defensible aquifer categorization.

The method demonstrates the need for regional discrimination: virtually identical rocks can have quite different aquifer characteristics in different areas, owing to differing geological and geomorphological histories.

Past and present colleagues in GSI are thanked for useful discussions. I am also grateful to many hydrogeologists, engineers and well drillers who contributed data, and to two anonymous reviewers for suggestions. The paper is published by permission of the Director of the Geological Survey of Ireland, Dr. Peadar McArdle.

References

DALY, D. & WARREN, W. P. 1998. Mapping groundwater vulnerability: the Irish perspective. *In*: ROBINS, N. S. (ed.) *Groundwater Pollution, Aquifer Recharge and Vulnerability*. Geological Society, London, Special Publications, **130**, 179–90.

DEPARTMENT OF ENVIRONMENT AND LOCAL GOVERNMENT, ENVIRONMENTAL PROTECTION AGENCY & GEOLOGICAL SURVEY OF IRELAND. 1999. Groundwater Protection Schemes. Department of Environment & Local Government, Environmental Protection Agency, and Geological Survey of Ireland, Dublin.

WRIGHT, G. R. 1997. *QSC Graphs*. GSI Groundwater Newsletter, Issue no. 32, December.

The Triassic Sherwood Sandstone aquifer in Northern Ireland: constraint of a groundwater flow model for resource management

G. W. McNEILL, A. A. CRONIN, Y. YANG, T. ELLIOT & R. M. KALIN

Environmental Engineering Research Centre, School of Civil Engineering, The Queen's University of Belfast, David Keir Building, Stranmillis Road, Belfast BT9 5AG, UK (e-mail: G.McNeill@qub.ac.uk)

Abstract: The most important aquifer in Northern Ireland is the Triassic Sherwood Sandstone formation that lies under the urban area of Belfast and the areas to the east (Newtownards) and south (Lagan Valley) of the city. The management of this groundwater resource is important in providing a sustainable supply for both public and industrial users. A groundwater flow model that collated relevant meteorological, hydrological and hydrogeological information for the Lagan Valley and Newtownards areas was developed. This conceptual model was coupled with geochemical and isotopic data (using inverse modelling) to constrain the groundwater flow parameters. The results of ^{14}C-dating suggest some of the groundwater in the aquifer may be up to 4000 years old, and that structural controls play a major role in both the flow rate and the spatial distribution of groundwater within the Sherwood Sandstone aquifer.

Study of integrated catchment management requires knowledge of groundwater flow parameters. These parameters can be estimated using a number of meteorological, hydrological and hydrogeological measurements but a model of groundwater flow requires validation against existing knowledge before it is useful for effective groundwater resource management. This can involve interpretation of the groundwater chemistry and an understanding of the often-complex geochemical reactions involved (from recharge to storage and/or abstraction). Traditional methods of geochemically modelling groundwater flow ('forward modelling') have relied on knowing the chemical (and isotopic) composition of the initial water at recharge and applying chemical reactions (of known stoichiometry, kinetics and mixing rates) to recharge water in order to predict the chemistry of the final water under study (Plummer *et al.* 1994). The alternative method of geochemically modelling groundwater evolution is 'inverse modelling' which takes known chemistries of both the initial water and the final water, and infers the chemical reactions (by predicting the stoichiometry, kinetics and mixing rates) the water may have undergone within the aquifer. Inverse hydrogeochemical modelling can be coupled with the overall model of groundwater flow to provide a well-constrained hydrologic and geochemical model of an aquifer, which is necessary for groundwater resource management.

A coupled inverse model of groundwater flow and geochemistry was generated for the Triassic Sherwood Sandstone aquifer of Northern Ireland and this paper presents the techniques used and data involved. Geochemical and isotopic information was incorporated into the hydrogeological understanding of the area so that the groundwater model provides a more complete representation of the catchments (both surface water and groundwater) for resource management purposes.

Description of study area

Northern Ireland's most important aquifer underlies Belfast and extends both to the east and south of the city (Robins 1996). The principal aquifer is the Triassic Sherwood Sandstone formation; however other formations, such as the Permian sandstone and the overlying Quaternary sands and gravel, can be locally important water supply sources (Bennett 1976). The study area can be subdivided into two main regions: (a) the Lagan Valley to the south of Belfast, and (b) the Newtownards area from the head of Strangford Lough to the east of Belfast.

The Sherwood Sandstone aquifer in Northern Ireland is described in detail by Robins (1996). Hydrogeological work in the area, which began with Hartley (1935), has more recently included a reconnaissance hydrogeological study by

From: ROBINS, N. S. & MISSTEAR, B. D. R. (eds) *Groundwater in the Celtic Regions: Studies in Hard Rock and Quaternary Hydrogeology*. Geological Society, London, Special Publications, **182**, 179–190. 1-86239-077-0/00/ $15.00

Bennett (1976), determination of recharge capture areas by Robins & Shearer (1994), analysis of the hydrogeochemistry by Kalin & Roberts (1997) and modelling of the chemical and carbon isotopic composition of the groundwater by Cronin *et al.* (2000).

Topography and hydrology

The valley of the River Lagan is approximately 6 km in width and 25 km in length (Fig. 1) with Belfast and its surrounding conurbation to the north. The river flows through a glacial valley carved into the Triassic Sandstone before reaching Belfast Lough. The Lagan Valley is flanked to the west and north by steep scarp slopes of Tertiary basalt overlying Cretaceous Chalk (the Antrim Plateau) and to the SE by the rolling hills of County Down, an area notable for its abundance of drumlins. Surface drainage in the Lagan Valley is by a network of small streams and the Ravernet River (Manning 1972).

An extension of the Triassic sandstone aquifer runs from Belfast eastwards to the towns of Comber and Newtownards. Flow in the aquifer east of the Lagan Valley is hydrogeologically distinct from the main Lagan Valley flow. The only major river in this area is the Enler River, which flows from Dundonald to Comber, and then into Strangford Lough (Fig. 1).

Geology

Smith (1985) updated the geological description of the study area (Fig. 2) to take into account the knowledge gained from the Lagan Valley Groundwater Development programme (1977–1984). A detailed geological description of the Lagan Valley can be found in the Belfast Memoir (Manning *et al.* 1970) and the geology of the Newtownards area is described in detail by Smith *et al.* (1991).

Groundwater usage

Although official statistics state that groundwater provides only about 15% of the public and private water supply in Northern Ireland (Robins 1996) the underground resources are nevertheless of great social importance and economic benefit to the community and industry alike. Historically, in the Lagan Valley, the Triassic Sherwood Sandstone supplied water primarily for agricultural and industrial processes (e.g. linen manufacturing, heavy engineering). During the last few decades, public supply boreholes have been the most dominant groundwater abstraction points in the valley (extracting 3000 to 10 000 m^3/d) with a number of industries also utilizing this resource.

The Triassic Sherwood Sandstone formation has been in use as an aquifer in the Newtownards area since the end of the nineteenth century

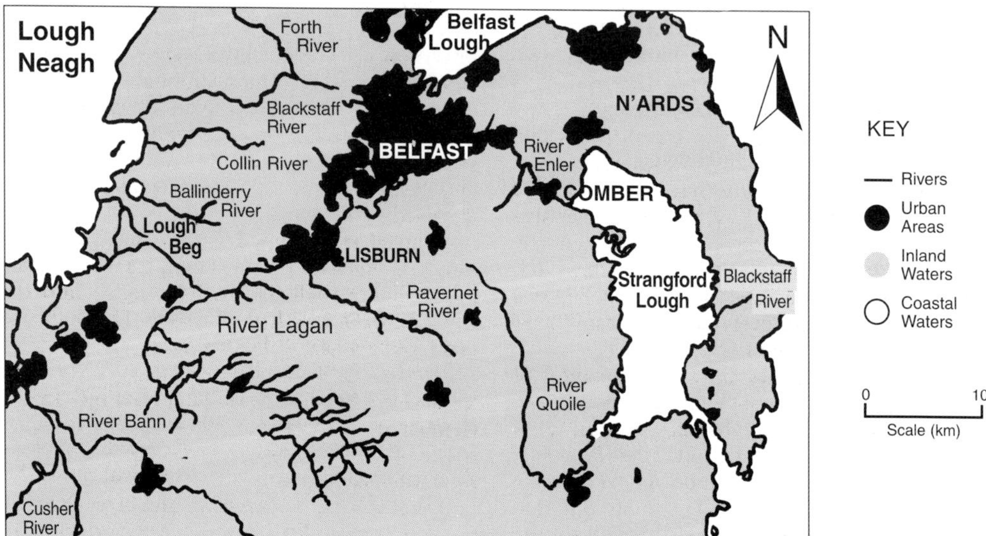

Fig. 1. Hydrology of the Lagan Valley and Newtownards areas.

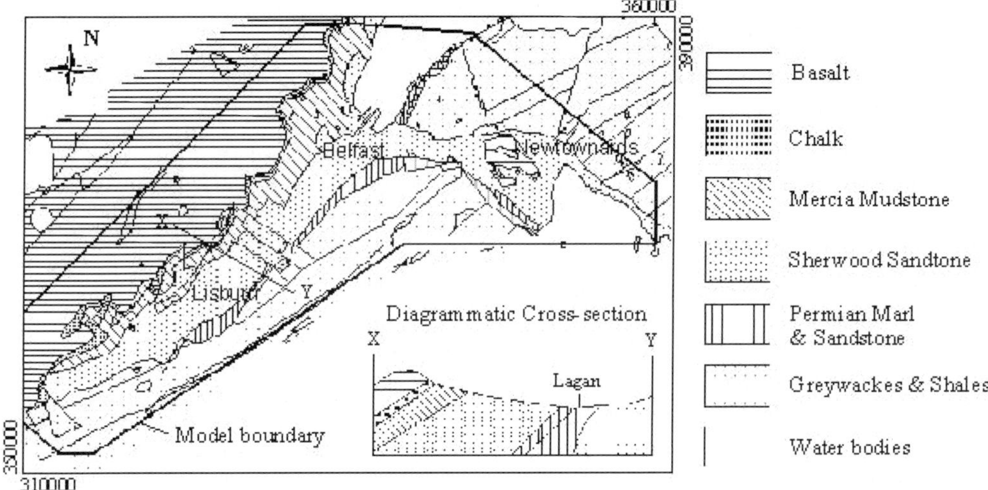

Fig. 2. Solid geology of the Lagan Valley and Newtownards areas (adapted from GSNI 1997).

(Manning 1972) and as a source of public supply since the 1950s (Bennett 1976). About 8000 m^3/d is currently withdrawn for public supply at Newtownards (as well as about 3500 m^3/d from the Permian sandstone and 1000 m^3/d from the sand and gravel aquifer at Comber).

Risk of groundwater pollution

It is important that groundwater resources are protected from contamination by both diffuse and point sources of pollution. Other UK cities overlying Triassic sandstone have found problems with organic pollutants (Lerner & Tellam 1992) or with sewer exfiltration (Misstear *et al.* 1996). The Triassic aquifer in Northern Ireland has potential water quality issues relating to agriculture, industrial processes and landfilling (Hydrogeological and Environmental Services Ltd. 1997; Kalin & Roberts 1997). For example, fertilizer application in the Enler catchment is extensive and in the River Enler itself between 1987 and 1997 nitrate levels increased from 5 mg/l to 40 mg/l (McConville 1999).

Not only water quality, but also the quantity of water available for sustainable abstraction, dictates whether an aquifer can be used as a groundwater resource. Despite the fact that abstractions in the area have decreased over the last sixty years, Robins (1996) points out that 'The Sherwood Sandstone is a critical aquifer unit in which the potential for overdrawing suggests that local licensing of major groundwater abstractions might be worthwhile but would require an adequate knowledge of the resources within the aquifer unit'. The importance of the Triassic aquifer will increase when one considers that further development of Quaternary aquifers is unlikely because of their high vulnerability to pollution (Robins 1995).

Objectives and methods of research

Coupled inverse groundwater flow and geochemistry modelling leads to a reduction of the uncertainty associated with a conceptual hydrogeological model. Identification and calibration of hydrogeological parameters of groundwater flow and geochemistry are fundamental for a strong understanding of the system, if sustainable management of groundwater is to be achieved.

Coupled inverse modelling may provide not only an efficient means of aquifer parameter identification, but also useful insights into the functioning of the physical system. The challenge is to create a useful representation of an aquifer system from necessarily limited data. These data are collected from a finite (and relatively small) number of observations made in a continuous spatial domain that will, in reality, exhibit considerable heterogeneity.

Use of inverse models facilitates assessment of prediction reliability because the results not only yield parameter estimates, heads and concentrations simulated for the stresses of interest, but also provide confidence levels, which are convenient for conveying the reliability of results to regulators (Poeter & McKenna 1995). The fundamental benefit of inverse modelling (over forward modelling) is its ability to automatically calculate parameter values that produce the best fit between observed and simulated hydraulic heads and solute/contaminant concentrations for a given conceptual model. A calibration obtained via a non-automated approach alone is not optimized, and numerous runs can be spent changing parameters whilst relatively few runs are spent actually adjusting the conceptual model. By coupling the groundwater flow and geochemistry in inverse modelling, the flows are constrained by geochemical and/or biochemical reactions along the flow paths, which makes the modelling of groundwater more robust and more effective. Another benefit is that coupled inverse modelling reveals issues that even an experienced modeller can easily overlook during non-automated calibration efforts. For example, parameter correlations (e.g. ratios of parameters rather than their individual values) are clearly delineated in an automated inverse calibration run (Poeter & Hill 1997). The impact both of available data and of anticipated (soft) data on the uncertainty can also be quantified by coupled inverse calibration. With the use of practical non-linear, least-square inverse software (such as PEST and UCODE) and stochastic uncertainty modelling (such as UNCERT), coupled inverse modelling for model calibration and alternative conceptual analysis is more readily accomplished with a reduction of the associated uncertainties (Wingle *et al.* 1994).

This study of the Sherwood Sandstone aquifer in Northern Ireland revealed that regional groundwater modelling was constrained by lack of both understanding of, and information on, the urban aquifer system. Thorough understanding of recharge, sound estimates of physical hydrogeological parameters and a reasonable representation of aquifer structure are crucial for confidence in the numerical simulation of flow rates and directions, and (of prime importance for contamination studies) advective transport. Much of the uncertainty for the Lagan Valley aquifer system was associated with the representation and interpretation of estimates of recharge and solute load, initial and boundary conditions, aquifer properties, and aquifer 2D/3D structure.

Approach

The popular computer modelling packages of MODFLOW, NETPATH, PhreeQC and PEST were employed as numerical simulation tools with proper modifications. From conceptual model, flow model and geochemical model, to inverse parameter optimization and uncertainty management, a series of comprehensive modelling techniques were carried out (Fig. 3). The identification of flow components and chemical/biochemical species (in solid, aqueous and gaseous phases) along typical flow paths, and the interpretation of uncertainty from parameter representation were completed as part of the research.

For the Sherwood Sandstone aquifer, recharge through the glacial till was found to be a critical factor (Cronin *et al.* 2000). Both the mechanism and the amount of recharge to the aquifer, largely covered by till, are still not fully known. However, the results of inverse modelling have placed upper and lower limits, and provide confidence levels in their values. The calculation of these limits began with a hydrologic balance using meteorological, hydrological and hydrogeological information.

Meteorology

The mean air temperature in Northern Ireland ranges from around 15°C in summer to approximately 5°C in winter, with a relative humidity often about 85%. In the Lagan Valley, the average rainfall is 953 mm a^{-1} and the average evapotranspiration is 404 mm a^{-1} (DoE 1995). The effective natural water resource available to the catchment therefore is on average 549 mm a^{-1}.

Hydrology

River flow stations are maintained by the Department of the Environment (N.I.) at various points on the Lagan, and its main tributary the Ravernet River. In the study area, the impermeability of the bedrock throughout much of the catchment means that flows in the Lagan respond rapidly to storm events, and limited seasonal groundwater transfer takes place in the catchment (Cruickshank 1997). The aquifer in the lower parts of the Lagan catchment is well below the river and not thought to be in direct connection with the river; however, groundwater possibly enters the river downstream of Lisburn (W. S. Atkins 1997). New geochemical and isotopic data suggest some connection in the southern part of the Lagan Valley (Cronin *et al.* 2000). There is no known hydraulic connection between

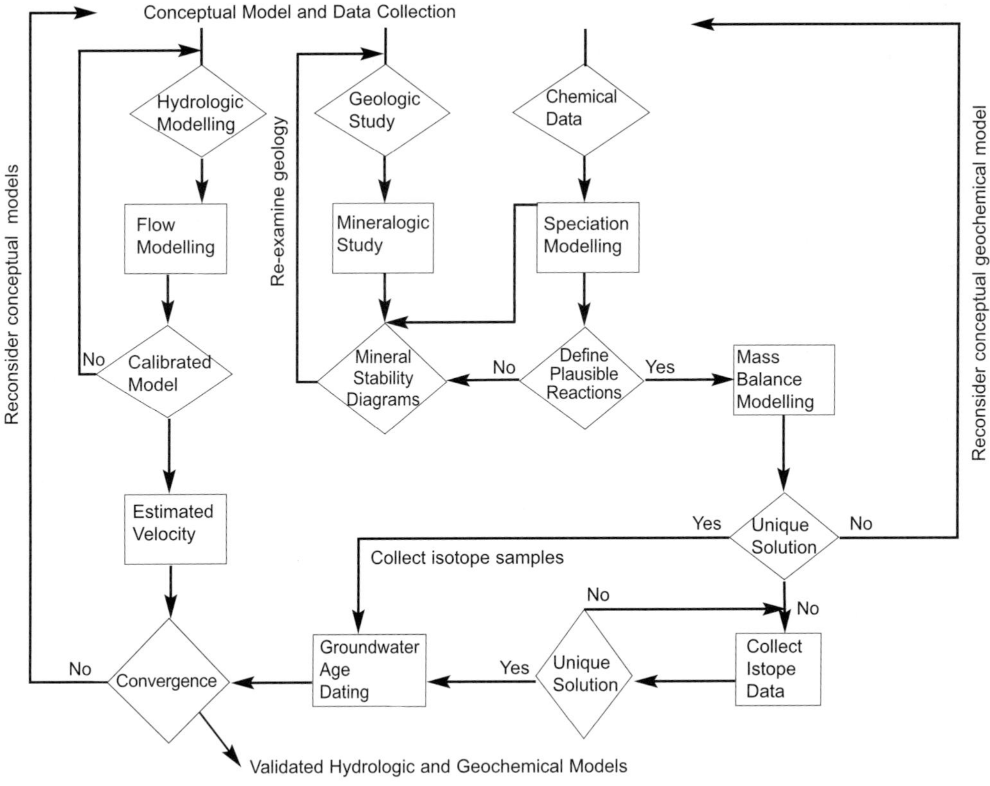

Fig. 3. Framework of coupled groundwater flow and geochemical modelling.

surface water and groundwater in the Newtownards area (Robins & Shearer 1994). However, an inferred link between high nitrate levels in the groundwater and agricultural practices in the area strongly suggests that there is some hydraulic connection (McConville 1999).

Aquifer discharge at Belfast has been estimated at 10.95 million $m^3 a^{-1}$ (Bennett 1976) a figure that is less than 3.5% of the average annual rainfall (Kalin & Roberts 1997). At present, given the uncertainties in all the available data, an accurate representation of the hydrologic balance is problematic (Kalin & Roberts 1997). On-going research at the Queen's University of Belfast is examining the salinity of Belfast and Strangford Loughs to try and calculate the amount of freshwater entering these water bodies via groundwater discharge.

Hydrogeology

Bennett (1976) produced a conceptual hydrogeological model that suggested groundwater flows through the sandstone towards the centre of the Lagan Valley from the NW side all along the length of the valley, and from the SE side at least in the area SW of Lisburn. The topography falls along the length of the aquifer towards Belfast and therefore Bennett (1976) suggested a component of flow through the sandstone is in a north-easterly direction. Flowing artesian conditions have only been encountered in a few locations but where the piezometric surface is above the river level, there may be an upward component of groundwater flow through the drift deposits to the surface along the central area of the valley.

The hydrogeological properties of the Triassic aquifer are not well defined because most of the exploratory drilling in the area was carried out during the late 1960s, before a systematic collection of hydrogeological data began (Robins & Shearer 1994). Nevertheless a substantial amount of pumping test data were obtained for the Lagan Valley, and some for the Newtownards area. The Sherwood Sandstone has an average transmissivity of 130 m^2/d and a storativity in the order of 2×10^{-3} (Kalin & Roberts 1997). This transmissivity may extend up to 200–250 m^2/d in the Newtownards area (Robins 1995). The glacial till (sandy clay) on the floor of the Lagan Valley has a permeability of $1.0–5.9 \times 10^{-6}$ m/d at 2 m depth, and $1.6–3.0 \times 10^{-6}$ m/d at 10 m depth (Kalin & Roberts 1997).

The Triassic Sherwood Sandstone aquifer overlies the Permian sandstone aquifer (Kennel Formation) in the vicinity of Newtownards and in the east of the Lagan Valley, but due to the intervening Permian marl they are reportedly not in hydraulic connection (Robins 1995). More important to groundwater movement is cross-basin geological faulting that has a significant influence on the groundwater flow between Lisburn and Belfast (P. Bennett pers. comm.).

Manning *et al.* (1970) estimated that average net rainfall in the Lagan catchment is 888 000 m^3/d. Subtracting an average river flow of 816 000 m^3/d results in a residual 72 000 m^3/d for groundwater infiltration. Assuming recharge is 40% of the residual infiltration results in a recharge rate of approximately 29 000 m^3/d (Manning *et al.* 1970). A useful estimate of the annual average recharge to the aquifer can be obtained by establishing the amplitude of the annual fluctuation of water level, and multiplying by its storage coefficient and its spatial extent. Bennett (1976) divided the Lagan Valley area into three separate zones. The first of these was the Lisburn dyke belt (10 km^2) with an annual fluctuation of 2 m and a storage coefficient of 1.5×10^{-4} to calculate an annual recharge of 3000 m^3. The second area was NE of Lisburn (33 km^2) with an annual fluctuation of 1 m and a storage coefficient of 2.5×10^{-3} giving an annual recharge of 82 500 m^3. The third area was SW of Lisburn (48 km^2) with an annual fluctuation of 1.5 m and a maximum storage coefficient of 8×10^{-4} that gave an annual recharge of up to 58 000 m^3. Bennett (1976) stated that these figures were absolute minima, and pointed out that it is academic to estimate the total storage of the aquifer, as only a tiny fraction of that storage can be exploited. Assuming absolute minimum values for all the necessary parameters (area of 91 km^2, specific yield of 0.1, and an effective aquifer thickness of 30 m) he produced a lower limit for the storage volume of the entire aquifer of approximately 270×10^6 m^3.

The previous conceptual model of the Sherwood Sandstone aquifer suggests that recharge is likely to occur along relatively high ground to the NW and SE sides of the valley, where the drift is absent or thin (Bennett 1976). The aquifer does not terminate against the NW side of the valley but continues at depth beneath the Antrim plateau. Its boundary with the Mercia Mudstone is gradational, with perhaps a continuous piezometric connection from one to the other. The Sherwood Sandstone therefore receives significant but undetermined recharge from the overlying formations along the NW side of the Lagan Valley.

The Sherwood Sandstone aquifer in the Enler Valley has a considerable outcrop to the NW of Newtownards through which direct recharge occurs. Elsewhere, the aquifer is concealed by till that is generally thought to be impermeable, and would therefore inhibit infiltration and promote runoff. However, both hydrochemical and available radiometric evidence, though not as yet conclusive, suggest that recharge through the till does occur (Robins 1995; McConville 1999). The till is certainly heterogeneous and may have preferential pathways for infiltration along discontinuities and sandy patches. The till and associated soils retain moisture and keep soils at, or near, field moisture capacity throughout most of the year. This counters the low permeability of the till and enables recharge to take place at virtually any time of the year.

Hydrochemistry

The major ion chemistry of groundwater in the Lagan Valley is dominated by calcium–magnesium–bicarbonate with some sulphate (Robins 1996). Cronin *et al.* (2000) highlighted the importance of dedolomitization in the geochemical evolution of the Lagan Valley groundwater, as measured by variable Ca–Mg ratios throughout the area. Kalin & Roberts (1997) found that groundwater in the Newtownards area had a lower calcium and sulphate concentration than in the Lagan Valley, which is overlain by the Mercia Mudstone. Kimblin (1995) studied the Permo-Triassic sandstone in NW England and found that where confined by Mercia Mudstone, dissolution of sulphate minerals is the dominant process controlling the major ion chemistry by contributing both calcium and sulphate to the groundwater. This process promotes Ca–Na, and possibly Ca–Mg, ion exchange between the water and the solid phases of the aquifer.

Dating of groundwater in the study area was found to be important, to indicate the different sources of the water, its residence time, and its travel path underground (Cronin *et al.* 2000). Groundwater samples from the mid-1970s with low tritium levels (Bennett 1976) indicated that the Sherwood Sandstone contained little post-1953 recharge. Higher tritium values at Lisburn were due to abstraction promoting the throughflow of water and drawing in relatively recent

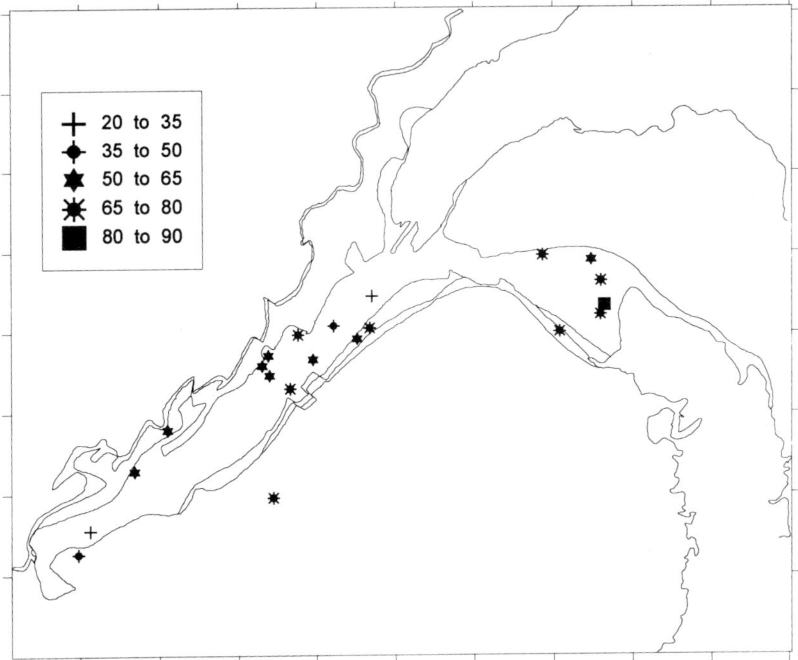

Fig. 4. Radiocarbon activity (expressed as percentage modern carbon, pmc) in groundwater from the Lagan Valley and Newtownards areas (adapted from Cronin *et al.* 2000).

water. However, it would appear that throughout the central area of the Lagan Valley aquifer, the groundwater is fairly old; the fact that there is no direct recharge to the aquifer in the central area of the valley supports this interpretation (Bennett 1976). Analysis of ^{14}C was carried out (Fig. 4) and a rough estimate may be made that the results around 55% modern carbon are between 500 and 1000 years old and those less than 35% modern carbon may be 3000 to 4000 years old (Cronin *et al.* 2000).

Results of modelling work

The conceptual model

The spatial distribution of groundwater flow systems is similar to surface water catchments in both the Lagan Valley and the Newtownards area. The initial conceptual model for the groundwater flow system from the piezometric data involves predominant flow in the Sherwood Sandstone towards the centre of the Lagan Valley, from the SE side at the top of the valley and from the NW side along the entire length of the valley. In the Newtownards area, the conceptual model involves groundwater flow from the east of Belfast towards Strangford Lough.

Borehole records obtained from the Geological Survey of Northern Ireland were examined to profile various stratigraphic units in the study area. Twelve distinct Quaternary deposits have been identified in the Lagan Valley that are variable in both thickness and extent. For conceptual representation of groundwater flow, these layers were reduced to (a) river alluvium, (b) fluvioglacial sands and gravels, and (c) glacial till. The solid geology of the area was digitized from a recent 1:250 000 edition of the Geological Map of Northern Ireland (GSNI, 1997). Six geological units (the Ordovician and Silurian greywackes, Permian sandstone and marl, Triassic Sherwood Sandstone, Mercia Mudstone, Cretaceous Chalk and Tertiary basalt) were defined for the geology in the conceptual model. Due to the complex topography, fracture development and hydrogeological significance of the rocks, a vertical profile was adopted. Two layers of solid geology were conceptualized to represent the depth profile of aquifer properties and groundwater movement. The lower layer was considered to have a reduced permeability due to the fact that the fractures will close with increasing overburden pressure.

Based on geochemical evidence (Cronin *et al.* 2000) the study area can be divided into four zones, namely the Upper Lagan Valley, Middle Lagan Valley, Lower Lagan Valley (separated by cross-basin faults) and the Newtownards areas. There is limited groundwater flow between the three Lagan Valley zones and the Newtownards zone is hydrogeologically separate.

Initial groundwater modelling

A steady state groundwater flow model was first developed to allow testing of the conceptual regional natural groundwater field prior to transient modelling. A regular model grid of 500 m × 500 m was used (Fig. 5) and inactive cells were placed all around the border of the model. So that no initial assumptions would bias the model, this inactive cell boundary was placed 5 km away from the geological and topographical divides in the study area. The whole of the Enler catchment and the majority of the Lagan hydrological catchment area were included. Thicknesses of 60–120 m and 50–80 m were used, respectively, for the solid geological layers (1 and 2) of the aquifer in most of the modelling area. Exceptional thickness was considered in topographically complex areas, for example in the high regions to the west of the study area. For solid geological layer 1, a thickness of approximately 100 m was used as this reflects the effective depth of the aquifer that is most productive.

All known meteorological data, river stage and flow data, borehole monitoring information, abstraction records and pump-test data were collated. A constant head of 0 m was assumed at Belfast and Strangford Loughs. An initial figure of 80 mm was assumed for the amount of net recharging precipitation reaching the water table. Model results showed this to be too high in the thick clay areas and so net precipitation was reduced to 30 mm in the steady state model for the centre of the valley. A higher net infiltration of 150 mm was used to simulate recharge areas (Bennett 1976; Robins & Shearer 1994), which was slightly higher than the value of 128 mm given by Foster (1969).

The rivers in the region that were assigned in the model were the Lagan, the Ravernet, and the Enler. River bed and river stage elevations were

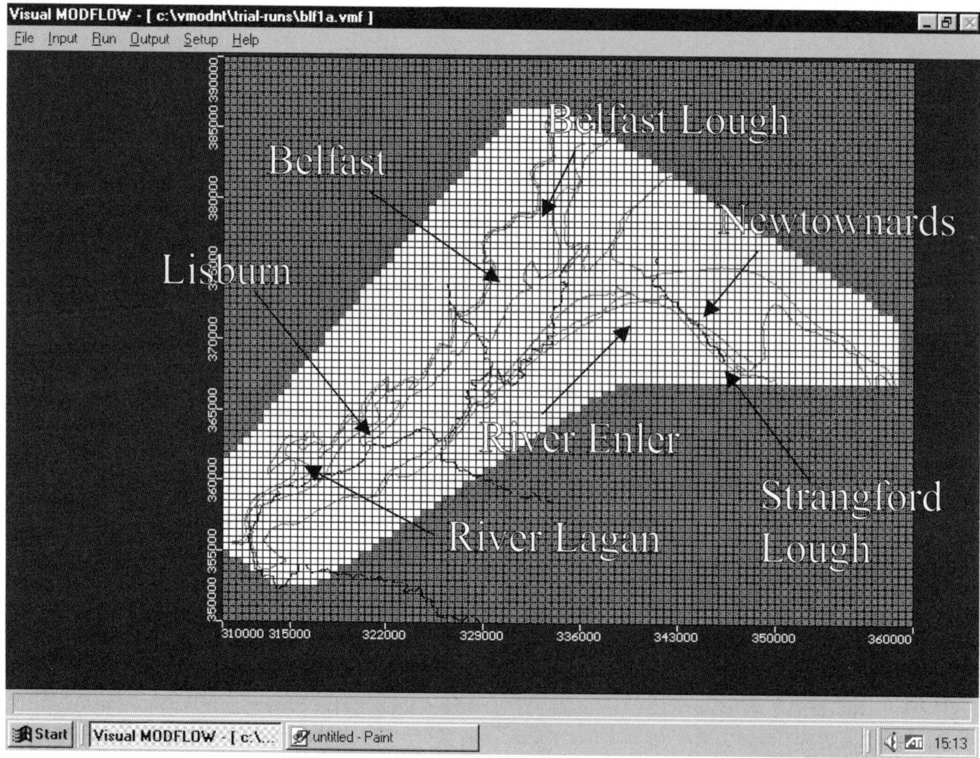

Fig. 5. MODFLOW-generated model grid of the Lagan Valley and Newtownards areas.

Table 1. *Aquifer property values*

Drift/solid layer	Kx (m/d)	Ky (m/d)	Kz (m/d)	Ss (1/m)	Sy	n
Alluvium	2.6	2.6	0.026	4×10^{-6}	0.2	0.4
Sand/Gravel	1.27	1.27	0.0127	6.6×10^{-6}	0.3	0.35
Glacial Till	0.06	0.06	0.0006	8×10^{-6}	0.1	0.45
Basalt	3.45	3.45	0.0345	1×10^{-10}	0.1	0.4
Chalk	2.0	2.0	0.02	1×10^{-5}	0.2	0.25
Mercia Mudstone	0.005	0.005	5×10^{-5}	1×10^{-10}	0.2	0.25
Sherwood Sandstone	0.2–6.0	0.2–6.0	1% of Kx	1×10^{-10}	0.2	0.24
Permian Sandstone	1.4	1.4	0.014	2.8×10^{-6}	0.1	0.16
Permian Marl	0.1	0.1	0.001	1×10^{-10}	0.2	0.25
Greywackes	0.6	0.6	0.006	1×10^{-10}	0.2	0.06

taken from maps and flow data supplied by the Rivers Agency. A nominal bed thickness of 1 m and a permeability of 3 m/d were used (Price & Foster 1974). The aquifer properties presented in Table 1 are from various pumping tests and reports carried out in the study area. Care has to be taken when using permeability values of sandstone calculated from cores, because it can result in an underestimation of water transport times through the media, which is essentially a dual-permeability system. Vertical permeability values were represented by a figure of 1% of horizontal permeability.

The steady-state model output predicts recharge at 40 000 m^3/d, which according to Bennett (1976) would be extremely low. A value of 21 000 m^3/d of groundwater flowed to Belfast Lough, again lower than the estimate of 30 000 m^3/d given by Hartley (1935). One feature of the region is the quite steep hydraulic gradients in the hills to the west of Belfast. This complicates the model as cells dry out quickly and are not easily re-wetted. Only cells with active heads were exported to an ASCII file as the initial starting point for the transient model. Pumping wells and observation wells were included in the transient model run. The period simulated was from August 1971 to September 1975 divided into 15 stress periods, one per season. Pumping records for this period are poor, apart from the public supply wells at Lisburn (abstracting around 2000 m^3/d) and Newtownards (7000 m^3/d). Most of the water levels predicted by the transient model were within 3 m of the observed values. The biggest variations occur in the hypothesized recharge areas at the Lagan Valley edges. Travel times in the sandstone layer were assumed to be about 0.1 m/d which, taking an approximate distance from the recharge point to the sea of 30 km, gives a calculated travel time of about 1000 years. Particle tracking carried out by MODPATH took similar lengths of time for particles to travel to the discharge points.

Utilization of PEST

The finite difference MODFLOW model of the area poses many important questions to examine with inverse modelling using PEST. These include the drying and re-wetting of cells, important because heads calculated during re-wetting are non-unique. Also convergence problems were experienced that could be overcome with a simple single layer model (Doherty 1994). However, in the regional groundwater flow model, the Quaternary sand and gravel aquifer layer is of great importance. Firstly, it is a water supply resource for the expanding towns of Newtownards and Comber. Secondly, it interacts with the underlying Sherwood Sandstone layer in the Lagan Valley (especially in the region around Lisburn) with leakage possible in both directions. The complex geology of the study area with its wide spatial variations in parameters, coupled with the topographical differences, makes it a worthy challenge for parameter estimation. It may well be that simplification of the regional model, or smaller models constructed with such processes as Telescopic Mesh Refinement, may ultimately prove more useful.

A detailed series of analyses with PEST was undertaken, in which a reasonable initial dataset of the hydrogeological parameters was used, for convergence of the inversion process and the overall iterations during optimization. The observation data were taken from the means of 1973–1974 at eighteen targets, and four recent supplementary measurements in the Newtownards area. Wide upper and lower boundaries of the adjustable parameters were set to give a wide range of uncertainty for the optimization procedure. The parameters were adjusted until the fits between model-generated and actual borehole groundwater heads were reduced to a minimum in the weighted least-squares result. PEST98 for Windows (WPEST) required about 30 minutes to run each groundwater flow model, therefore the optimization procedure took between two and three days for each computer run (Yang & Kalin 2000).

System uncertainty arose due to the limited observation data in the region and the different data measurements. To smooth the parameter estimation problem and reduce the uncertainty, lower weights were assigned to the four observations in the Newtownards area. The optimal objective function, the sum of squared weighted residuals of groundwater heads, was achieved after five inversion iteration runs. The sensitivity of each parameter with respect to all the weighted observations during five optimization iterations is shown in Fig. 6. The inversion of the regional flow model was most sensitive to the parameter values of sand and gravel, glacial till and Sherwood Sandstone in the middle part of the Lagan Valley, and then to those of basalt and greywackes. Zonation of hydraulic conductivity was determined upon the basis of geology, hydrology, topography and land use. Based on the spatial zonation of the adjustable parameters, the automated calibration of the parameters was conducted for the groundwater flow model. The estimated hydraulic conductivity (K) values for the solid geological strata were nearly identical to the initial estimations. The much lower value for the Mercia Mudstone, conformably lying on the Sherwood Sandstone, suggests a slow velocity and long travel time of groundwater recharge through the escarpment and hills in the western part of Belfast (Bennett 1976). For the Sherwood Sandstone aquifer, the spatial distribution of K is characterized by a general decrease, as expected, from shallow to deep in the vertical direction. Horizontally, K is relatively greater in the north and middle (around Lisburn) of the Lagan Valley than in the rest of the area. This confirms the high anisotropy of the Sherwood Sandstone aquifer as a result of the geological structures present (Kalin & Roberts 1997).

The overall 95% confidence limits of the inverse solution are wide. The limits in the inversion provide only an indication of parameter uncertainty and indicate that they probably extend further into parameter space than the linear assumption itself and may be exaggerated in logarithmic transformation (Doherty 1994). The contours of automated and measured groundwater heads (Fig. 6) show an overall good match even with the uncertainty on the sparse distribution of observations. After the model automation, the groundwater mass balance was within a reasonable range. Even so, there are still too few observations and relatively too many parameters in the automation process.

Inverse geochemical modelling based on groundwater Ca–Mg ratios and carbon isotopes (Cronin *et al.* 2000) results in groundwater flow paths that differ from those from hydrogeological

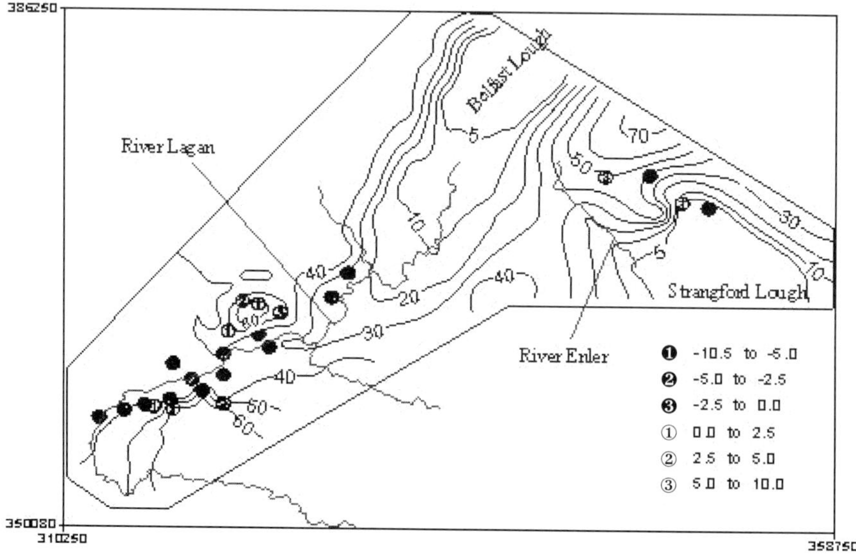

Fig. 6. Groundwater head contours and weighted residuals of the calibration.

modelling alone. For instance, the presence of 'older' water in the Upper Lagan Valley compared with that in the Middle Lagan Valley contradicts the initial assumption that predominant flow is from valley top to bottom (Bennett 1976). This implies that continuing work that will couple geochemical data with the hydrogeological data by inverse modelling is essential to constrain the true distribution of these flow paths.

Conclusions

Inverse modelling and geochemical modelling of groundwater are not new techniques. However, combining inverse groundwater flow and geochemical modelling is a novel area. The temporal and spatial variations of natural isotopes and geochemistry in groundwater systems were used in Northern Ireland, to constrain the flow systems and improve our understanding of system boundaries, sources and sinks, and other aquifer properties for the Triassic Sherwood Sandstone system. Current models were refined with available geochemical and isotope hydrology datasets. Estimates of recharge are of a similar magnitude to previous predictions, but the overall understanding of the system is currently limited by a lack of detailed data for the entire study area.

We would like to thank the Department of the Environment (N.I.), The Department of Education (N.I.), and the Queen's University of Belfast for supporting this work. The Geological Survey of Northern Ireland and the Department of Agriculture (N.I.) are also acknowledged for their discussions and provision of information.

References

BENNETT, J. R. P. 1976. *The Lagan Valley Hydrological Study*. Geological Survey of Northern Ireland, Open File Report **57**.

CRONIN, A. A., ELLIOT, T., YANG, Y. & KALIN, R. M. 2000. Geochemical modelling and isotope studies in the Sherwood Sandstone aquifer, Lagan Valley, Northern Ireland. *In*: DASSARGUES, A. (ed.) *Tracers and Modelling in Hydrogeology* (Proceedings of the TraM'2000 Conference held in Liege, Belgium, May 2000). IAHS Publication, **262**, 425–431.

CRUICKSHANK,, J. G. 1997. *Soil and Environment, Northern Ireland*. Department of Agriculture (NI) & Queen's University of Belfast, Belfast.

DEPARTMENT OF ENVIRONMENT (UK). 1995. *Industry profiles. 1995 Environment Act: Statutory Guidance and Regulations on Contaminated Land*. HMSO, London.

DOHERTY, J. 1994. PEST: A unique computer program for model-independent parameter optimisation. *In*: *Water Down Under 94*, Adelaide, Australia, 551–554.

FOSTER, S. S. D. 1969. Newtownards Borough Council Enler Gravels Abstraction Scheme, Further Report on Groundwater Conditions in the Kennel Bridge Area with particular reference to the relation between the Permian Sandstone Aquifer and the Overlying Drift Formations. Institute of Geological Sciences, Open File Report **22**, Wallingford.

GEOLOGICAL SURVEY OF NORTHERN IRELAND. 1997. Northern Ireland Solid Geology (Second Edition) 1:250 000, British Geological Survey, Keyworth, Nottingham.

HARTLEY, J. J. 1935. *The Underground Water Resources of Northern Ireland*. The Institution of Civil Engineers, Belfast and District Association Pamphlet.

HYDROGEOLOGICAL AND ENVIRONMENTAL SERVICES LIMITED. 1997. *White Mountain to Black Mountain, Groundwater Survey: Landfill impact study*. DoE (N. I.), Belfast.

KALIN, R. M. & ROBERTS, C. 1997. Groundwater resources in the Lagan Valley sandstone aquifer, Northern Ireland. *Journal of Chartered Institution of Water and Environmental Management*, **11**, 133–139.

KIMBLIN, R. T. 1995. The chemistry and origin of groundwater in Triassic sandstone and Quaternary deposits, northwest England and some UK comparisons. *Journal of Hydrology*, **172**, 293–311.

LERNER, D. N. & TELLAM, J. H. 1992. The protection of groundwater from pollution. *Journal of the Institution of Water & Environmental Management*, **6**, 28–37.

MANNING, P. I. 1972. *The Development of the Groundwater Resources in Northern Ireland*. Institution of Civil Engineers, N. Ireland.

——, ROBBIE, J. A. & WILSON, H. E. 1970. *Geology of Belfast and the Lagan Valley*. Memoir of the Geological Survey of Northern Ireland, Belfast.

MCCONVILLE, C. 1999. Using stable isotopes to estimate groundwater recharge in a temperate zone. Unpublished PhD Thesis, The Queen's University of Belfast, Belfast.

MISSTEAR, B. D. R., WHITE, M., BISHOP, P. K. & ANDERSON, G. 1996. Reliability of sewers in environmentally vulnerable areas. *CIRIA Project Report No*. **44**.

PLUMMER, L. N., PRESTEMON, E. C. & PARKHURST, D. L. 1994. An interactive code (NETPATH) for modelling net geochemical reactions along a flow path. *USGS Water-resources Investigation Report*, **94-4169**.

POETER, E. P. & HILL, M. 1997. Inverse models: a necessary next step in groundwater modelling. *Groundwater*, **35**, 250–260.

—— & MCKENNA, S. A. 1995. Reducing uncertainty associated with groundwater flow and transport predictions. *Groundwater*, **33**, 899–904.

PRICE, M. & FOSTER, S. S. D. 1974. Water Supplies from Ulster Gravels. *Proceedings of the Institution of Civil Engineers*, **57**, 451–466.

ROBINS, N. S. 1995. Groundwater in Scotland and Northern Ireland: similarities and differences. *Quarterly Journal of Engineering Geology*, **28**, 163–169.

——1996. *Hydrogeology of Northern Ireland.* British Geological Survey, HMSO, London.

—— & SHEARER, T. R. 1994. The Permian and Triassic Aquifers at Comber, determination of recharge capture areas for aquifers at Bellaghy and Comber. British Geological Survey Report **WD/94/39**.

SMITH, A. 1985. *Geology of the Lagan Valley.* Geological Survey of Northern Ireland, Belfast.

SMITH, R. A., JOHNSTON, T. P. & LEGG, I. C. 1991. *The Geology of the Country Around Newtownards.* Memoir of the Geological Survey of Northern Ireland, Sheet 37 and part of Sheet 38, HMSO, London.

WINGLE, W. S., MCKENNA, S. A. & POETER, E. P. 1994. *UNCERT, a geostatistical tool for evaluating, modelling and designing remediation facilities.* Colorado School of Mines, USA.

W. S. ATKINS. 1997. Lagan Catchment – Draft Water Quality Management Strategy for Northern Ireland. DoE (N. I.), Belfast.

YANG, Y. & KALIN, R. M. 2000. Multi-objective optimisation for sustainable groundwater resource management in a semi-arid catchment. *Hydrological Sciences Journal* (in press).

Modelling the potential impact of climate change on a shallow coastal aquifer in northern Scotland

R. MALCOLM & C. SOULSBY

Department of Geography, University of Aberdeen, Aberdeen AB24 3UF, UK (e-mail: r.malcolm@abdn.ac.uk)

Abstract: The potential impact of climate change on a freshwater wetland in a coastal dune system at St. Fergus, north-east Scotland, was assessed using hydrogeological modelling techniques. A range of climate change scenarios were input into a numerical model (MODFLOW) that had been calibrated for the site using contemporary field data. The combination of parameter and hydrogeological uncertainty in models, together with uncertainty over likely site-specific climate change in the future, dictates that such modelling studies are often inconclusive. Nevertheless, a range of scenarios was forecast which indicates that future average recharge and flooding of the wetland is likely to fall within the range of interannual variation observed at the site, regardless of whether recharge increases or decreases. It is therefore likely that any resulting ecological impacts will be limited and reflect subtle hydrological changes that will affect different species in different ways. The utility of such a modelling approach is in providing environmental managers with a range of likely scenarios that can be incorporated into future management plans and policies to maximize flexibility of response.

Coastal dunes fringe much of the NW coastline in the Celtic countries from Galway in Ireland to the machair regions of NW Scotland (Haes & Wolters 1992). These sites are often rich reservoirs of biodiversity and thus have a high conservation value. They are however, increasingly being exploited, primarily for recreational, but also for industrial use (Viles & Spencer 1995). Recent research has examined the possible impact of climate change on coastal regions with respect to sea level rise and saltwater inundation; however, the impact of climate change on the groundwater behaviour of coastal sand dunes and associated wetlands has been rarely considered (Soulsby *et al.* 1997). The recharge of freshwater aquifers and the behaviour of groundwater are important controls on the structure and function of coastal dune and wetland ecosystems (van der Meulen 1990). The high biodiversity is partly explained by the marked hydrological gradients that occur, with xerophytic species dominating the arid dune ridges whilst organisms tolerant of waterlogging are found in low-lying slacks. Thus, understanding the possible impact of climate change on the groundwater behaviour is a fundamental prerequisite to the conservation of soft coastlines.

Coastal dunes extend over 7% of the British coastline; more specifically, they occupy over 50% of the 350 km coastline between Aberdeen and Inverness in NE Scotland (Ritchie 1983). The St. Fergus area, north of Peterhead, is the site of the UK's largest North-Sea gas-receiving terminal and is also the site of a major freshwater wetland within an extensive dune system, which is important for bird breeding and migration (Paterson 1997). The annual flooding of the wetland is controlled by the rise and fall of the water table in response to seasonal recharge patterns and is an important influence on the distribution of vegetation communities and ornithological habitats in the coastal dune system (Soulsby *et al.* 1997). The development of the gas terminal in close proximity to the wetland led to the initiation of several monitoring programmes to investigate what impact, if any, the complex was having on the hydrology, ecology and geomorphology of the region. This large data resource permitted a detailed investigation of the hydrology of this coastal wetland, which included an assessment of the possible impact of climate change on the wetland. The aims of this paper are three-fold: (i) to describe the development of a numerical model of groundwater behaviour in the freshwater aquifer; (ii) to describe the simulation of the impact of possible climate change scenarios on the groundwater levels; and (iii) to highlight the potential utility of groundwater modelling in forecasting the impact of climate change on coastal habitats and in aiding management of these environments.

From: ROBINS, N. S. & MISSTEAR, B. D. R. (eds) *Groundwater in the Celtic Regions: Studies in Hard Rock and Quaternary Hydrogeology*. Geological Society, London, Special Publications, **182**, 191–204. 1-86239-077-0/00/ $15.00

Study area

Physiography of St. Fergus

The St. Fergus coastline comprises a dynamic beach and dune system that extends over 6 km (Fig. 1). A detailed description of the site is provided elsewhere (Ritchie 1990) and only the main physical features are reported here. The main dune ridge is approximately 16 m high and to the west of this lies the major physiographic feature, the 'Winter Loch', a coastal dune slack which extends over 2 km in length and floods for on average five months of the year due to the rise and fall of the water table (Soulsby & Malcolm 1997). The base of the Winter Loch is some 1.5 m above sea level. A smaller and more stable inner dune ridge runs along the adjacent western margin of the dune slack rising to 5–7 m Above Ordnance Datum (AOD). An escarpment marks the position of a former post-glacial cliff-line and rises steeply to a plateau approximately 12–16 m AOD. The site has been crossed in seven places by the gas pipelines that have been brought onshore here, which were carefully restored to minimize any impact of installation on the site hydrology or dune stability.

Hydrogeology

The solid geology of the Peterhead area comprises roughly equal proportions of metamorphosed sedimentary and granitic rocks with a subordinate volume of metavolcanic rocks. The dominant underlying igneous rocks are

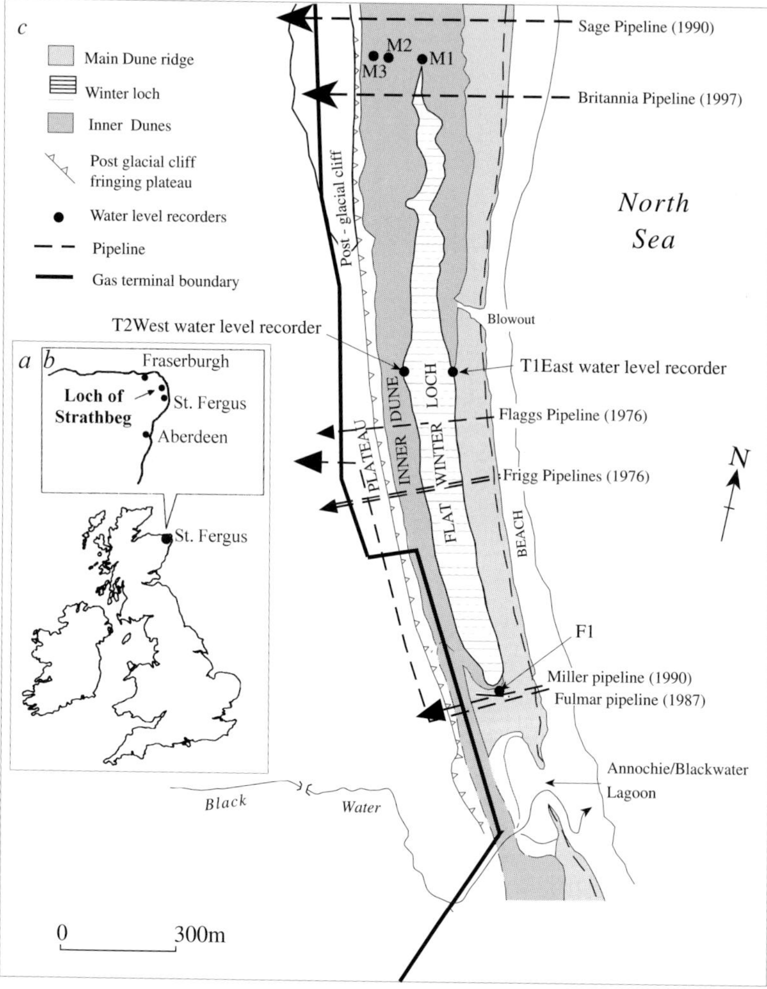

Fig. 1. Location map of St. Fergus and the main physiographic elements.

the Peterhead and St. Fergus granites, which are considered to be mineralogically similar and have been described as a regularly jointed pink leucocratic adamellite (British Geological Survey 1992).

The drift geology at St. Fergus is more complex, and represents the result of ice movement from differing source regions and sea level fluctuations during the Quaternary. Two differing ice-masses converged in this region and resulted in interdigitated tills of differing sedimentary origin (Gemmell 1990). Above the tills are layers of sand/silt/clay beds of glaciolacustrine origin (Fig. 2). Once the ice melted, the tills and other sediments were eroded by the sea creating the current 'fossil' cliffline. Thus, beach sands and gravels were deposited at the foot of the cliff. During the Flandrian transgression, which culminated at St. Fergus approximately 7000 years before present, sea level rose again to a few metres above its present position. Relative sea level subsequently fell due to isostatic rebound and allowed the development of the contemporary main dune ridge resulting in the present morphology of the dune system.

Bedrock was encountered in few of the boreholes which were drilled prior to the terminal and reported by Gemmell (1990); typically, weathered pink granite was observed which descended steeply towards the Blackwater basin to the south. A high proportion of the unconsolidated sediment of the St. Fergus aquifer is derived from this granitic parent material.

The till, which directly overlies the bedrock, is described as 'boulder clay' or in borehole terminology a 'sandy silty clay'. Particle size analysis was completed on samples from selected boreholes (Gemmell 1990). The results showed a sediment composed of a high percentage of clay–silt particles (*c.* 40%) with a predominant sand component. The composition of the till results in low hydraulic conductivity (Britannia, unpublished report) thus it may be defined as the lower impermeable boundary to groundwater flow (Soulsby *et al.* 1997). The till rarely occurs far above sea level and, from the borehole data,

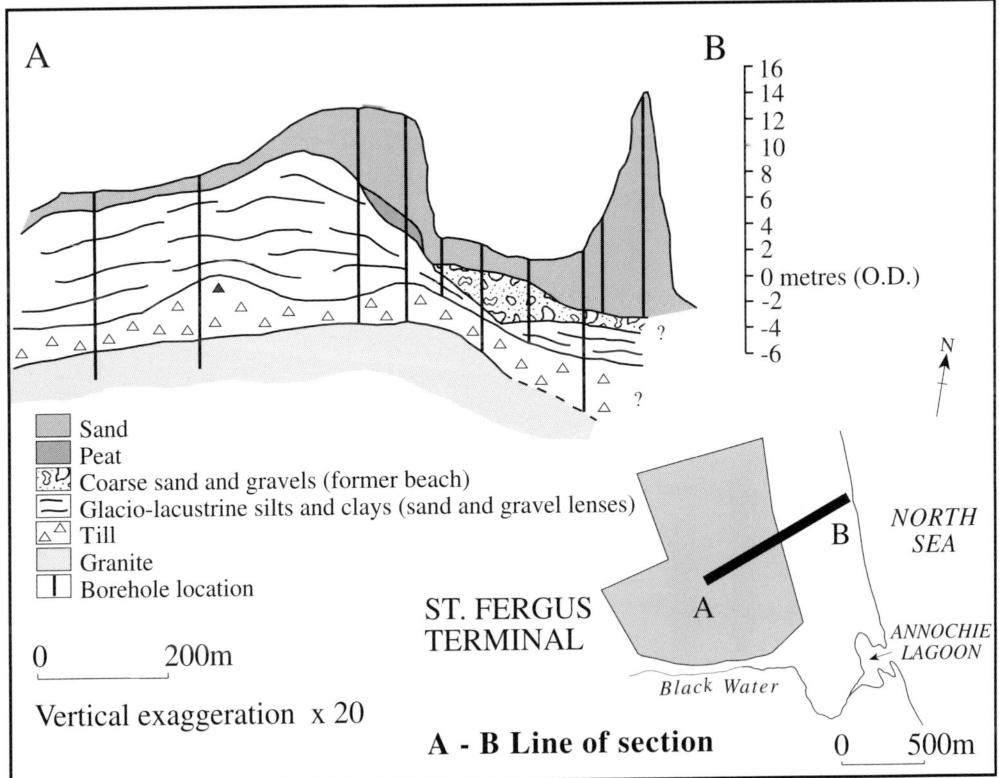

Fig. 2. Drift stratigraphy of St. Fergus dune system (after Gemmell 1990).

appears to dip gently to the east and south. However, this is in contrast with the borehole information which describes a silty sand with occasional gravel lenses. Gemmell (1990) states that the sediment described as till in the seismic profiles is, in part, glaciolacustrine sands and silts. It is important to note the complex stratigraphy of the plateau and fossil cliff. There is an intermingling of sands, silts and clays with gravel lenses and tills. Engineers on site have noted that within a 100 metre range in the north of the site two deep (5–10 m) cuttings in the escarpment revealed differing sedimentary profiles (C. Caulfield pers. comm.). For example, a trench made for the 1990 Sage pipeline (Fig. 1) cut through a predominantly clayey sediment. For the most recent development (1997), approximately 100 m to the south, the sediment cut consisted of sands and gravels. The complex interdigitation of these sediments with differing hydraulic properties (published literature suggests that the hydraulic conductivities of these sediments would differ by greater than three orders of magnitude (Dingman 1994)) has implications for groundwater modelling of the field system. A spatially averaged value for individual hydrogeological parameters is required for each model grid cell which masks the field variability. These complex beds are often 8–10 m thick but to the west of the fossil cliff their presence is much less substantial (Gemmell 1990).

The upper layer of the majority of boreholes consists of aeolian sand deposits. These are often of considerable thickness to the east of the fossil cliffline (borehole data shows 7–8 m thickness is common). Sediment size ranged from fine- to coarse-sands with a discontinuous layer of coarse-sand and gravel near the contact with the underlying tills and glaciolacustrine sediments. This layer represents the principal unconfined aquifer of the field site and is less common to the south and west of the St. Fergus region particularly above the fossil cliffline.

Hydrology

The mean annual precipitation at St. Fergus is approximately 673 mm and this is fairly evenly spread over the year, though the winter months tend to be wetter (Fig. 3). The majority of groundwater recharge also occurs at this time when evaporation rates are at their lowest and an annual average of 250 mm recharge has been estimated (Malcolm 1998). This was achieved by applying the Thornthwaite method of calculating monthly potential evapotranspiration values from monthly rainfall and sunshine hours (subsequently more fully explained) (Shaw 1983). Thornthwaite was applied as it was the most appropriate method given the data available. Analysis of the water table and stratigraphy data suggest that groundwater flows down a steep hydraulic gradient generated by the inclination of the base of the aquifer and flows across the dune system at a much shallower gradient before discharging into the sea to the east of the main dune ridge. The recharge of the water table results in an annual fluctuation of approximately 0.7 m. This leads to the flooding of the dune slack wetland generally from October/November through to April/May.

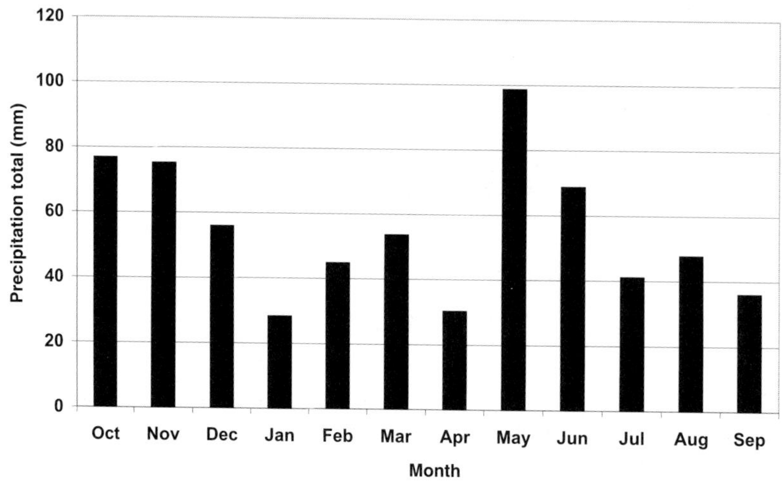

Fig. 3. Mean monthly precipitation values for St. Fergus (source: Transco gas terminal).

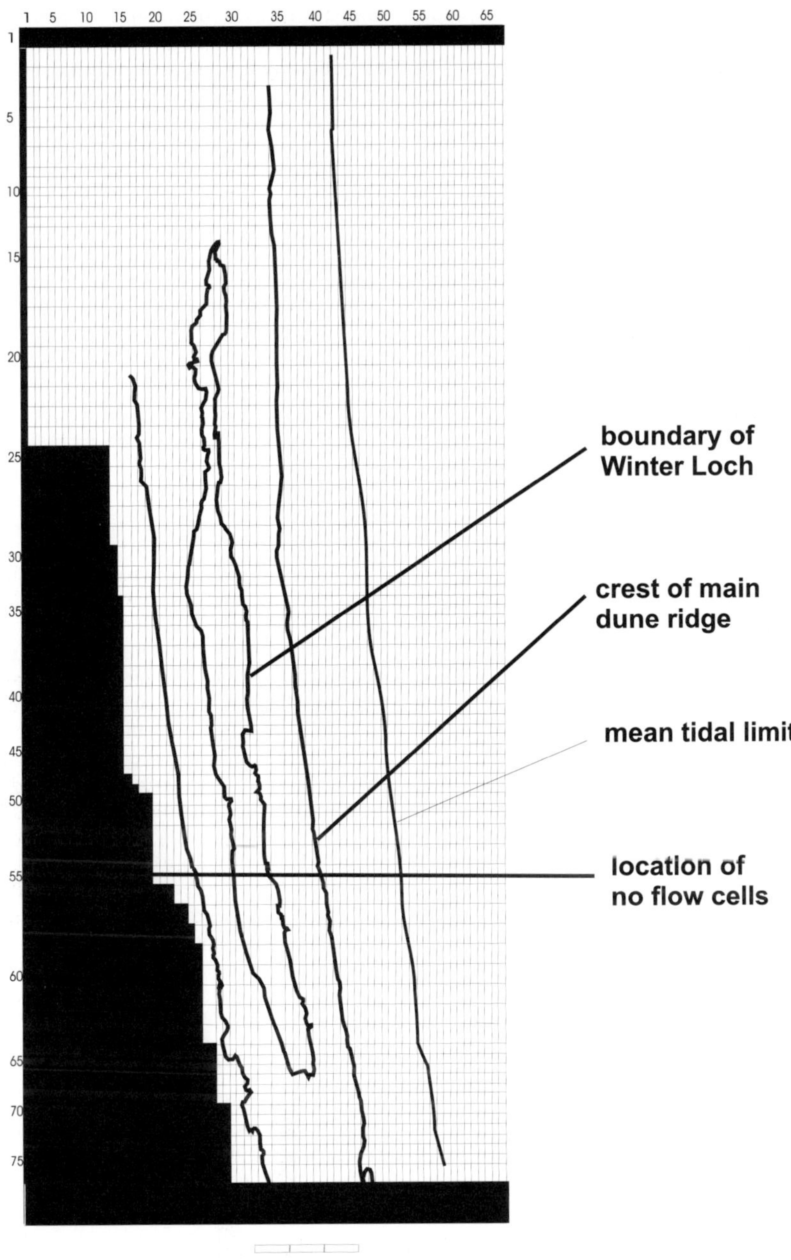

Fig. 4. Grid design of the St. Fergus MODFLOW model.

Methodology

The use of groundwater modelling techniques to simulate water table fluctuations in wetlands has been widespread (Gilvear *et al.* 1993; Gerla & Matheney 1996). In this study it provides the means for quantifying the behaviour of the groundwater regime which can subsequently be used to infer the possible impact of climate change on the freshwater aquifer. To achieve the aim of examining the impact of climate change on the aquifer water table, the method advocated by Arnell & Reynard (1996) in conjunction with that of Anderson & Cheng (1998) was adopted. The principle of this method is as follows: (i) the numerical groundwater model of the hydrologic system under consideration is firstly conceptualized, then defined and calibrated; (ii) the climate change scenarios obtained from published literature are used to alter the climate data input to the model and simulations run. In determining which climate change parameters to use in the model it is apparent that difficulties exist in predicting future estimates of them all. Particular problems surround the selection of appropriate timesteps for model calculation and simulation using global circulation model data; generally, there is greater uncertainty associated with calculating data on daily than monthly time steps. Currently, there appears to be less difficulty in estimating changes in future precipitation figures than evaporation due to the number and complex interaction of variables involved in calculating evaporation and their associated uncertainty. Thus, most published findings provide precipitation forecasts as percentage changes in monthly levels from current values and hence such data were used in the model.

The US Geological Survey code MODFLOW, a three-dimensional finite-difference groundwater model developed by McDonald & Harbaugh (1988), was used in quantifying changes in the groundwater regime (Fig. 4). MODELCAD (Rumbaugh 1993), a pre-processor package was used to design the model which was then configured into files for MODFLOW execution. The main topographic features were digitized using the ARC/INFO package, which provided a base map for the model to be built upon. To calibrate the model, a trial and error approach was adopted for both steady state and transient simulations. These used target data taken from five observation boreholes located within the field system. The water table elevations from the model were recorded at monthly time steps in consideration of the available literature on future climate estimates. From the model output, the water table contours created were drawn using a SURFER kriging interpolation (Golden Software 1995). From this and the output file of head values, the groundwater regime could be quantified and compared under differing scenarios.

Various hydraulic parameter values and information on recharge and stratigraphy were required for the model. A few hydraulic conductivity values were derived from site field data though the majority were derived from published data, which introduced further uncertainty in to the model, which is not fully quantified (Beven 1989; Bizzari & Ferraresi 1990). To a certain extent, such uncertainty is inevitable given the difficulty in measuring hydraulic variables at suitable scales for model estimates. Nevertheless, a detailed field investigation of the behaviour of the water table and of the process of recharge in coastal sand dunes (Malcolm 1998) resulted in the development of a good conceptual model of groundwater movement within the site.

Sea-level rise was not considered in the model as the rate of uplift due to isostatic rebound in the north-east of Scotland currently cancels the rate of sea-level rise occurring. Thus, there was no need to consider this in the modelling procedure (Shennan 1993; Kay *et al.* 1996).

Climate change scenarios

A range of recent climate change scenarios was chosen: these scenarios were considered representative of the wide variability of findings of recent research on climate change predictions for Britain (Table 1). Those scenarios chosen reflected the diversity of estimates being published. This provides a basis for predicting the range of likely responses. As previously outlined, there are issues to be considered in applying the precipitation estimates to the model and in interpreting the results. Principally, the precipitation data used are normally disaggregated from the output of Global Circulation Models (GCM's) which are at a much greater spatial scale than the approximate 1 km^2 being considered here. Thus, uncertainty in the estimates is amplified in consideration of local climate conditions which are nested within coarse regional GCM estimates. Furthermore, literature relating to contemporary climate change for the NE of Scotland is sparse with a few notable exceptions (Duncan 1992; Faulkner *et al.* 1997; Foster *et al.* 1997).

Precipitation data have been collected at one location by the gas companies at St. Fergus for 17 years, which was for model calibration and as baseline conditions with which to compare the predictive simulations. From this record, the lowest annual precipitation total recorded (October 1988–September 1989) was used as an historical example for modelling. From the published literature, the precipitation estimates published by Arnell & Reynard (1996), based on the UK Climate Change Research Group GCM models (CCIRG 1991) were used to provide a comprehensive range of scenarios (Table 1). Estimates of the change in precipitation values

Table 1. *Percentage change in precipitation values for individual months for each climate change scenario from contemporary values*

	Oct	Nov	Dec	Jan	Feb	Mar	Apr	May	Jun	Jul	Aug	Sep
1	−25	−25	−25	−25	−25	−25	−25	−25	−25	−25	−25	−25
2	−8	−8	0	0	0	−8	−8	−8	−16	−16	−16	−8
3	−5	−5	0	0	0	−5	−5	−5	−10	−10	−10	−5
4	0	0	0	0	0	0	0	0	0	0	0	0
5	+2	+2	+5	+5	+5	+2	+2	+2	0	0	0	+2
6	0	0	+10	+10	+10	0	0	0	+6	+6	+6	0
7	+4	+4	+8	+8	+8	+4	+4	+4	0	0	0	+4
8	+10	+10	+10	+10	+10	+10	+10	+10	+10	+10	+10	+10
9	+16	+16	+16	+16	+16	+16	+16	+16	+16	+16	+16	+16
10	0	0	+40	+300	+140	0	0	0	0	0	0	0

1 Contemporary driest scenario.
2 Arnell driest scenario year 2050.
3 Arnell driest scenario year 2030.
4 Current conditions.
5 Arnell best guess scenario year 2030.
6 Hulme scenario year 2050.
7 Arnell best guess scenario year 2050.
8 Arnell wettest scenario year 2030.
9 Arnell wettest scenario year 2050.
10 Rowntree scenario year 2050.

from current conditions were provided for the years 2030 (mid range) and 2050 (long range) under 'driest', 'best guess' and 'wettest' conditions. Hulme *et al.* (1993) provided data from the ESCAPE project (CRU 1992): however, the majority of scenarios were similar to those of Arnell & Reynard (1996) and thus were not considered. One scenario of Hulme *et al.* was used where precipitation was predicted to increase only in the winter months (defined as December to February) and the summer months (defined as June to August) by 1 mm per day. Finally, Rowntree (1990) provides a comprehensive review of climate change data for the United Kingdom, which supplied one further set of precipitation estimates.

Conceptual model

The conceptual model for the St. Fergus numerical groundwater model was relatively simple, consisting of one hydrostratigraphic unit (layer one, unconfined type). The upper layer of the underlying glaciolacustrine sediments marked the position of the confining unit as the majority of these sediments had hydraulic conductivities two orders of magnitude lower than the overlying sands and gravels (Anderson & Woessner 1992). A 2-dimensional areal model was proposed for groundwater simulation. Inflow to the aquifer system predominantly consists of groundwater recharge from precipitation (Fig. 5). Overland flow is considered negligible at a scale considered significant for modelling. No surface flow occurs and the main outflow is groundwater discharge to the seaward margin. The Winter Loch feature is an ephemeral dune slack which occurs solely in response to the rise and fall of the water table and is not considered as a separate entity in the modelling procedure. For the majority of the duration when the water table rises above the ground surface and the

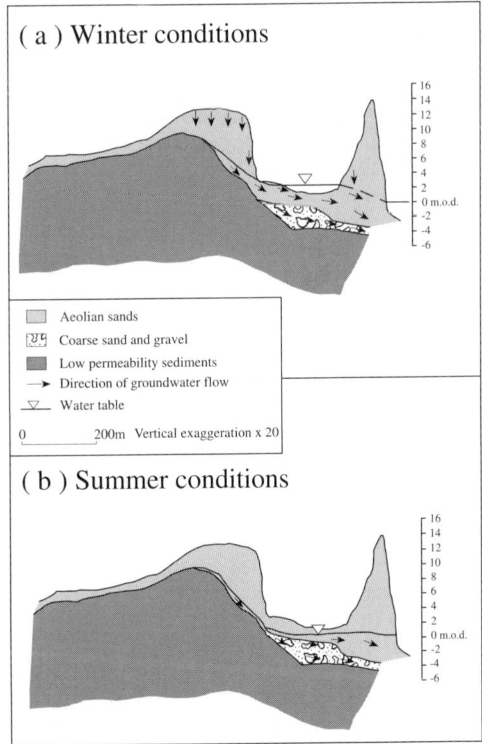

Fig. 5. Model of the groundwater flow at St. Fergus.

'loch' occurs, the total area of flooding is relatively small in comparison to the total area of dune system.

To the north and south, groundwater streamlines identified from field water table maps were chosen to mark no-flow cell boundaries to contain the model. The thinning of the hydrostratigraphic unit to the west on the plateau also marked the approximate location of the groundwater divide; thus a no-flow boundary was set. Finally, to the east, constant head cells (set at 0.0 m) were set to mark the seaward margin.

The modelled area was discretized using a variably spaced, block-centred, finite difference grid of 68 columns and 79 rows; giving a total of 5372 cells including 77 constant head cells (Fig. 4). The grid itself was orientated subparallel to the principal hydraulic gradient of the site. Finer grid spacing was specifically designed around: (a) the catchment divide to the west, where water would flow east either into the dune system or to inland streams to the west; (b) the target borehole locations and; (c) the steep hydraulic gradient of the cliff region.

Data input and model calibration

Geological data were available from 16 boreholes which were spatially distributed in 4 transects generally concentrated in the north of the field site. The hydrogeological characteristics were determined predominantly from published literature and available log data. Slug tests were attempted to determine hydraulic conductivity; however, values were not obtained due to the high permeability of the sediment. Thus, for initial model development, a range of published values of hydraulic conductivity and specific yield were used, the hydraulic conductivity data relating to the local area (Robins 1986; Dingman 1994). Values of hydraulic conductivity in the calibrated model ranged from 0.01–50 m per day relating to differing regions of the aquifer. Near the western margin, lower values of hydraulic conductivity were initially chosen for model set up and were only slightly altered during calibration (K values of between 0.01–2 m per day). In the majority of the dune system, calibrated values of hydraulic conductivity ranged from 20–50 m per day, reflecting the highly permeable sands of the dune aquifer. The specific yield of the sediment was also obtained from published literature and ranged from 0.15 (silts and silty sands) to 0.30 (medium sands). All of the hydrogeological and geological information was input to the model by formulation of zones of average parameters which were computed by MODELCAD using SURFER gridded files. Steady state and transient calibration occurred over the hydrological year October 1995 to September 1996 using the target data.

Calibration could only be undertaken on one year of data (October 1995 to September 1996) as this was the only year water table elevations were recorded at five points across the dune system. Water table variation had been recorded previously at two boreholes which were restricted to a small area of the field site and were subsequently found to poorly represent the behaviour of the water table throughout the dune system (Malcolm 1998). Nevertheless the data from the two boreholes provided the opportunity to determine how representative the calibration period was of previous years. The groundwater hydrograph over the calibration year followed a similar sinusoidal pattern to previous years and monthly values of water table elevations in the two boreholes were within one standard deviation of average conditions over the previous ten years of observations. For the transient model, the initial and final head values of the observation wells were also within one standard deviation of the longer term average conditions.

To calculate monthly effective recharge values, potential evapotranspiration figures [calculated by the Thornthwaite method as no other method was possible (Shaw 1983)] were subtracted from the monthly contemporary precipitation totals. This was the most appropriate method in consideration of the data available and resulted in approximately 250 mm of effective recharge from the annual rainfall totals, which gives an average rate for inclusion in the steady state model of 0.68 mm d^{-1}; this was uniformly distributed across the modelled area. Twelve monthly time steps were chosen with the simulated water table elevation compared to the mean monthly water table recorded in the field. The recharge data for the calibration year was divided in to individual monthly values of recharge. These monthly time steps were chosen in consideration of the constraints of the regional estimates of climate change.

Model calibration allowed for refinement of the model's representation of the hydraulic properties, hydrogeologic framework and boundary conditions which were included in sensitivity analysis. Thus a desired level of agreement between the model simulations and the observed groundwater field data could be achieved (Chen *et al.* 1997). Sensitivity analysis was undertaken to quantify the uncertainty in the calibrated model parameters (Table 2). The north and south boundaries represented groundwater stream lines and thus were not considered

Table 2. *Results of the sensitivity analysis for the steady state model*

Parameter	Zone	Multiplier	Residual sum of sources
Base			0.042
K1 (m/d)	0.1	2	0.044
K2 (m/d)	1	2	0.044
K3 (m/d)	2.5	2	0.042
K4 (m/d)	5	2	0.059
K5 (m/d)	7.5	2	0.042
K6 (m/d)	10	2	0.096
K7 (m/d)	15	2	0.051
K8 (m/d)	20	2	0.082
K9 (m/d)	30	1.5	0.668
K10 (m/d)	40	1.5	0.209
K11 (m/d)	50	1.5	0.331
recharge m/d	0.00068	0.5	4.367
recharge m/d	0.00068	1.5	2.520

in sensitivity analysis. The western boundary, representing the groundwater divide was relatively inflexible as it was controlled by known locations of the aquifer elevation.

Simulation results

The results of the calibration procedure for the steady state model are shown in Table 3. The residual sum of mean squares was 0.042 with a maximum residual found at M_3 of 0.19. The largest residuals occurred in the north-eastern end of the field site, reflecting the lack of geological data in this vicinity. The resulting water table contour map generally agreed with the conceptual model for the field system as represented by a drawn contour map of field head data (Fig. 6). A high hydraulic head is generated from the cliff and plateau region which drives the groundwater across the dune system to the seaward margin according to the conceptual model shown in Fig. 5. The gradient of the water table is relatively shallow in the dune system.

Trial and error calibration was also undertaken with the transient model but additionally, hydrographs of simulated v. observed water table elevations at each target point were also created. Overall, simulated water table contours agreed with the observed values over the monthly time steps though the greatest difference occurred in the summer months, particularly in the northern section (Figs 7 and 8). Similarly to the steady state calibration, the largest divergence between model and field hydrographs occurred in the most north-easterly section of the field site, where the recession limb of the model hydrograph could not accurately simulate the observed.

Table 3. *Final simulated water table values and associated statistics for calibrated steady state model*

Well	Target head	Model head	Residual
M3	2.45	2.26	0.19
M2	1.98	2.00	−0.02
M1	1.67	1.60	0.07
T2	1.55	1.56	−0.01
F1	1.59	1.55	0.04
Residual mean		0.0523	
Residual standard deviation		0.0756	
Residual sum of squares		0.0416	
Absolute residual mean		0.0659	
Minimum residual		−0.0193	
Maximum residual		0.1864	
Observed range in head		0.9000	
Residual standard deviation range		0.0829	

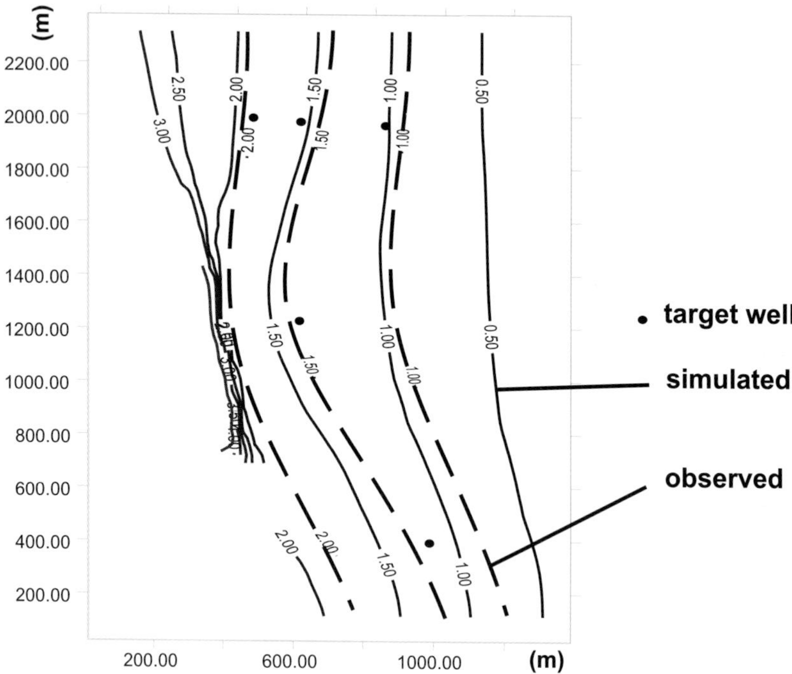

Fig. 6. Steady state water table contour map showing observed and simulated contours.

Climate change predictions

Using the calibrated model, climatic parameters were varied according to each climate change scenario. The mean annual water table elevation for each borehole under the differing climate change scenarios is shown in Fig. 9. The range in mean annual groundwater levels under each climate change scenario for the individual target boreholes is between 20–45 cm (with the exception of the 1st and 10th scenarios). Model simulations 2 through to 7 range from roughly −0.08 m below the current mean annual water table of each borehole to +0.11 m above. The inclusion of the Arnell driest scenario, year 2050 (1st) and the Rowntree scenario 2050 (10th) greatly increases the range in mean water table elevation. Nevertheless, the mean values do not reflect the dynamic nature of the physical environment.

These changes in water table level were translated to the equivalent spatial extent and temporal duration of flooding (Table 4). This was achieved by delineating on a detailed topographic site map the regions where the simulated monthly water tables rose above ground surface. The duration of flooding is relatively similar for most simulations (with the exception of 1, 8 and 10). Flooded areas generally emerge during November/December and remain until early summer. However, the duration of flooding is significantly different for each scenario. The wettest prediction (Arnell 2050) shows that, on average, a 100% increase in the flooded area for each month might be expected, despite the apparent small increase in effective recharge. The more extreme predictions show possibilities ranging from no flooding at all (1st scenario) to 300% increase in flooded area in December (10th scenario).

Implications

As with all model prediction, uncertainty in terms of the structure and parameterization of the model and the realism of the scenarios used bedevils interpretation (Wheater & Beck 1995). It is widely recognized that disaggregating outputs from GCM's is a major constraint on attempts to develop site specific climate change predictions, as evidenced by the range of scenarios considered in this study. Despite this, the medium term data record from St. Fergus implies that a likely shift in average conditions over the long term, of precipitation, evaporation and recharge, will fall within the range

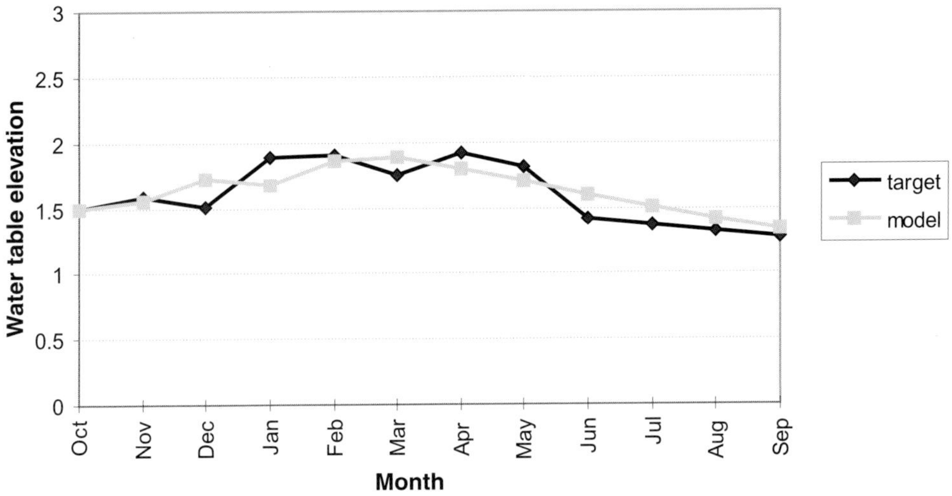

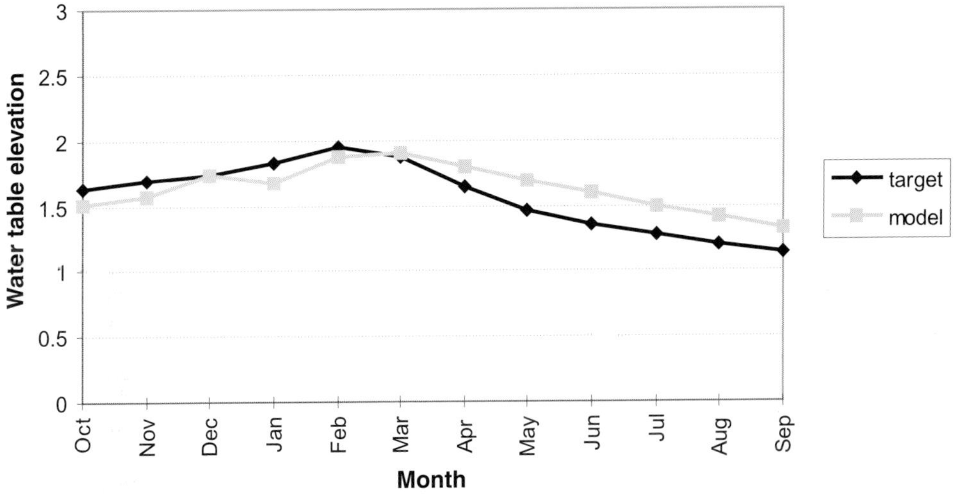

Fig. 7. Comparison of monthly target and simulated water table elevations at two target points.

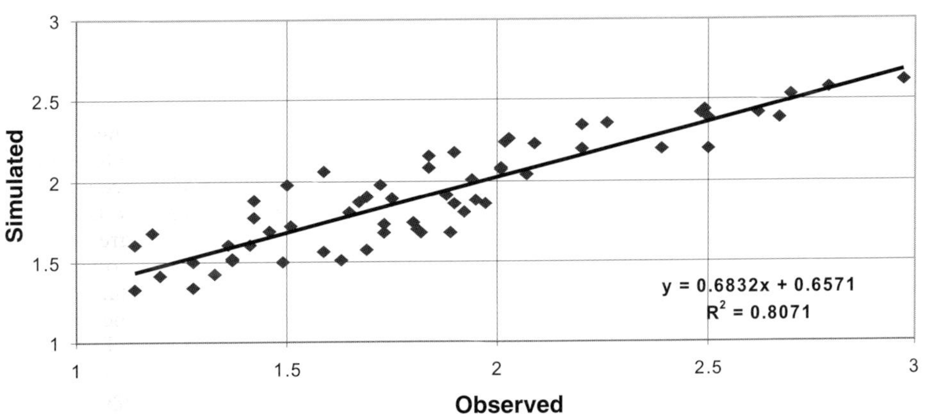

Fig. 8. Correlation of observed versus simulated water table elevations for all target points.

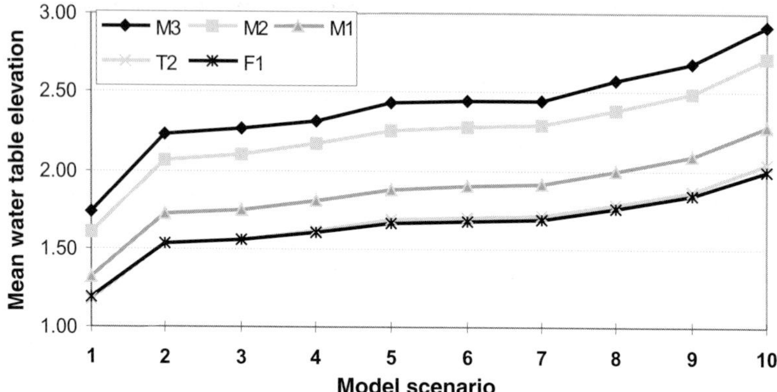

Fig. 9. Mean annual water table elevation for individual target boreholes under each climate change scenario.

Table 4. *Areas (in 1000 m^2) of dune slack flooded at monthly intervals under differing climate change scenarios*

	Oct	Nov	Dec	Jan	Feb	Mar	Apr	May	Jun	Jul	Aug	Sep
1	0	0	0	0	0	0	0	0	0	0	0	0
2	0	0	14.9	11.8	62.9	59.3	27.6	11.8	0	0	0	0
3	0	0	20.5	14.2	67.8	64.9	39.3	15.8	0	0	0	0
4	0	0	27.6	22.9	77.6	76.6	52.6	23.7	0	0	0	0
5	0	0	52.6	39.3	94.4	95.6	72.7	50.0	19.7	0	0	0
6	0	0	56.0	44.3	101.3	9.0	74.7	49.3	22.1	0	0	0
7	0	0	62.0	52.6	102.5	101.3	76.6	52.6	23.7	0	0	0
8	0	25.2	76.6	69.8	110.0	109.2	94.4	70.8	42.6	18.1	0	0
9	0	46.0	95.6	86.3	117.2	117.2	107.6	87.5	64.9	30.9	13.4	0
10	22.1	63.9	111.6	114.8	140.8	137.2	124.4	112.0	103.7	75.7	35.9	23.7

Areas in 1000's of m^2.
1 Contemporary driest scenario.
2 Arnell driest scenario year 2050.
3 Arnell driest scenario year 2030.
4 Current conditions.
5 Arnell best guess scenario year 2030.
6 Hulme scenario year 2050.
7 Arnell best guess scenario year 2050.
8 Arnell wettest scenario year 2030.
9 Arnell wettest scenario year 2050.
10 Rowntree scenario year 2050.

of extreme years recorded over the last two decades. Nevertheless, the ecological impacts of this are difficult to predict. A small shift in the average level of the water table may have a great impact on vegetation composition and diversity. This requires ecological modelling for such questions to be answered.

The possible impact of climate change on coastal wetlands in the Celtic regions, is a complex issue which is currently surrounded by a great deal of uncertainty. To date, there has been limited discussion on changes in the groundwater regime of coastal aquifers, despite their importance as ecological resources and natural coastal defences. Resource managers often require detailed information on the nature and magnitude of likely impacts yet this is often difficult or impossible to obtain. If the direction of climate change is also uncertain, as in this case, then a range of possible scenarios provides managers with the basis for options which can be acted upon when clearer evidence of the direction and magnitude of climate change becomes apparent. Possible action plans can be developed or initiated to preserve biodiversity if required. This paper provides a first approximation of the possible impact of climate change on a typical freshwater wetland as a pre-cursor to improved integration of climate change information and ecological modelling. The application of groundwater models, integrated with good field data, and used in conjunction with ecological information will result in better informed guidelines and estimates of the impact of climate and environmental change on the wetlands occupying dune systems that characterises much of the Celtic coastline.

This work was funded by the St. Fergus Dunes Management Committee and the University of Aberdeen. The support and assistance of L. Kingham and B. Ritchie is gratefully acknowledged. Supplementary resources were generously provided by Britannia Operator and the help given by C. Caulfield and H. Bradshaw is greatly appreciated. R. Gard provided much needed assistance with field instrumentation. The advice of M. Chen in the intricacies of groundwater modelling was invaluable. Thanks are also given to two anonymous referees who provided valuable comments on the original manuscript.

References

ANDERSON, M. P. & WOESSNER, W. W. 1992. *Applied Groundwater Modelling simulation of flow and advective transport*. Academic Press, London.

ANDERSON, M. P. & CHENG, X. 1998. Sensitivity of groundwater/lake systems in the Upper Mississippi River basin, Wisconsin, USA, to possible effects of climate change. *In*: *Hydrology, Water Resources and Ecology in Headwaters* (Proceedings of the HeadWater'98 Conference, Italy). International Association of Hydrological Sciences Publication Number, **246**, 3–8.

ARNELL, N. W. & REYNARD, N. S. 1996. The effects of climate change due to global warming on river flows in Great Britain. *Journal of Hydrology*, **183**, 397–424.

BEVEN, K. 1989. Changing ideas in hydrology – the case of physically based models. *Journal of Hydrology*, **105**, 157 172.

BIZZARI, A. & FERRARESI, M. 1990. Model uncertainties in the simulation of the river Marecchia alluvial fan northern Italy. *In*: *ModelCARE 90: Calibration and reliability in groundwater modelling*, Proceedings of the conference held in The Hague, September 1990. IAHS Publication no. **195**, 363–372.

BRITISH GEOLOGICAL SURVEY. 1992. *Solid and drift geology of the Peterhead region*. Sheet 87E, 1:50 000.

CHEN, M., SOULSBY, C. & WILLETTS, B. 1997. Modelling river-aquifer interactions at the Spey Abstraction scheme, north-east Scotland: implications for aquifer protection. *Quarterly Journal of Engineering Geology*, **30**, 123–136.

CCIRG. 1991. *The potential effects of climate change in the United Kingdom*. HMSO, London.

CRU (ed). 1992. *A scientific description of the ESCAPE model*. Norwich: Climate Research Unit, University of East Anglia.

DINGMAN, S. L. 1994. *Physical Hydrology*. Prentice-Hall, New Jersey.

DUNCAN, K. 1992. The implications of global warming in south-east Scotland: an historical analogue approach. *Scottish Geographical Magazine*, **108**, 172–178.

FAULKNER, D. S., ARNELL, N. W. & REYNARD, N. S. 1997. Everyday aspects of climate change in Europe. *Proceedings of the Sixth National Hydrology Symposium*, Salford 1997, 8.1–8.8.

FOSTER, M., WERRITY, A. & SMITH, K. 1997. The nature, causes and impacts of recent hydroclimatic variability in Scotland and Northern Ireland. *Proceedings of the Sixth National Hydrology Symposium*, Salford 1997, 8.9–8.17.

GATES, P. 1994. Climate change: implications for conservation management. *Ecos*, **15**, 29–34.

GERLA, P. J., & MATHENEY, R. K. 1996. Seasonal variability and simulation of ground-water flow in a prairie wetland. *Hydrological Processes*, **10**, 903–920.

GEMMELL, A. M. D. 1990. Stratigraphy and Drift Geology. *In*: RITCHIE, W. & KINGHAM, L. *The St. Fergus Coastal Environment: The physical and biological characteristics*. CEMP April 1990, 7–15.

GILVEAR, D. , ANDREWS, R., TELLAM, J. H., LLOYD, J. W. & LERNER, D. N. 1993. Quantification of the water balance and hydrogeological processes in the vicinity of a small groundwater-fed wetland; East Anglia, UK. *Journal of Hydrology*, **144**, 311–334.

GOLDEN SOFTWARE. 1995. *SURFER for Windows: a user's guide, version 6* Golden Software Inc.

HAES. H. DE. & WOLTERS, A. R. 1992. The Golden Fringe of Europe: Ideas for a European Coastal Conservation Strategy and action plan *In*: CARTER, R. W. G., CURTIS, T. G. F. & SHEEHY-SKEFFINGTON, M. J. (eds) *Coastal Dunes*. Proceedings of the Third European Dune Congress, Galway, Ireland. Balkema, Rotterdam.

HULME, M., HOSSELL, J. E. & PARRY, M. L. 1993. Future climate change and land use in the United Kingdom. *The Geographical Journal*, **159**, 131–147.

KAY, R. C., ELIOT, I., CATON, B., MORVELL, G. & WATERMAN, P. 1996. A review of the Intergovernmental Panel on climate change's Common Methodology for assessing the Vulnerability of Coastal Areas to Sea-Level Rise. *Coastal Management*, **24**, 165–188.

MALCOLM, R. 1998. *The hydrology of a restored coastal dune slack, St. Fergus, north-east Scotland: an integrated field and modelling approach*. Unpublished PhD thesis, University of Aberdeen.

MCDONALD, M. G. & HARBAUGH, A. W. 1988. *A modular three-dimensional finite-difference groundwater model*, US Geological Survey, Techniques of Water-resources Investigations Book 6, chapter A1.

PATERSON, I. 1997. The birds of the St. Fergus dunes: 1977–1996. *In*: RITCHIE, W. & KINGHAM, L. (eds) *The St. Fergus Coastal Environment: monitoring and assessment of the coastal dune at the Northsea gas terminals, St. Fergus*. Aberdeen University, Aberdeen, 115–126.

RITCHIE, W. 1983. Coastal land use in North East Scotland. *In*: RITCHIE, W. (ed.) *North East Scotland coastal field guide and geographical essays*. Department of Geography, University of Aberdeen.

——1990. General Physical setting of the St. Fergus terminal. *In*: RITCHIE, W. & KINGHAM, L. (eds) *The St. Fergus Coastal Environment: The physical and biological characteristics*. CEMP April 1990, 1–6.

ROBINS, N. S. 1986. *Hydrological observations on the borrow pit at North Kirkton, St.Fergus*. Hydrogeology Research Group, British Geological Survey.

ROWNTREE, P. R. 1990. Estimates of future climate change over Britain. Part 2: results. *Geographical Magazine,* 79–89.

RUMBAUGH, J. O. 1993. *MODELCAD386: Computer aided design for groundwater modelling, Version 2.20,* Geraghty and Miller Modelling Group Software.

SHAW, E. M. 1983. *Hydrology in practice*. Van Nostrand Reinhold International.

SHENNAN, I. 1993. Sea-level changes and the threat of coastal inundation. *The Geographical Journal,* **159**, 148–156.

SOULSBY, C. & MALCOLM, R. 1997. Hydrology of St. Fergus dune system. *In*: RITCHIE, W. & KINGHAM, L. (eds) *The St. Fergus Coastal Environment: monitoring and assessment of the coastal dune at the northsea gas terminals, St. Fergus*. Aberdeen University, Aberdeen, 75–90.

——, HANNAH, D., MALCOLM, R., MAIZELS, J. K. M. & GARD, R. 1997. Hydrogeology of a restored coastal dune system in northeastern Scotland. *Journal of Coastal Conservation,* **3**, 143–154.

VAN DER MEULEN, F. 1990. European Dunes: consequences of climate change and sea-level rise. *In:* BAKKER, T. W., JUNGERIUS, P. D. & KLIJN, J. A. (eds) *Dunes of the European Coasts – Geomorphology, Hydrology and Soils*. Catena Supplement 18, New York, 205–233.

VILES, H. & SPENCER, T. 1995. *Coastal Problems: Geomorphology, ecology and society at the coast.* Edward Arnold, London.

WHEATER, H. S. & BECK, M. B. 1995. Modelling upland stream water quality: process identification and prediction uncertainty. *In*: TRUDGILL, S. T. (ed.) *Solute Modelling and Catchment Systems.* John Wiley and Sons, Chichester, 306–369

Groundwater exploration in rural Scotland using geophysical techniques

A. M. MACDONALD[1], D. F. BALL[1] & D. M. McCANN[2]

[1] *British Geological Survey, Murchison House, West Mains Road, Edinburgh EH9 3LA, UK (e-mail: amm@bgs.ac.uk)*

[2] *Department of Civil and Environmental Engineering, University of Edinburgh, Kings Buildings, West Mains Road, Edinburgh EH9 3JN, UK*

Abstract: Identifying the location, size, and characteristics of small aquifers is becoming increasingly important in Scotland with the growing demand for groundwater to supply isolated rural communities. These small, localized aquifers are found within superficial deposits, such as alluvium, blown sands and raised beach deposits, and in fracture zones in the underlying bedrock. Geophysical techniques offer a rapid and inexpensive method of characterizing these aquifers. Electromagnetic techniques using EM34 and EM31 instruments have proved useful in identifying variations in the thickness of superficial deposits and detecting buried channels, for example, at Palnure, SW Scotland. Ground penetrating radar has been used at several locations, including the island of Coll, to detect the boundary between bedrock and alluvium or blown sand. More detailed techniques, such as resistivity soundings, seismic refraction and resistivity tomography have been used to identify fractures in basement rocks and help calibrate the other methods, for example at Foyers, near Loch Ness. Magnetic profiling has also been used to locate dykes (acting as hydraulic barriers) within Permian aquifers in the west of Scotland.

Groundwater is increasingly important for water supply in rural Scotland, as surface water, although plentiful, can sometimes be strongly coloured, variable in quality or unreliable during drought. This increased interest in groundwater has driven exploration in a variety of hydrogeological environments throughout Scotland – many of which were dismissed in the past as being of little potential. Because the water demands of rural Scotland are low (a village of 100 people may require only 25 m^3/d), even small amounts of groundwater can have a strategic importance in supplying the dispersed population of rural areas.

Many of the minor aquifers in Scotland are heterogeneous. Aquifer properties and geometry can vary significantly within a few metres, which can have a large impact on borehole yield and sustainability. It is necessary, therefore, to explore these minor aquifers thoroughly, to ensure that production boreholes are sited in optimum locations. Surface geophysics, when combined with exploration drilling, is a cost-effective way of defining the geometry of minor aquifers.

Hydrogeological environments in Scotland

Groundwater occurs in a variety of hydrogeological environments throughout Scotland. Major aquifers are limited and comprise Devonian, Permian and Carboniferous sediments (Fig. 1). Surface geophysics can add little to the process of exploration in these major aquifers, except perhaps in identifying dykes which may act as barriers to groundwater flow. Borehole geophysical techniques, however, have proved useful within the major aquifers, identifying fractures and major inflow zones (Buckley this volume).

Surface geophysical techniques have been found to be most useful in characterizing small shallow aquifers, such as drift deposits, or fractured bedrock. These small aquifers are often thin and discontinuous – therefore, the potential for finding a sustainable water supply can vary significantly over short distances. The nature and geometry of the aquifers directly affect the applicability and effectiveness of geophysical methods. The characteristics of the minor aquifers of rural Scotland are summarized in Table 1 and discussed briefly below.

From: ROBINS, N. S. & MISSTEAR, B. D. R. (eds) *Groundwarer in the Celtic Regions: Studies in Hard Rock and Quaternary Hydrogeology*. Geological Society, London, Special Publications, **182**, 205–217. 1-86239-077-0/00/

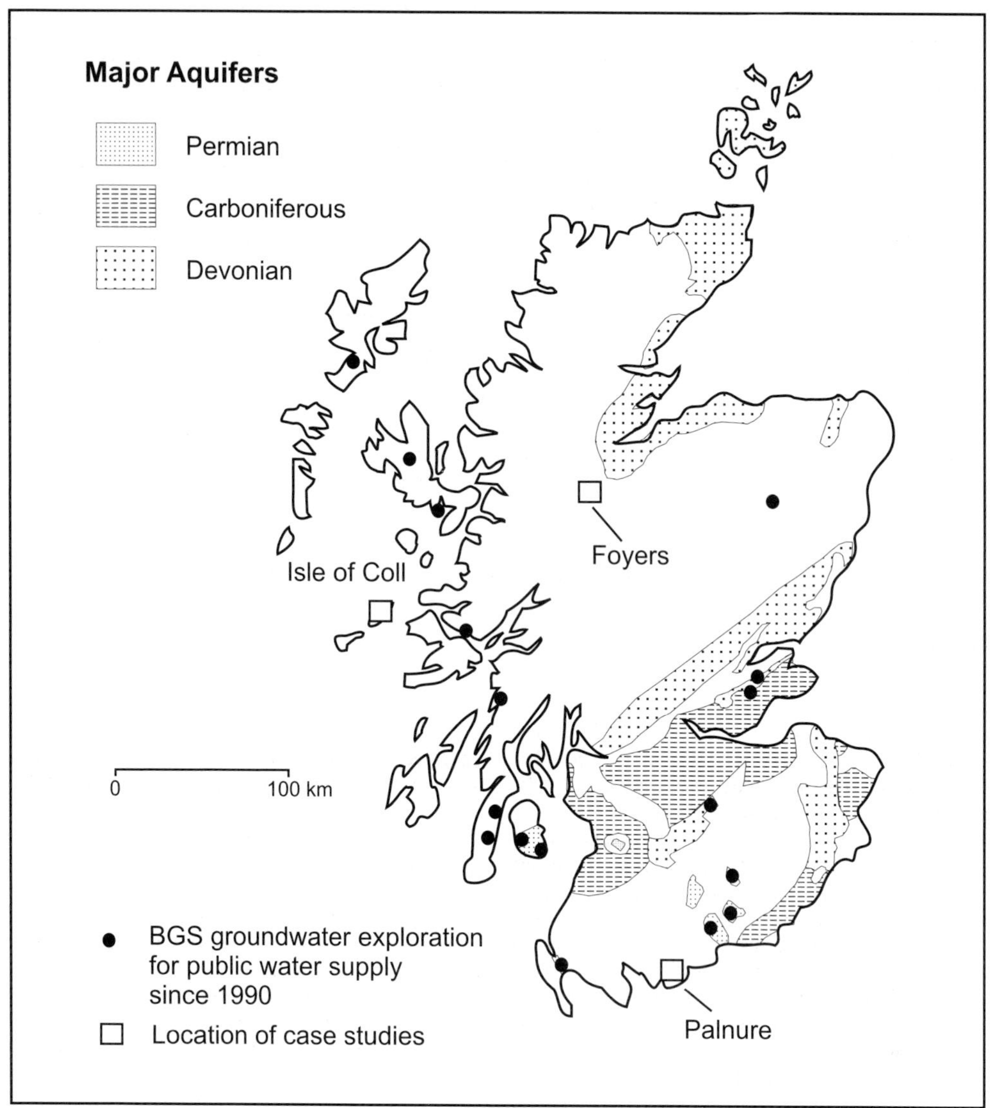

Fig. 1. Major aquifers in Scotland and locations of groundwater investigations for public supply by BGS in Scotland since 1990. [Major aquifers based on British Geological Survey (1988)].

Fluvioglacial deposits are often highly heterogeneous, varying from coarse gravel to fine, clayey sands. It is common for large blocks of boulder clay to be present within the deposit or for lenses of lacustrine clays to occur adjacent to thick beds of sand. Where buried gravel-filled channels are present, fluvioglacial deposits can offer useful groundwater potential.

Riverine alluvium commonly contains surface layers of silty clay, interbedded with thin beds of well-sorted gravel. In many cases, river gravel constitutes glacial material reworked into more uniform beds with a greater degree of sorting. Buried channels are common features beneath river floodplains in valleys once occupied by glaciers. There are currently several rural water supply schemes from river gravels planned or in operation, the highest yielding being from the River Spey in Morayshire (Watt *et al.* 1987; Jones & Singleton this volume) and the River Lochy near Fort William (Johnstone & Rennie 1991).

Raised beach deposits are variable in composition, ranging from coarse beach shingle containing no fine material, to thick, silty clays formed in

Table 1. *Summary of aquifer conditions and geometry of small dispersed aquifers in Scotland*

Hydrogeological environment	Groundwater potential	Groundwater target	Aquifer geometry	Depth to target	Groundwater flow
Fluvioglacial sand and gravel	Moderate	Sand and gravel within buried channels or lenses	Thin sand lenses, and long buried channels	0–40 m	Intergranular
Alluvial deposits	High	Medium–coarse-grained sand and gravel	Variable thickness, often homogeneous	0–20 m	Intergranular
Raised beach deposits	Moderate/low	Beach gravel, particularly where in small basins	Thin (<2 m) lenticular	0–20 m	Intergranular
Blown sand	Moderate/high	Saturated sand away from sea	Thin, homogeneous	0–10 m	Intergranular
Glacial till	Low	Thin gravel layers within the till	Thin (<2 m) and lenticular	0–30 m	Intergranular/ fracture
Basement rocks	Low	Fault zones and areas of intense fracturing	Narrow sub-vertical fractures zones	0–100 m	Fractures

lagoons. Groundwater occurrence within raised beach deposits is sporadic. The main targets for groundwater development are beach gravels, especially where the underlying bedrock surface is depressed and the gravel thick. However, high concentrations of sulphate and organic material may be present; and groundwater can be brackish where deep deposits occur below sea level.

Blown sands are generally well-sorted, comprising fine and medium-grained permeable sand. Although permeable, these deposits can be thin, and conventional boreholes may not perform satisfactorily. On Tiree, an infiltration gallery has been constructed within blown sand, which takes advantage of the good lateral permeability of the deposit (Robins 1990).

Glacial till, or boulder clay, is a characteristically poorly-sorted material composed predominantly of clay and stones. It is a widespread deposit and covers much of Scotland, apart from the highest peaks. Many old shallow wells that supply individual houses or farms are present within the till, but the till is generally unsuitable for large scale groundwater abstraction.

Much of Scotland is underlain by basement rocks with little intergranular porosity or permeability. Unlike the basement rocks of tropical Africa, any weathered material has been largely removed by glaciation and productive boreholes have to penetrate fractures associated with bedding planes or faults. Yields are highly variable, depending on the number and size of fractures intersected, but up to several litres per second have been achieved (Ball & MacDonald 1997).

Exploring for groundwater in Scotland

Scottish aquifers are small in extent, and yields are low and variable. Groundwater exploration relies heavily on investigating the aquifer geometry at a particular site, and arranging the abstraction regime to optimize the long term yield from the aquifer (Table 1). Collecting as much information as possible about a site prior to drilling can prove cost-effective. The British Geological Survey (BGS) is a major source of geological information and much of the country has been mapped at approximately 1:10000 scale. Data are also available from over 250000 mineral and site investigation boreholes, and approximately 4000 water boreholes.

Satellite data or aerial photographs can be used to locate potential aquifers. Both have been used extensively in arid areas to find fractures or sedimentary features (e.g. Marsh & Greenbaum 1995; Teeuw 1995; Isiorho & Nkereuwem 1996). Landsat Thematic Mapper (TM) data are the satellite data most commonly used in groundwater exploration. In Scotland the interpretation of satellite data is complicated by the presence of thick glacial drift and dense vegetation. Fracture

Table 2. *Appropriate geophysical methods for groundwater studies in Scotland*

Geophysical technique	What it measures	Output	Approximate maximum depth of penetration	Appropriate hydrogeological environments	Comments
Frequency domain EM	Apparent terrain electrical conductivity (calculated from the ratio of secondary to primary EM fields)	Single traverse lines or 2-D contoured surfaces of bulk ground conductivity	50 m	Alluvium Raised beach Blown Sand Fluvioglacial sand and gravel Basement	Quick and easy method for determining changes in thickness of superficial deposits, or locating buried channels. Interpretation is non-unique and requires careful geological control. Can also be used in basement rocks to help identify fracture zones.
Transient EM	Apparent electrical resistance of ground (calculated from the transient decay of induced secondary EM fields)	Output generally interpreted to give 1-D resistivity profile	100 m	Alluvium Raised beach Blown Sand Fluvioglacial sand and gravel Basement	Better than FDEM at locating targets through conductive overburden; also better depth of penetration.
Ground penetrating radar	Reflections from boundaries between bodies of different dielectric constant	2-D section showing time for EM waves to reach reflectors	10 m	Blown sand Alluvium	Accurate method for determining thickness of sand and gravel. The technique will not penetrate clay, however, and has a depth of penetration of about 10 m in saturated sand or gravel.

Resistivity	Apparent electrical resistivity of ground	1-D vertical geoelectric section; more complex equipment gives 2-D or even 3-D geoelectric sections	50 m	Basement Major aquifers Alluvium Raised beach Blown Sand Fluvioglacial sand and gravel	Can locate fracture zones in basement rocks or major aquifers. Also useful for identifying thickness of sand or gravel within superficial deposits. Often used to calibrate EM surveys.
Seismic refraction	P-wave velocity through the ground	2-D vertical section of P wave velocity	100 m	Basement Thickness of drift deposits	Can locate fracture zones in basement rock and also thickness of drift deposits. Not particularly suited to measuring variations in composition of drift.
Magnetic	Intensity (and sometimes direction) of earth's magnetic field	Variations in the earth's magnetic field either along a traverse or on a contoured grid	30 m	Major aquifers	Can locate magnetic bodies such as dykes or sills which can act as barriers in major aquifers. Susceptible to noise from any metallic objects or power cables.
VLF	Secondary magnetic fields induced in the ground by military communications transmitters	Single traverse lines, or 2-D contoured surfaces.	40 m	Basement and major aquifers	Can locate vertical fracture zones and dykes within basement rocks or major aquifers.

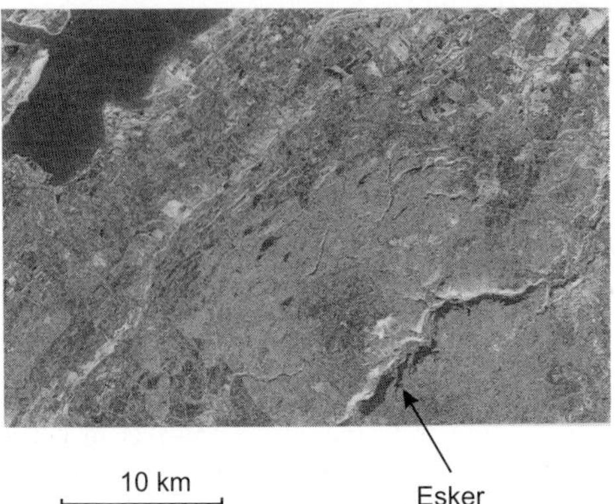

Fig. 2. Landsat TM image showing large esker in the Inverness area (Marsh & Greenbaum 1995).

mapping is only possible in upland areas where the physiography is well defined. Faults and fractures can then be identified and standard statistical methods applied to find areas with a higher density of fractures and higher groundwater potential. In low-lying areas the glacial drift masks the fracture patterns of the bedrock, but satellite data can be used to identify small aquifers within the superficial deposits. Satellite data acquired in winter, when the sun angle is low, can identify subtle geomorphological features such as moraines, eskers, in-filled glacial channel fans and fan deltas. Figure 2 shows a major esker system identified in the Inverness area from satellite data.

Investigations in rural areas, away from the main aquifers (Fig. 1), have increased substantially in the last few years. This has been driven by the increasing stringency of water quality standards, which has placed many surface water sources derived from peaty uplands in disfavour. Many of these investigations have been for small communities with a water requirement of less than one litre per second. Different hydrogeological environments have been investigated, including Permian sediments in Arran, basalt lava flows in Skye, raised beach deposits in Luing and alluvial deposits in Kintyre. The small costs of investigation are significantly outweighed by the financial benefits of finding a few successful groundwater supplies that do not require the same level of water treatment as many surface water sources.

The use of geophysics in exploration

Prospecting for groundwater supplies using geophysics is well established. In the United States and Canada, geophysical techniques have been applied to locate groundwater within fluvioglacial environments (e.g. Heigold *et al.* 1979; Kosinski & Kelly 1981; Frohlich & Kelly 1985; Kelly & Frohlich 1985). In Africa and India, geophysics is used routinely to locate groundwater in fractured basement rocks (e.g. Beeson & Jones 1988; Olayinka & Barker 1990; McNeill 1991; Barker *et al.* 1992; Carruthers & Smith 1992). These same geophysical techniques are also applicable to Scotland and other Celtic countries with heterogeneous low-yielding aquifers.

General chracteristics for each of the various Scottish aquifers are given in Table 1. Small shallow targets, with contrasts in electrical conductivity, or vertical electrical or magnetic features such as dykes or faults, are generally most easily located using geophysical techniques. Once the nature of the target is known, the most effective technique for identifying that particular target can be chosen. Table 2 summarizes the main geophysical survey techniques and the targets they can most effectively locate. Results from these surveys can be used to locate further exploration boreholes, or to determine the extent of the aquifer. If aquifer geometry is straightforward, some geophysical data can be analysed to estimate aquifer parameters such as porosity or transmissivity (e.g. Kelly & Frohlich 1985; Mbonu *et al.* 1991).

Case studies

Three different examples of using geophysics in groundwater exploration are presented below. They have been selected to illustrate the variety of techniques that are appropriate to rural Scotland. Other examples are: magnetic profiling to locate dykes (acting as barriers) within Permian sandstone; electromagnetic methods to identify gravel within raised beach deposits; and electromagnetic and electrical resistivity tomography to identify sandy horizons within fluvioglacial deposits.

Ground penetrating radar studies in the island of Coll

The Island of Coll lies off the west coast of Scotland (Fig. 1). It is a generally flat island, about 22 km long and 6 km wide. The resident population on the island is approximately 140, rising to about 300 during the summer. Due to the unreliability of the surface water supplies, West of Scotland Water Authority (WSWA) investigated the possibility of using groundwater to supply the 80 m^3/d required by the island. The greatest potential for groundwater is within the drift deposits, particularly blown sand. The underlying bedrock comprises Precambrian Lewisian Gneiss, which has no primary porosity, and unless heavily fractured has little groundwater potential.

Ground penetrating radar (GPR) was chosen to characterize the thin blown sand aquifer in Coll. GPR is a geophysical technique related to seismic reflection (Reynolds 1997). Short pulses of electromagnetic radiation are propagated into the ground and reflections recorded on a receiver. Reflections are produced at interfaces where there is a change in electrical impedance, which is dominated by the dielectric constant of the ground. The larger the contrast in dielectric properties at the interface, the stronger the signal reflected back to the surface. A subsurface profile is built up by moving the antennae over the ground surface keeping a fixed distance between the transmitter and receiver. Data are then collated to give a graphical plot of reflectors against depth (the depth being determined from the two-way travel time). The depth of penetration is governed by the electrical conductivity of the ground. Penetration is low where electrical conductivity is high (such as in clay); in dry sands and gravel however, penetration can be several tens of metres. Examples of GPR applied to characterizing sandy drift deposits are given by Beres & Haeni (1991),

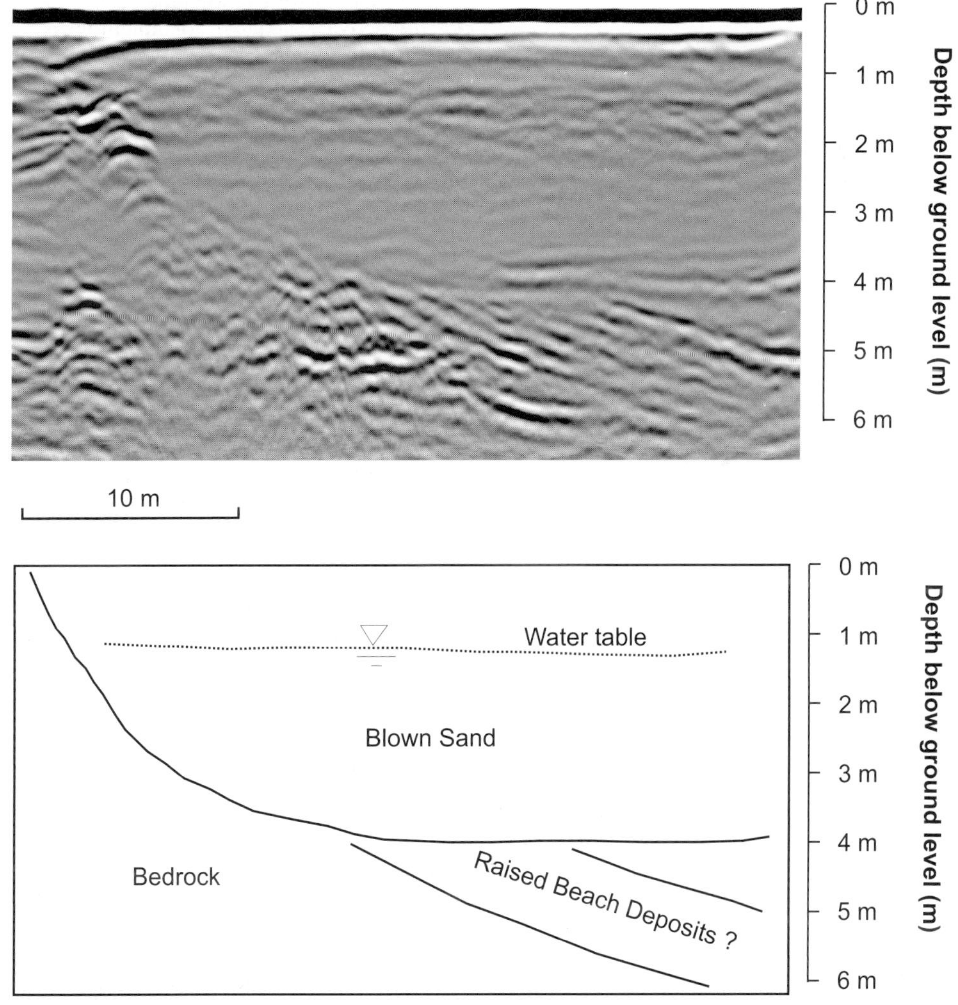

Fig. 3. GPR profile and interpretation for a small section of the surveys undertaken in Cliad Bay, Island of Coll.

Gawthorpe *et al.* (1993), Overmeeran (1994), Smith & Jol (1995) and Bridge *et al.* (1995).

Approximately 800 m of GPR surveys were undertaken across the largest blown sand deposit on the island in Cliad Bay. The results of the survey were confirmed by augering. The interface between the bedrock and the blown sand was clearly identified in the surveys (Fig. 3) and was generally from 3 m to 4 m in depth throughout the bay. Raised beach deposits beneath the blown sand could also be identified, as could the water-table at about 1 m depth. Water samples showed the groundwater to be of good quality, with total dissolved solids of less than 200 mg/l. Therefore, the surveys proved a relatively extensive aquifer comprising up to 3 m of saturated sand, with a 1 m unsaturated zone. The investigations and interpretation took less than one week and cost less than the price of one exploration borehole. In the end, WSWA decided not to develop the groundwater within the blown sands, due to concerns about protecting the groundwater from a nearby landfill.

Identifying a buried channel in alluvium

An alluvial aquifer was investigated at Palnure, in Dumfries and Galloway, as part of a scheme by WSWA to augment the water supply for the area (Fig. 1). The site chosen for investigation was an area of alluvium in a river meander of the Palnure Burn. Ground surface elevation varies by approximately 1.5 m from the road to the river. Several test boreholes had been drilled previously which identified two thin aquifers separated by clayey silt (boreholes 196, 296 and 496 on Fig. 4). The shallow aquifer comprises several metres of gravel with a transmissivity of $100\,m^2/d$; a deeper sandy layer has a transmissivity of $20\,m^2/d$. Approximately $400\,m^3/d$ was required from the site, which could be supplied from two or three production boreholes penetrating the shallow gravel. Finance for the investigation was limited, therefore an EM34 survey was chosen as a rapid means of mapping the bulk electrical conductivity of the alluvium and of identifying sites for additional exploratory boreholes. Vertical electrical sounding was used to help calibrate the EM34 survey.

EM34 measures the bulk electrical conductivity of the ground by passing an alternating electromagnetic field over and through the ground and measuring the secondary electromagnetic fields produced. By making several assumptions, an estimate of the apparent electrical conductivity can be made from the ratio of

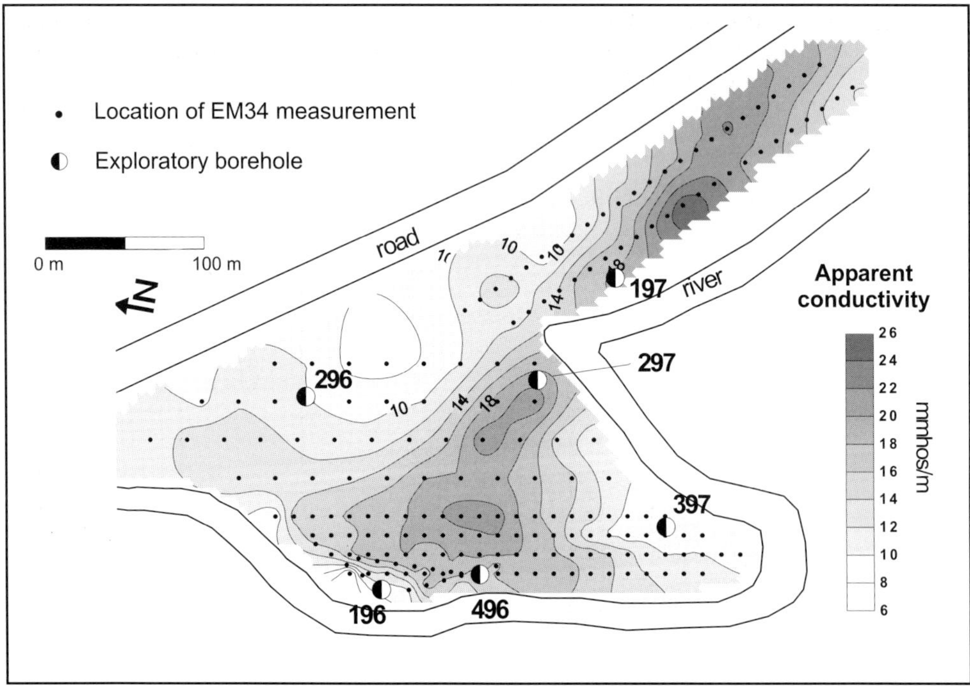

Fig. 4. Apparent electrical conductivity for Palnure measured with EM34 using horizontal dipoles and 20 m coil separation.

the primary and secondary EM fields (McNeill 1980). The measured electrical conductivity depends on various factors, such as porosity, groundwater chemistry and clay content. As such, the technique is inherently ambiguous: increased electrical conductivity may be caused by increased porosity (and therefore high groundwater potential) or increased clay content (and therefore low groundwater potential). Despite the ambiguity, with good geological control, EM34 is an inexpensive and rapid method of extrapolating information between boreholes or of characterizing areas of similar geophysical properties. The technique can give information to about 50 m below subsurface – depending on the configuration of the transmitter and receiver, and the frequency of signal used.

Figure 4 shows the bulk apparent electrical conductivity for the alluvial deposit at Palnure measured using a 1600 Hz signal with 20 m transmitter-receiver separation in a horizontal dipole configuration. With a horizontal dipole, much of the signal is from the top 15 m (McNeill 1989). The horizontal dipole survey found low conductivity close to the road; subsequent resistivity surveys indicated that the underlying bedrock had very low conductivity and was close to the surface by the road. A more complex pattern of electrical conductivity is observed within the main body of alluvium. A band of higher conductivity crosses the field, with low conductivity closer to the river. Resistivity surveys suggested that the high conductivity was caused by a layer of high conductivity material at shallow depth. Three additional exploratory boreholes were drilled as a result of the geophysical survey (boreholes 197, 297 and 397). Two were drilled within the high conductivity band and one within the low conductivity area next to the river.

Lithological logs for the area are shown in Fig. 5. The area of low conductivity close to the road is clearly due to the alluvium being thin at this point, and the low conductivity bedrock being close to surface (borehole 296). The high conductivity band can be attributed to a clayey silt layer from 0 to 6 m (boreholes 197, 297 and 496). The lower conductivity areas close to the river correspond to a gravel layer at shallow depth (borehole 397 and 196). Groundwater within the clayey silt channel is of poor quality, with high manganese and iron. This suggests a reducing environment for the groundwater, which may be brought about by the thick layer of clayey silt overlying the thin gravel aquifer. Towards the river, where the shallow clayey silt is replaced by gravel, the manganese and iron are low, and yields were found to be high. Therefore production boreholes can be positioned in the areas of low conductivity close to the river, where the alluvium is thick and there are no shallow clayey silt layers.

The EM34 survey proved an inexpensive and rapid method of characterizing the alluvium aquifer at Palnure. Exploration boreholes could be accurately targeted to provide information that could be extrapolated over the rest of the area. The survey enabled a silty channel with poor quality water to be identified and therefore

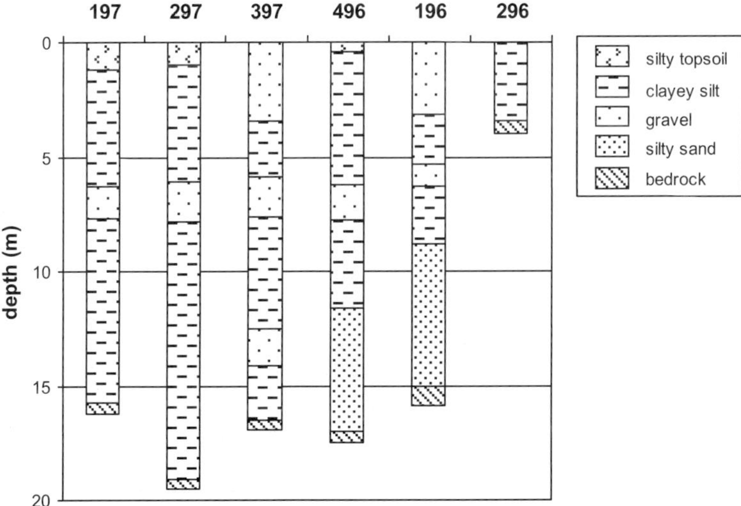

Fig. 5. Lithological logs for the boreholes drilled at Palnure. Borehole locations are shown on Fig. 4.

avoided. Due to external factors however, WSWA has not yet proceeded with developing groundwater at this site.

Locating a fracture zone in granite

An example of the use of geophysical methods to characterize basement rocks is given as part of a study made of the low-pressure tunnel sections at the Foyers pumped storage hydroelectric scheme, Loch Ness (Cratchley *et al.* 1976). While the original purpose of this survey was to evaluate the degree of fracturing in the rock mass and was thus an engineering rather than a hydrogeological study, nevertheless it is an excellent example of locating water-bearing features in basement rocks. The location of the survey area is shown in Fig. 1.

The main geophysical survey was carried out on the ground surface along the line of the Glen Lia to Loch Mhor section of the tunnel, which runs through tonalite and granodiorite facies of the Foyers granite complex. As tunneling progressed, a heavily fractured fault zone was encountered and additional support was required for the rock mass. Blasting operations slowed considerably as the tunnel progressed through it, incurring significant additional costs. After the tunnel had been constructed, a geophysical survey was carried out to establish whether prior knowledge of the extent of this zone of fractured rock could have been obtained from surface electrical resistivity and seismic refraction surveys.

The seismic refraction survey was carried out across part of the tunnel line using a 24 geophone spread 220 m long, with shot points at either end of the spread, and in lines at distances up to 150 m from either end. Charges of up to 0.45 kg were used. The seismic profile was reversed so that the reciprocal method due to Hawkins (1961) could be used to compute the depth to the various seismic interfaces below each geophone.

The final interpretation of the seismic refraction survey indicated the presence of three layers (Fig. 6). Seismic layer I (velocity 1100 m/s) was identified as glacial drift, composed of silty sand with some clay and boulders. Seismic layer II is variable in both thickness and P-wave velocity and identified as the weathered zone of the granodiorite. Values for P-wave velocity in layer II generally range between 3600 and 4500 m/s – suggesting that the granodiorite is in the completely to highly weathered category. Between 0 and 50S the velocity in layer II falls to between 3000 and 3300 m/s, which can be interpreted as the completely weathered and fractured material

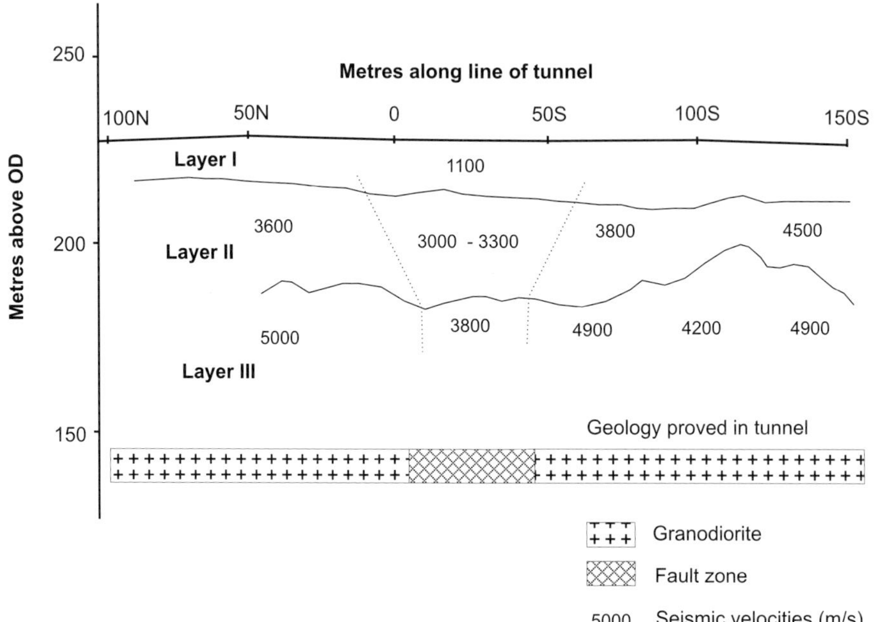

Fig. 6. Interpretation of seismic refraction data across a fracture zone in the Foyers granite complex. The geology proved by tunnelling is also shown (adapted from Cratchley *et al.* 1976).

of a fault zone. Similarly, in layer III (identified as fresh to slightly weathered granodiorite) the low-velocity zone of 3800 m/s centred on 25S also indicates the downward continuation of the fault zone.

The resistivity interpretation of results obtained from expanding probes carried out at the surface at right angles to the line of the tunnel comprises four main resistivity layers (Fig. 7). Layers 1 and 2 correlate in depth with seismic layer I and were identified as glacial drift. Samples from the drift encountered in the borehole investigations show a change at about 5 m depth from a silty sand with boulders, to a non-cohesive sand with clay and boulders. The resistivity method is thus capable of detecting variations in superficial layers that cannot be accurately resolved by the seismic method.

Resistivity layer 4 correlates with fresh to slightly weathered granodiorite (seismic layer III) south of the fault zone, but lies at much shallower depth to the north of the fault zone. The fault zone is again clearly shown, centred at 25S, as a break or depression in the bedrock surface. Resistivity layer 3 correlates reasonably well with the weathered zone of the granodiorite (seismic layer II), but shows a range of resistivity values. The low values of 400 ohm-m may be associated with the main faulting and possibly indicate another zone of highly weathered granodiorite.

From a hydrogeological standpoint this heavily fractured and weathered zone in the granodiorite represented a significant source of groundwater. During the tunnelling operations a large quantity of water entered the tunnel as it passed through this zone. Although no aquifer properties were determined, two exploratory boreholes drilled within the fault zone showed rapid recovery of water levels to within 0.5 and 3 m of ground level after the cessation of pumping. It is likely that water supply boreholes within this fault could have provided significant quantities of groundwater.

Conclusions

Geophysical methods can make a significant contribution to the location of rural water supplies in Scotland. Many of the Scottish aquifers have complex geometry and are difficult and

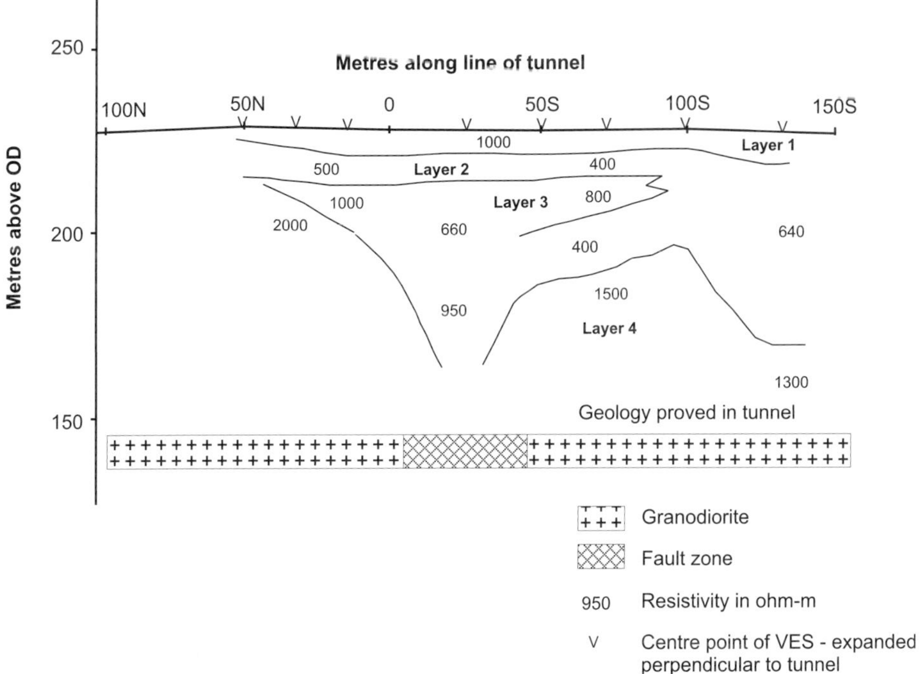

Fig. 7. Interpretation of Sclumberger resistivity data across a fracture zone in the Foyers granite complex. Approximate resistivity for each interpreted layer is shown along with the geology proved by tunnelling (adapted from Cratchley *et al.* 1976).

expensive to characterize using only conventional approaches, such as drilling boreholes. The experience of using geophysics to characterize small heterogeneous aquifers in Africa and north America/Canada can readily be applied to small, but strategically important, aquifers in the rural Celtic regions. Geophysical methods are best used to extrapolate information away from exploratory boreholes, and indicate areas for further test drilling. In this way geophysics can improve hydrogeological understanding and reduce the cost of groundwater investigations.

Electromagnetic techniques using EM34 and EM31 equipment have proved useful in identifying variations in the thickness of superficial deposits and for detecting buried channels. These techniques are rapid, but require good geological control to interpret the geophysical data correctly. Ground penetrating radar has been used at several locations to detect the boundary between bedrock and alluvium or blown sand. The results from the surveys have been confirmed with hand augering and drilling. Detailed resistivity and seismic refraction clearly identified a fracture zone in granite. The fracture zone was proved by tunnelling. A more rapid method, such as time domain EM or resistivity tomography using a multi-electrode array, would give similar information in less time. Magnetic profiling has also proved useful in Scotland to locate dykes (acting as hydraulic barriers) within Permian aquifers in the west of Scotland.

As water authorities look increasingly to groundwater to provide sustainable good quality water supplies to rural communities, the use of geophysical methods is likely to increase. These methods offer a rapid means of characterizing the complex small aquifers that occur across Scotland, and with careful interpretation can increase the success and decrease the costs of hydrogeological investigations.

This paper is published by the permission of the Director, British Geological Survey (NERC). Many of the studies for groundwater have been in collaboration with West of Scotland Water Authority and North of Scotland Water Authority. The help of G. Lloyd and N. Brennan with the case examples shown here is much appreciated.

References

BALL, D. F. & MACDONALD, A. M. 1997. *Scottish rural water supplies: the role of groundwater*. British Geological Survey Technical Report **WD/97/41**.

BARKER, R. D., WHITE, C. C. & HOUSTON, J. F. T. 1992. Borehole siting in an African accelerated drought relief project. *In*: WRIGHT, E. P. & BURGESS, W. G. (eds) *The Hydrogeology of Crystalline Basement Aquifers in Africa*. Geological Society, London, Special Publications, **66**, 183–201.

BEESON, S. & JONES, C. R. C. 1988. The combined EMT/VES geophysical method for siting boreholes. *Ground Water*, **26**, 54–63.

BERES, M. & HAENI, F. P. 1991. Application of ground-penetrating-radar methods in hydrogeological studies. *Ground Water*, **29**, 375–386.

BRIDGE, J. S., ALEXANDER, J., COLLIER, R. E. L., GAWTHORPES, R. L. & JARVIS, J. 1995. Ground-penetrating radar and coring used to study the large-scale structure of point-bar deposits in three dimensions. *Sedimentology*, **42**, 839–852.

BRITISH GEOLOGICAL SURVEY. 1988. *Hydrogeological Map of Scotland. 1:625 000*. British Geological Survey.

BUCKLEY, D. K. 2000. Some case histories of geophysical downhole logging to examine borehole site and regional groundwater movement in Celtic regions. *This volume*.

CARRUTHERS, R. M. & SMITH, I. F. 1992. The use of ground electrical methods for siting water supply boreholes in shallow crystalline basement terrains. *In*: WRIGHT, E. P. & BURGESS, W. G. (eds) *The Hydrogeology of Crystalline Basement Aquifers in Africa*. Geological Society, London, Special Publications, **66**, 203–220.

CRATCHLEY, C. R., MCCANN, D. M. & ATES, M. 1976. Application of geophysical techniques to the location of weak tunnelling ground with an example from the Foyers hydroelectric scheme, Loch Ness. *Transactions of the Institution of Mining and Metallurgy*, Section A, **85**, A127-A135.

FROHLICH, R. K. & KELLY, W. E. 1985. The relation between hydraulic transmissivity and transverse resistance in a complicated aquifer of glacial outwash deposits. *Journal of Hydrology*, **79**, 215–219.

GAWTHORPE, R. L., COLLIER, R. E. L., ALEXANDER, J., BRIDGE, J. S. & LEEDER, M. R. 1993. Ground penetrating radar: application to sandbody geometry and heterogeneity studies. *In*: North, C. P. & PROSSER, D. J. (eds) *Characterization of Fluvial and Aeolian Reservoirs*. Geological Society, London, Special Publications, **73**, 421–432.

HAWKINS, L. V. 1961. The reciprocal method of routine shallow seismic refraction investigations. *Geophysics*, **26**, 806–819.

HEIGOLD, P. C., GILKESON, R. H., CARTWRIGHT, K. & REED, P. C. 1979. Aquifer transmissivity from surficial electrical methods. *Ground Water*, **17**, 338–345.

ISIORHO, S. A. & NKEREUWEM, T. O. 1996. Reconnaissance study of the relationship between lineaments and fractures in the southwest portion of the Lake Chad Basin. *Journal of Environmental and Engineering Geophysics*, **1**, 41–54.

JOHNSTONE, J. M. C. & RENNIE, G. M. 1991. Development of a new water supply for Fort William. *Journal of the Institution of Water and Environmental Management*, **5**, 503–512.

JONES, C. R. C. & SINGLETON, A. 2000. Public water supplies from alluvial and glacial deposits in northern Scotland. *This volume*.

KELLY, W. E. & FROHLICH, R. K. 1985. Relations between aquifer electrical and hydraulic properties. *Ground Water*, **23**, 182–189.

KOSINSKI, W. K. & KELLY, W. E. 1981. Geolectric soundings for predicting aquifer properties. *Ground Water*, **19**, 163–171.

MARSH, S. H. & GREENBAUM, D. 1995. *Unconsolidated sedimentary aquifers: review No. 7 remote sensing*. British Geological Survey Technical Report **WC/95/71.**

MBONU, P. D. C., EBENIRO, J. O., OFOEGBU, C. O. & EKINE, A. S. 1991. Geolectric sounding for the determination of aquifer characteristics in parts of the Umuahia area of Nigeria. *Geophysics*, **56**, 284–291.

MCNEILL, J. D. 1980. *EM34-3 survey interpretation techniques*. Technical Note TN-8, Geonics Ltd., Mississauga, Canada.

——1989. Advances in electromagnetic methods for groundwater studies. *In:* GARLAND G. D. (ed.) *Proceedings of Exploration '87: Third Decennial International Conference on Geophysical and Geochemical Exploration for Minerals and Groundwater*. Ontario Geological Survey Special Volume 3, 678–702.

——1991. Advances in electromagnetic methods for groundwater studies. *Geoexploration*, **27**, 65–80.

OLAYINKA, A. & BARKER, R. 1990. Borehole siting in crystalline basement areas of Nigeria with a microprocessor controlled resistivity traversing system. *Ground Water*, **28**, 178–183.

OVERMEERAN, R. A. 1994. Georadar for hydrogeology. *First Break*, **12**, 401–408.

REYNOLDS, J. M. 1997. *An introduction to applied and environmental geophysics*. John Wiley & Sons, Chichester.

ROBINS, N. S. 1990. *Hydrogeology of Scotland*. HMSO for the British Geological Survey, London.

SMITH, D. G. & JOL, H. M. 1995. Ground penetrating radar antenna frequencies and maximum probable depths of penetration in Quaternary sediments. *Journal of Applied Geophysics*, **33**, 93–100.

TEEUW, R. M. 1995. Groundwater exploration using remote sensing and a low-cost geographical information system. *Hydrogeology Journal*, **3**, 21–30.

WATT, G. D., MELLANBY, J. F., VAN WONDEREN, J. J. & BURLEY, M. J. 1987. Groundwater investigations in the Lower Spey valley near Fochabers. *Journal of the Institution of Water & Environmental Management*, **1**, 89–103.

Some case histories of geophysical downhole logging to examine borehole site and regional groundwater movement in Celtic regions

DAVID K. BUCKLEY

British Geological Survey, Maclean Buildings, Wallingford, Oxfordshire OX10 8BB, UK
(e-mail: dkb@bgs.ac.uk)

Abstract: Many Celtic regions comprise hard fractured rock aquifers and hilly terrain. They are generally regions which have also been affected by repeated glaciation and deglaciation during the Pleistocene period. This has influenced their hydraulic properties and the groundwater flow systems which have developed. Geophysical borehole logging provides a useful method to examine the occurrence of groundwater, water inflows and the movement of water in local and regional groundwater flow systems. The downhole measurements show that wellbore flow, aquifer layering and separate groundwater circulations are common features of the hydrogeology.

Celtic regions generally have more rainfall than the national average, and a high proportion of this is underlain by ancient hard rock. These regions were repeatedly covered by ice sheets, often in excess of 1 km thick, during the Pleistocene period. The effects of repeated glaciations, the extensive cover of glacial deposits and the rise in land level due to deglaciation rebound, are important influences on the hydrogeology. Large areas of these regions are underlain by igneous, metamorphic and hard sedimentary rocks, where the extent and distribution of weathering and fracturing are also important controls on the hydrogeology. Thus, groundwater exploration and development strategies akin to those employed in basement aquifers in other regions of the world (Wright & Burgess 1992) can be appropriate. A challenge facing hydrogeologists working in Celtic areas is relating aquifer development with Pleistocene history, in particular the injection of glacial meltwater under high hydraulic head during deglaciation phases (Boulton *et al.* 1995), and flow route development by groundwater circulation to lower base levels during the glacial periods. In northern regions, which were under considerable ice cover for long periods, glacial rebound is significant so that some former base levels and groundwater flow routes may now be found well above current sea level. Modification of the near-surface layers by glacial processes, deposits and permafrost is also an important component of aquifer development in Celtic areas.

Geophysical logging provides basic information which can be used to examine the aquifers in terms of their development processes. This includes information on the aquifer layering, the glacial deposits, the distribution of permeability and the water inflows at depth, as well as the local and regional groundwater flow systems which have developed in response to the Pleistocene and Holocene processes. This paper describes some of the results of downhole geophysical logging surveys undertaken in boreholes in Scotland, Northern Ireland, Wales and Cornwall over the last 15 years, to examine the occurrence and movement of groundwater. The examples are selected to illustrate particular features of hydrogeology. Figure 1 shows some of the sites where downhole investigations have been carried out.

Groundwater flow systems

Groundwater flow systems develop within rock masses when recharge can take place at high elevation and a flow route can develop to discharge the groundwater at a lower elevation, ultimately down to base level (sea level) in coastal regions, (Hubbert 1940; Tóth 1963; Freeze & Cherry 1979). Over time, hydraulic circulation by groundwater establishes preferential flow routes through physical and chemical interaction. The different circulation routes and rates of flow govern the residence time of the groundwater which in turn directly influences its chemistry and age. Hence the groundwater body evolves a chemical stratification because of geological

From: ROBINS, N. S. & MISSTEAR, B. D. R. (eds) *Groundwater in the Celtic Regions: Studies in Hard Rock and Quaternary Hydrogeology*. Geological Society, London, Special Publications, **182**, 219–237. 1-86239-077-0/00/

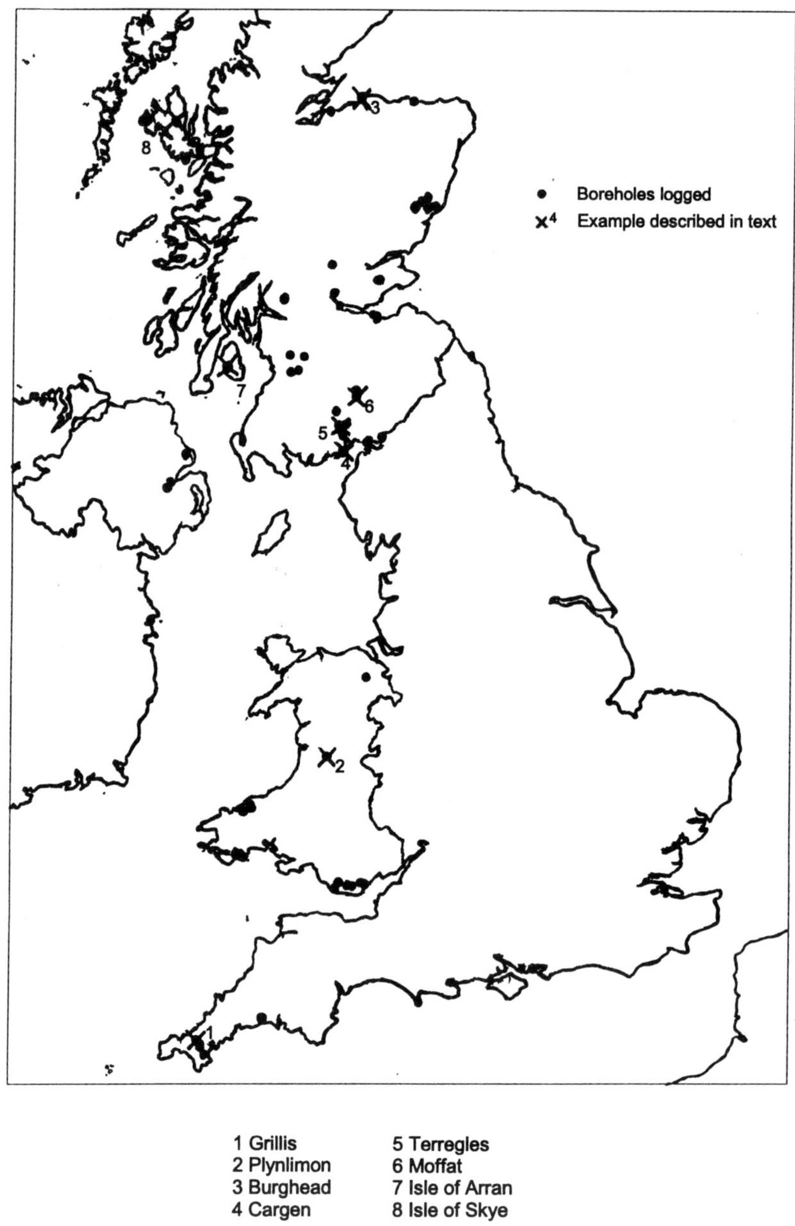

Fig. 1. Location map of logged boreholes.

layering and hydraulic circulation. Alley (1993) describes some of the chemical quality changes that take place in regional groundwater flow systems. The lithological and chemical stratification can be examined by geophysical formation and fluid logging of boreholes drilled.

Individual aquifer layers recharged at different elevations and separated by different permeability strata tend to display different hydraulic heads. As a consequence there is a potential for groundwater to flow vertically from horizons of higher head to those of lower head through available routes in the rock to equilibrate the head differences. Such water movement through openings in the rock mass is termed *aquifer crossflow*; in porous rocks this takes place through pore space, while in fractured rocks water follows fracture and joint openings. The crossflow waters that invade horizons of low head may have quite different chemistry.

When boreholes are drilled into separate water-bearing layers and are constructed open-hole or with long screens, ie. open to several layers having different hydraulic heads, water will flow vertically up or down the borehole to equalize the hydraulic heads. The movement within the borehole, upwards or downwards, can be relatively rapid and substantial and is termed *wellbore flow*. This is an indication of the potential for aquifer crossflow to occur. It can be detected by fluid logging techniques. Flowing (artesian) boreholes are an obvious visible example of upward wellbore flow where it overflows at the surface (Fig. 2). Most boreholes exhibit wellbore flow that does not overflow at the surface but is nevertheless taking place inside the borehole and causing invasion of horizons of lower head. The overall effect of aquifer crossflow and wellbore flow is to create a mixed groundwater and this has implications for representative water sampling from boreholes. Schmidt (1977) and Reilly *et al.* (1989) showed that pumped water sampled from boreholes represented a mixture of the component

Fig. 2. Geophysical logging of an overflowing borehole in Permian sandstone, Isle of Arran.

inflows. Wellbore flow can be inferred from fluid electrical conductivity and fluid temperature log profiles in boreholes, and borehole flowmeter measurements can be used to quantify the inflow proportions and also determine the velocity and direction of vertical fluid movement taking place.

The geophysical logging measurements

Geophysical logging is a remote sensing method, used in boreholes, that characterizes the rock mass and its contained fluids. Each set of measurements is referred to as a 'log' and by comparing several logs, each measuring a different property, the physical and chemical properties of the rock mass can be characterized. Methods and interpretations described in the literature relate mainly to practice in the petroleum industry (e.g. Dewan 1983; Bateman 1985; Desbrandes 1985). Texts relating to its application in hydrogeology include Keys & MacCary (1971), Repsold (1989), and Chapellier (1992). One of the principal advantages of geophysical logging over visual description of the rock cuttings is that the log measurements are not restricted to the narrow visible part of the electromagnetic spectrum (400–700 nm), traditionally used by field or wellsite geologists to describe rock samples, but encompass a much wider range of imaging frequencies, typically 1 kHz–10^{20} Hz (ELF to gamma rays), which provides more effective characterization of the rock mass (Doveton 1999). Furthermore, the measurements are continuous so that bed boundaries and depths are defined more accurately.

The detailed log profiles obtained can be compared with those from adjacent boreholes where it may be possible to recognize the same (or different) vertical sequence of layers, thus indicating the dip and continuity of the strata and providing a correlation. When combined with fluid logging identification of water inflow and outflow, the horizons of water movement within the lithological sequence can be determined, and the influence of lithological layering and geological structure on groundwater movement can be appreciated.

Geophysical log identification of fluid movement at depth

Fluid logging comprises fluid electrical conductivity (EC) and fluid temperature (TEMP) measurement of the borehole fluid, plus measurement of the velocity of wellbore flow using borehole flowmeters. Waters circulating in the groundwater flow system, following different routes, tend to have different residence times and groundwater chemistry. These differences can be recognized by recording the small changes in temperature and EC of the borehole fluid, where it enters the borehole. Sometimes the differences are only slight and the logs must be considerably amplified, or a differential measurement is necessary to detect them. The fluid EC and temperature logs are normally run under the ambient hydraulic conditions, and then repeated under changed hydraulic conditions, usually by pumping. Overlay plotting of the curves then usually identifies the positions where water entry is taking place.

The fluid temperature and EC profiles obtained give an indication of any wellbore flow taking place. In isotropic granular aquifers fluid temperature normally increases with depth according to the geothermal gradient. However, where natural or induced wellbore flow is taking place, the vertical fluid movement disturbs the natural geothermal gradient and straight, vertical sections on the fluid temperature (and often fluid EC) profiles usually signify intervals of wellbore flow, which may be up or down, depending upon the relative hydraulic head (Figs 3 and 4).

The fluid EC and temperature measurements themselves also give an indication of the nature of the groundwater flow system in the rock mass. Shallow circulating (younger) waters normally have a lower EC (unless polluted) due to shorter residence time, and are generally cooler, than waters found at depth. Relatively low EC waters found at depth, that are also relatively cool, may represent more rapidly moving groundwaters circulating in fractures or fissures which are bypassing slower matrix pore routes. The presence of relatively warm and elevated conductivity borehole fluids at shallow levels may indicate upward wellbore flow of older waters from depth, or pollution from the surface.

Borehole flowmeter measurements, used to quantify the separate inflows and gauge their relative contribution to the total flow are normally made with a spinner (impeller) flowmeter, but where the fluid velocity is low, either because of low yield or large diameter, a more sensitive heat-pulse (thermal) method or injection of a chemical tracer may be used. The heat-pulse flowmeter is a convenient technique and injects a pulse of heat at a heating grid placed centrally between two equidistant thermistors (Dudgeon *et al.* 1975). The timing of the arrival of the injected heat-pulse, and the particular thermistor

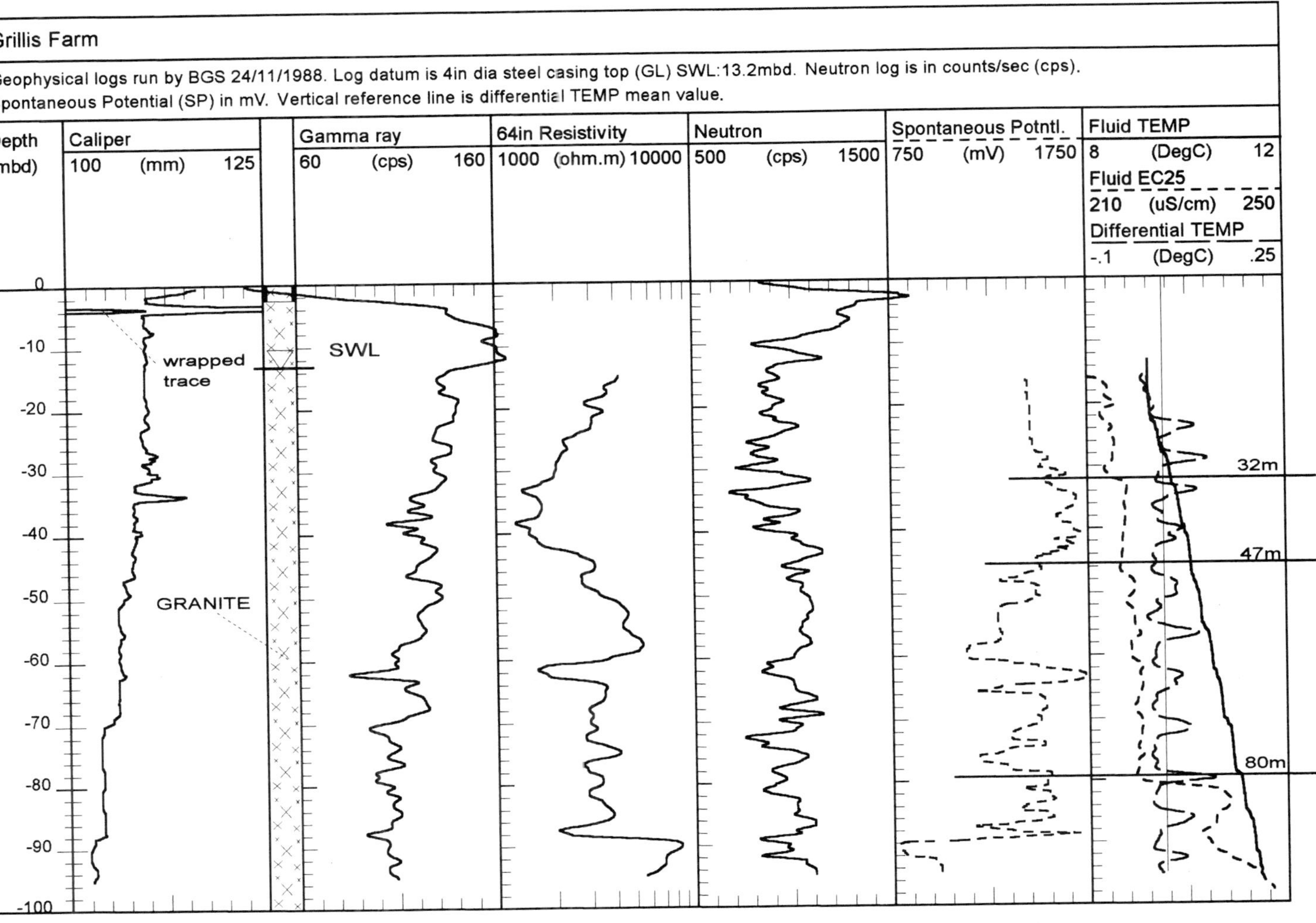

Fig. 3. Geophysical logs, Grillis Farm borehole, Carnmenellis Granite, Cornwall.

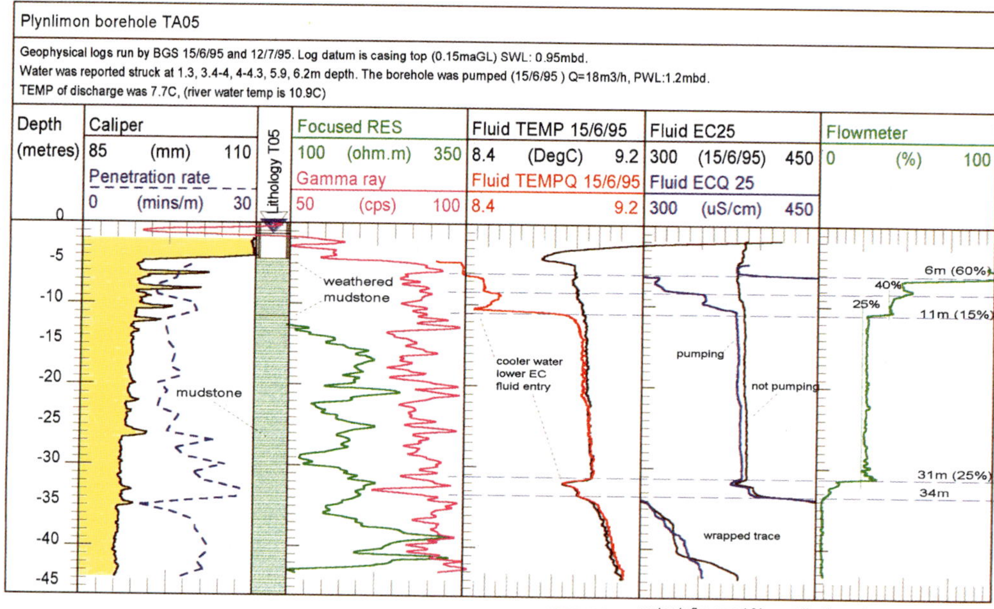

Fig. 4. Geophysical logs of Ordovician mudstone, borehole TA05, Plynlimon, Wales.

responding, provides the velocity and direction of the fluid movement. Examples of flowmeter applications are given by Molz *et al.* (1989), Molz & Young (1993), Molz *et al.* (1994), Boman *et al.* (1997) and Gossell *et al.* (1999).

For short screen boreholes, and piezometers only open to a single aquifer horizon, water pumped is usually representative of the target layer, provided the construction is sound. However, water pumped from open hole or long screen boreholes, especially in hard rock terrains, represents a mixture of the groundwaters from all the contributing permeable layers penetrated, in proportion to their relative contributions, each possibly having different circulation histories, transit times and chemical evolutions (Church & Gramato 1996). Sometimes, where there are several boreholes on a site, these provide pathways for water to move between them, and this can lead to further mixing of waters. Very often pumped water is a mixture of modern recharged water obtained from shallow level with a smaller proportion of deeper circulating waters that are older, and might include palaeowaters. Pumping of palaeowater (that is not recharged) has implications for water balance and model calibration estimates, so that it is important to resolve and characterize the individual layer contributions at depth by downhole logging.

Case histories

Formation and fluid logs in granite, Carnmenellis, Cornwall

Figure 3 shows geophysical logs run in one of several boreholes penetrating the Carnmenellis Granite in Cornwall. The boreholes were logged as part of a groundwater chemistry study. Analogue logging equipment was used and the curves have been digitized from the paper logs.

One of the features evident from logging is that there is little or no development of a weathered zone at the surface which would normally store groundwater. The boreholes entered fresh rock just below the surface and were completed with a minimum of blank casing. The caliper log (track 1) shows only 3 m of steel casing present, below which the borehole is open hole and becoming harder with depth (closer to gauge). Static water level was at 13.2 metres below datum (mbd). Enlargements (probably fractures) are indicated at 28–40 m depth. The gamma ray activity is high in some holes due to the rock being granite.

Fluid EC and TEMP logs were run in the borehole without pumping. These logs and a differential temperature curve (Differential TEMP) derived from the measurements, are presented in track 7. The fluid logs suggest inflows at 22, 32,

47 and 80 m depth where changes are displayed on the curves, and indicated on Fig. 3 by the horizontal lines. The profiles suggest natural circulation is taking place between horizons in the borehole, which may be fractures, although confirmation is needed from pumped fluid logging and CCTV or image logging. The cooler water and lower temperature gradient above 22 m suggests that a more rapid circulation of groundwater is taking place in the rock mass above that depth, as part of a groundwater flow system that is currently recharged from the surface. The deepest circulation encountered in the holes logged in the area was 136 m depth.

Fluid movement at depth in mudstone of an upland catchment, Plynlimon, Wales

Figure 4 is an example of geophysical formation and fluid log responses in a borehole penetrating Ordovician mudstone in the upland catchment of the River Severn. The borehole is at approximately 300 m above Ordnance Datum (m AOD), near Plynlimon in Wales. The borehole is on the edge of a stream and the logs show groundwater inflows from fractures and cleavage of the mudstone at depth. Mudstones are not normally a groundwater exploration target and the logging was part of an examination of groundwater and stream interaction, for a catchment water balance. The logging identified a significant groundwater occurrence and movement within the mudstones at shallow levels, feeding local streamflow, and also at depth.

In Fig. 4 formation logs (caliper, drilling penetration rate, focused resistivity and gamma ray) are shown in tracks 1 and 3 and a lithological column, showing casing interval and static and pumped water levels, is illustrated in track 2. Fluid logs are shown on the right hand side in tracks 4–6.

The caliper log shows the base of casing at approximately 4.5 m depth, below which the borehole is open hole with some enlargements in diameter to about 13 m depth. The lithological description, not listed in detail, indicated up to 2 m of peat and weathered mudstone overlying fresh mudstone to 13 m. The gamma ray log deflects to the right alongside more clayey layers, and to the left against decreasing clay and increasing sand content. A prominent sandier horizon is inferred at 32 m where water inflow is taking place.

The fluid temperature and conductivity logs shown in tracks 4 and 5 were run prior to pumping and then repeated whilst pumping (ECQ/TEMPQ). The flowmeter log, run downwards whilst pumping, is shown in track 6. The fluid logs represent the upward moving mixture moving towards the pump suction which was placed at 4.5m depth. Cooler lower EC waters enter at 6, 8.5 and 11 m on pumping and represent 75% of the total. The flowmeter shows that the remaining 25% enters at 31 and 34 m depth. The inflow positions are represented on the plot by the horizontal dashed lines.

The fluid temperature log run prior to pumping (track 4) is virtually identical to that when pumping, except above 11 m. The low temperature (and similar EC) gradient between 32 m and 11 m is an indication that natural *upflow* is probably taking place over this interval, without any pumping. Below 34 m depth, the increased temperature gradient and increased fluid EC both suggest very little groundwater movement and this was confirmed by the flowmeter logging. It can be expected that the natural upflow between 31 m and 11 m invades the horizons at 11 m and above, and, on pumping, some of the invaded higher EC water is returned. Hence care needs to be taken when interpreting the chemistry of samples collected when pumping the borehole. The flowmeter profile is a stepped response rather than a gradual incremental change in flow rate, and this is an indication that water entry is from discrete fractures or cleavage rather than intergranular contribution. The possibility of breakaway on yield–drawdown relationships might, therefore, be expected. Logging of nearby boreholes showed a similar picture of concentrated shallow inflow of cool, low EC water, but with some deeper inflow at 30–40 m depth. In one nearby borehole, inflow at 32 m represented 70% of the total pumped.

The groundwater flow system of the mudstone can be examined by incorporating the log analysis into scale cross-sections. In this particular example downhole acoustic and optical scanning was recommended to determine the relationship and orientation of the fracture and cleavage system to the water inflows identified.

Water inflow through wellscreen and identification of sand pumping

Geophysical logs, which illustrate water and sand entry through wellscreen from a sandstone aquifer in NE Scotland, are shown in Fig. 5. The borehole was drilled in 1986, not far from the Moray Firth, for supply purposes and penetrated Permian sandstones overlying Devonian sandstones in a sequence dipping north.

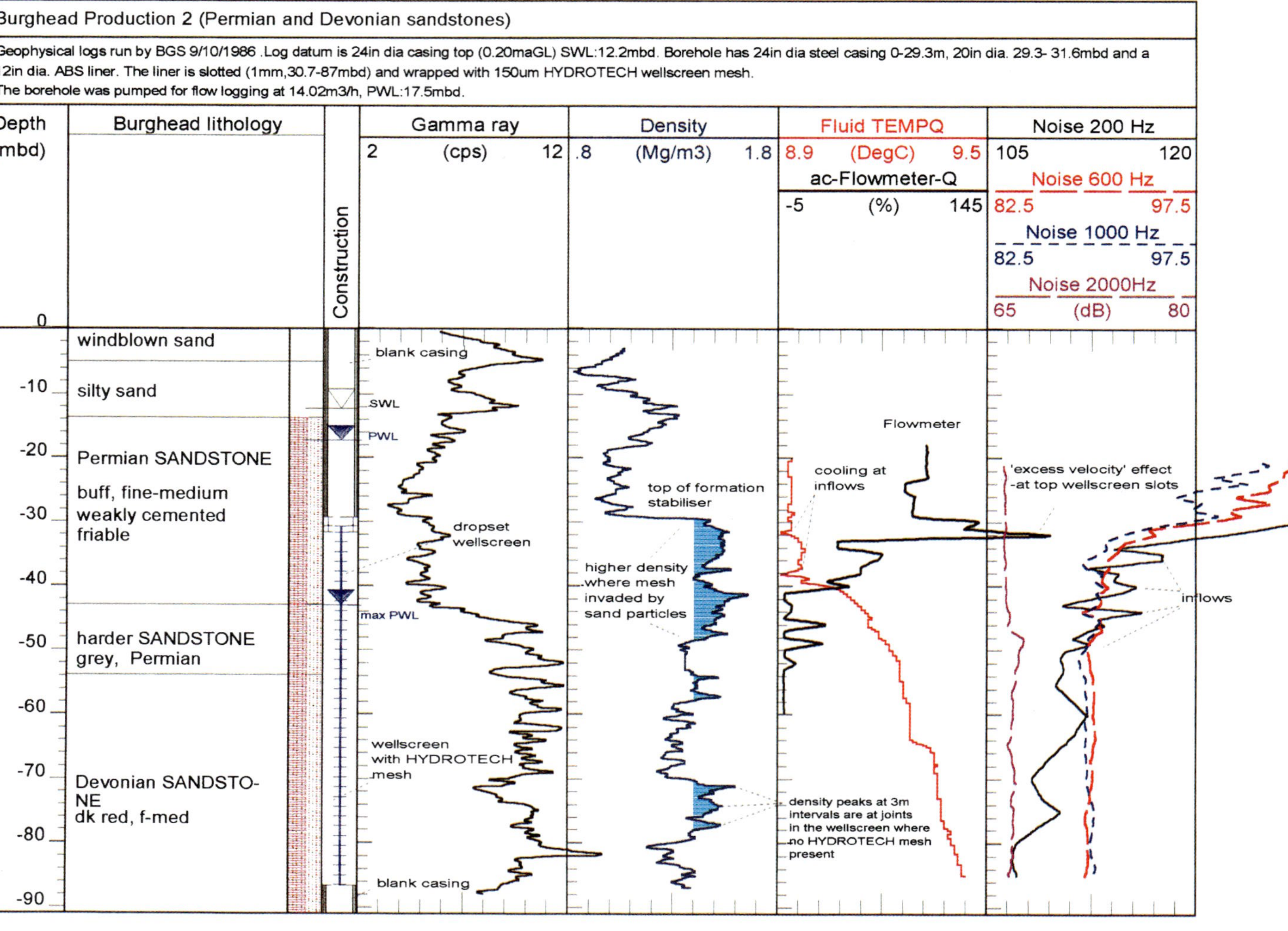

Fig. 5. Geophysical logging to identify sand entry, Burghead Production borehole, Moray, Scotland.

The borehole has a fairly complicated construction which influences some of the log measurements made. It is cased at 610 mm diameter down to 29 m depth, inside of which is a dropset 508 mm diameter steel casing to 31.6 m depth. Inside and below the casings is a 305 mm diameter plastic (ABS) wellscreen which is slotted (30.7 to 87 m 1 mm slots). The slotted sections are wrapped with 0.15 mm HYDROTEC plastic mesh which serves as a substitute for a gravel pack. The annular space outside the mesh was additionally filled with formation stabilizer gravel (3–5 mm grains). Steel centralizer guides were used to locate the screen inside the casings and their position at 4, 15 and 24 m depth was detected by casing collar locator (CCL) log measurements. The top of the formation stabilizer was detected at 29 m depth by density logging. During initial step drawdown test pumping in 1986, fine (0.25–0.1 mm) to very fine-sand (<0.1 mm) passed through the wellscreen slots at higher rates of pumping, preventing use of the borehole. Geophysical logging was undertaken to examine the situation and, if possible, locate the sand entry.

The formation and fluid logs run in the plastic liner are shown in Fig. 5. The caliper log showed no obvious break or damage to the liner. The lithology is summarized in tracks 1 and 2 and the borehole construction is depicted alongside. The static water level (SWL) on logging was 12.2 mbd and the pumping water level (PWL) during fluid logging was 17.5 mbd. The gamma ray log, although influenced by the formation and gravel pack, shows a higher activity below 43 m depth. The windblown sand at the surface and weakly cemented Permian sandstone described at approximately 33 m depth, are potential sources of the sand pumped.

The density log is a short spacing measurement and its limited depth of penetration means that little if any of its signal is contributed by the formation. Its profile largely represents the variation in density of the gravel pack and plastic mesh. Some confirmation of this is given by the low density ($\sim$1.0 Mg/m^3) indicated above the top of the pack at 29 m depth where the tool is measuring the density of the liner. Where the wellscreen sections are joined the mesh wrap is absent and the gravel formation stabilizer comes in direct contact with the base pipe. This locally increases the bulk density which is seen as regular peaks (at 3 m intervals) on the log.

Overall, the log shows an increased bulk density from 1.30 Mg/m^3 at 70 m to 1.50 Mg/m^3 at 30 m. Assuming the pack is of uniform density, the zones of highest density within this interval (infilled for emphasis) are interpreted as being due to fine-sand and formation stabilizer invasion of the plastic mesh wrapping and fluid pore space. The density is highest between 30 and 50 m depth, where flowmeter and fluid temperature logging also indicates the greatest concentration of fluid inflow. A similar high density is also shown at 70–78 m depth.

A fluid temperature log run whilst pumping (track 6) revealed inflow commencing at 78 m depth, with cooling of the upward moving mixture taking place at 64 m, between 38 and 51 m and at 31.7 m depth (top slot). Analysis of the flowmeter log, shown in the same track, shows that approximately 35% enters from 38–51 m and approximately 65% enters at the top of the wellscreen.

The flowmeter log displays three features typical of impeller measurements. Because of the large diameter (305 mm) and the relatively low pumping rate (14 m^3/h), the log is unable to resolve the velocity over the full depth. The fluid velocity falls below rotation threshold (about 1–2 m/min) below 41 m depth whilst the temperature log indicates that deeper inflows exist, but are below the spinner threshold velocity. A more sensitive impeller or a heat-pulse flowmeter, or use of a higher pumping rate, would identify them. The spinner flowmeter log reveals the inflows between 35–41 m to have a smooth profile, characteristic of intergranular flow contribution, rather than the stepped incremental profile typical of fracture or fissure flow entry. The area-corrected flow curve indicates a flowrate at the top of the wellscreen approaching 200% of that pumped. This apparent anomaly is on account of high velocity at the top wellscreen slots where there is concentrated inflow which physically narrows the cross-section area for vertical flow to take place. When multiplied by the borehole area (from the caliper log) the 'excess velocity' is apparent. In this particular example, because of the slotted lining, the impeller flowmeter does not necessarily reflect the positions of inflow from the aquifer but rather shows the positions of entry through the wellscreen slots, although it can be said that the bulk of the inflow into the borehole takes place between 31.7 and 41 m depth.

The four curves presented in track 7 are acoustic noise recorded downhole (in dB) in different frequency bands by a piezoelectric transducer coupled to the temperature tool, whilst pumping. Low frequency noise below 100 kHz associated with the pump and generator is filtered out. It was thought that sand entry accompanying water inflow might cause 'pinging' and be recognized by the measurements.

The technique can be of value in sand pumping boreholes where conventional impeller flowmeters become blocked by fine sand. The curves are generally similar to the flowmeter profile, especially the 200 Hz curve, and suggest that inflows may take place at 44, 60 and 75 m depth, as also suggested by the fluid temperature log. The relatively high density recorded over the interval 70–77 m is of interest, therefore, because it might be a reflection of sand invasion of the plastic mesh surrounding the wellscreen. During the final step of step drawdown testing, maximum pumping water level was at approximately 43 m depth (shown in track 3) and water production must then have come from below; the evidence of water inflows at 60 m and 75 m from the fluid TEMP and 200 Hz noise log profiles suggests where the production may have come from. The inferred sand invasion at 70–77 m depth may have occurred then. Thus, the short-spacing density log provides an indication of the probable sand entry positions. Video logging of the borehole whilst pumping might reveal the lower boundary of the sand entry. The example is a fairly unusual application of logging measurements to identify sand entry.

Lithological layering and water inflows, Permian breccias with thin sandstone layers, Cargen borehole, Dumfries basin, Scotland

Figure 6 is an example illustrating wellbore flow (both upwards and downwards) and typical formation and fluid logs obtained in Permian deposits comprising thick breccia layers and thin sandstone horizons from a borehole in the Dumfries basin, Scotland. The pronounced layering is a common and important hydrogeological feature of the Permian sediments (see Gaus & Ó Dochartaigh this volume). Eleven thin sandstone layers can be identified by a combination of their lower gamma radiation, lower resistivity and low magnetic susceptibility. The breccias are harder, have a higher resistivity, and contain a high proportion of Palaeozoic mudstone which increases their gamma ray activity sharply. The caliper log can also identify the sandstone layers because they tend to be softer than the breccias.

Fluid EC and TEMP logs run prior to pumping showed relatively cool (9.8°C) and relatively high EC (approximately 750 μS/cm) water in the borehole from the water level (16 m) to 54 m depth. The relatively cool temperature and the form of the logs both suggest wellbore downflow taking place over this interval, presumably induced by adjacent pumping. Upflow from 92 m to 72 m and 54 m is also suggested by the profiles.

When the borehole was pumped (pump inlet at 18 m) the pumped fluid logs showed a marked change in profile above 54 m as the direction of fluid movement was reversed. Below 96 m the temperature logs were unchanged but cooling of the upward moving mixture at 96, 72, 55 and 19 m reveals the inflows. At 54 m and above, the cooling is accompanied by an increase in fluid EC.

The area-corrected impeller flowmeter log shown in the right hand track resolved inflow at 96 m (17%), 72 m (23%), 65 m (50%) and 54 m. A certain amount of processing has been performed on the flow log data to remove the background linespeed effect and correct for the borehole area. A heat-pulse flowmeter log run whilst pumping, also shown in track 8, showed a similar, though not identical, profile.

The fluid logging data therefore suggest *upflow* of warm, low EC water from near the base of the hole to sandstone layer 6 at 64 m and sandstone 7 at 54 m depth, and *downflow* of cool and higher EC water from near the SWL to sandstone 7 at 54 m depth (without pumping). Adjacent pumping is believed to be responsible for reducing the head of thin sandstone layer 7 at 54 m. Some of the cool, higher EC water that moves down from shallow depth, invading the shallow sandstone layers, appears to be returned at 32 m and 20 m on pumping. Given time, the invading shallow water will also appear in the abstraction responsible. The effect is particularly marked and laterally extensive because the aquifer comprises thin layers of high permeability sandstone separated by thick low permeability breccia layers (Tate 1968). The example illustrates upward and downward wellbore flow taking place in an openhole borehole penetrating a strongly layered aquifer where the producing horizons are thin sandstone beds separated by thick, low permeability breccia wedges. The relatively high permeability of the thin sandstones means that the effects of abstraction are spread over a wide area. Video logging showed that the large inflow identified at 72 m was from fractured sandstone.

Water inflows and wellbore flows, Permian sandstone/breccia sequence, Dumfries basin, Scotland

A similar example of aquifer layering, inflow from thin sandstones and fractures and wellbore

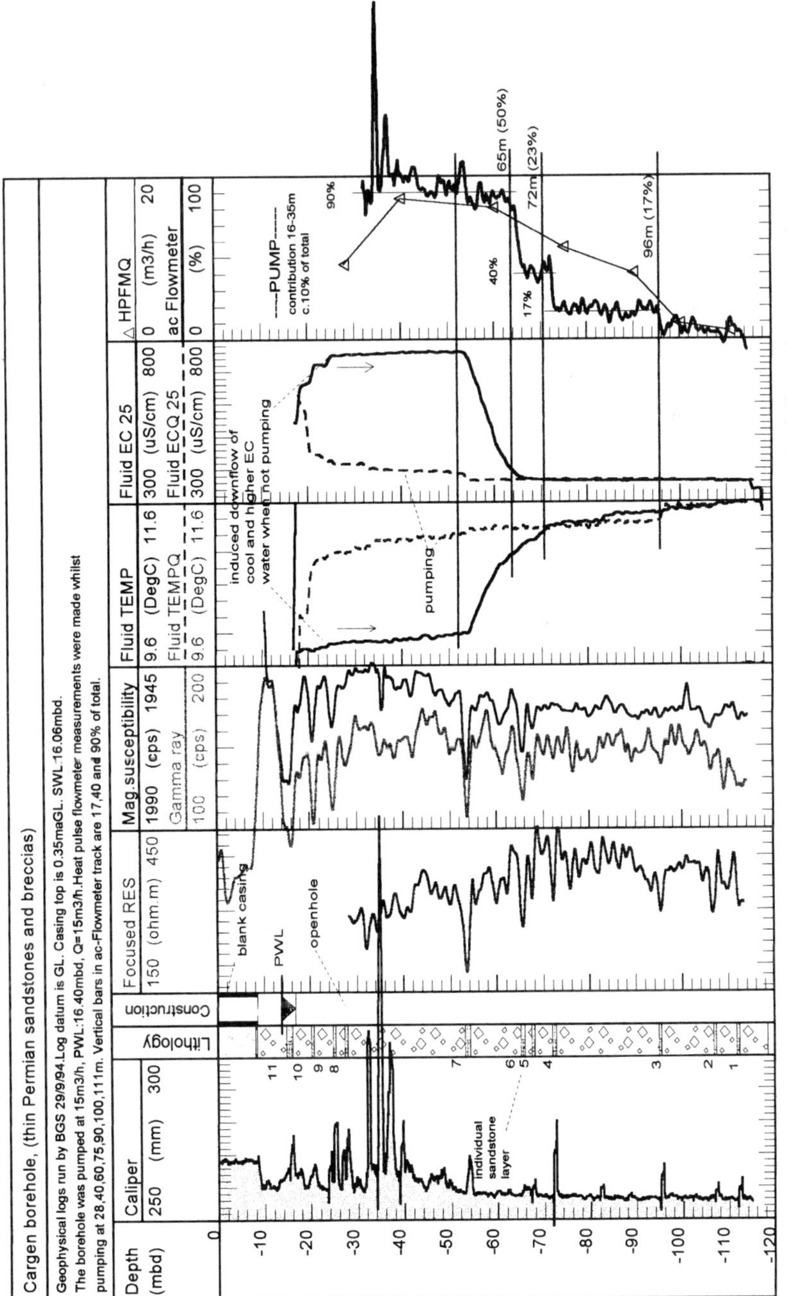

Fig. 6. Upward and downward wellbore flow, Permian sandstones and breccias, Cargen borehole, Dumfries basin.

flow comes from the same aquifer on the NW side of the basin, approximately 6 km to the north where three production boreholes were drilled on a site and penetrated Permian sandstones and breccias.

Caliper logging showed that the sandstones were softer than the breccias, and laboratory tests of the material (Lovelock 1969) showed the breccias to have very low intergranular hydraulic conductivity ($<1 \times 10^{-6}$ cm s^{-1}) whilst the sandstones have much higher values, up to 6.6×10^{-4} cm s^{-1}). Gamma ray and resistivity logs similarly resolved the layering, and breccias below 65 m depth showed higher gamma ray activity on account of a higher Palaeozoic mudstone content, which must have been exposed nearby at the time of deposition.

The pumping fluid logs of the three boreholes are shown in Fig. 7. The boreholes are approximately 50 m apart. Boreholes 2 and 3 are 130 m deep and penetrate three sandstone layers separated by thicker breccia layers, whilst borehole 1 is 146 m deep and penetrates a fourth sandstone layer as shown by the lithological column in Figs 7 and 8. A heat-pulse flowmeter log run in borehole 1 prior to pumping, identified strong upflow (approximately 25 m^3/h, 23.6 m/min) taking place from sandstone layer 4 at 146 m depth, with loss of flow at 129 m and at 60–65 m depth where it invades sandstone units 3 and 2 (Fig. 7).

Fluid logging again identified wellbore *downflow* prior to pumping from approximately 38 to 65 m depth, and *upward* wellbore flow was identified from the base of borehole 1. Five separate inflows were identified, representing 20% moving down to the pump at 38 m from approximately 31 m depth, and 80% moving up to the pump from below. Reference to their positions and the caliper logs is presented in Fig. 8 and shows that the main inflows generally correspond to enlargements representing the sandstone layers and fracturing, but several are inflows from breccias in proximity to the sandstone layers.

Figures 7 and 8 show that water in the deeper borehole is significantly warmer (approximately 0.4°C) due to the deeper circulation in sandstone unit 4 and unusual peak warming is displayed on the temperature logs of boreholes 2 and 3 at approximately 65 m depth alongside sandstone 2 horizon, flanked above and below by cooler water. The most likely origin for this log feature is warmer water invasion of the sandstone 2 horizon from the upflow in borehole 1. The fluid EC logging showed that whilst the pumped water from the three boreholes is very similar ($ECQ_{25} = 370$–$380\ \mu$S/cm), the fluid EC at depth in each borehole when pumping is quite different.

The impeller flowmeter profiles for the three holes, shown to the right of the EC curves (Fig. 7), are each similar and represent cumulative flowrates whilst pumping. They indicate the proportion of total flow moving up the boreholes and reveal *stepwise* changes in flow rate indicative of inflows from fractures at 65 and 105 m depth. The largest inflow in each borehole is at approximately 65 m depth where 40–45% of the total flow enters. The invasion of sandstone unit 2 and 3 by sandstone 4 waters in borehole 1 is also of interest for sampling purposes because pumped samples taken from boreholes 2 and 3 will contain some groundwater from sandstone 4 horizon, derived from the wellbore flow invasion, even though they do not themselves penetrate that horizon. It is also evident that pumped samples from boreholes 1 and 3 will also contain a proportion of the high EC wellbore downflow currently invading sandstone 2 horizon from shallow level in borehole 2. The abstraction responsible for lowering of the head in sandstone 2 horizon will eventually receive these waters as well. The hydraulic situation described is typical of thin layered aquifers within a discharge area of a sedimentary basin, having recharge at higher elevation on its margins, and influenced by adjacent abstractions.

Assembly of the logging data for the three boreholes into a scale cross-section of the site (Fig. 8) illustrates the vertical lithological layering, and the vertical variation in water quality and fluid temperature in relation to the geological layering. It reveals the strata in this example to be almost horizontal. Elsewhere in this region the strata may be dipping, as illustrated by the logging data in Fig. 9 which shows the relationship of water inflows and a steep strata dip recorded in some water exploration boreholes near Moffat. In this example the dip is equivalent to approximately 75 m/km, which could be very significant in terms of different sources of groundwater, and chemical characteristics within the capture zone. These examples illustrate how the combination of formation and fluid logging data on scale cross-sections allows the relationship between water inflow and lithostratigraphy and geological structure to be examined. They also provide a view of how the water inflows identified on a particular site relate to the local and regional groundwater flow systems, and how water may be derived from several aquifers and move between individual boreholes on a site when pumped.

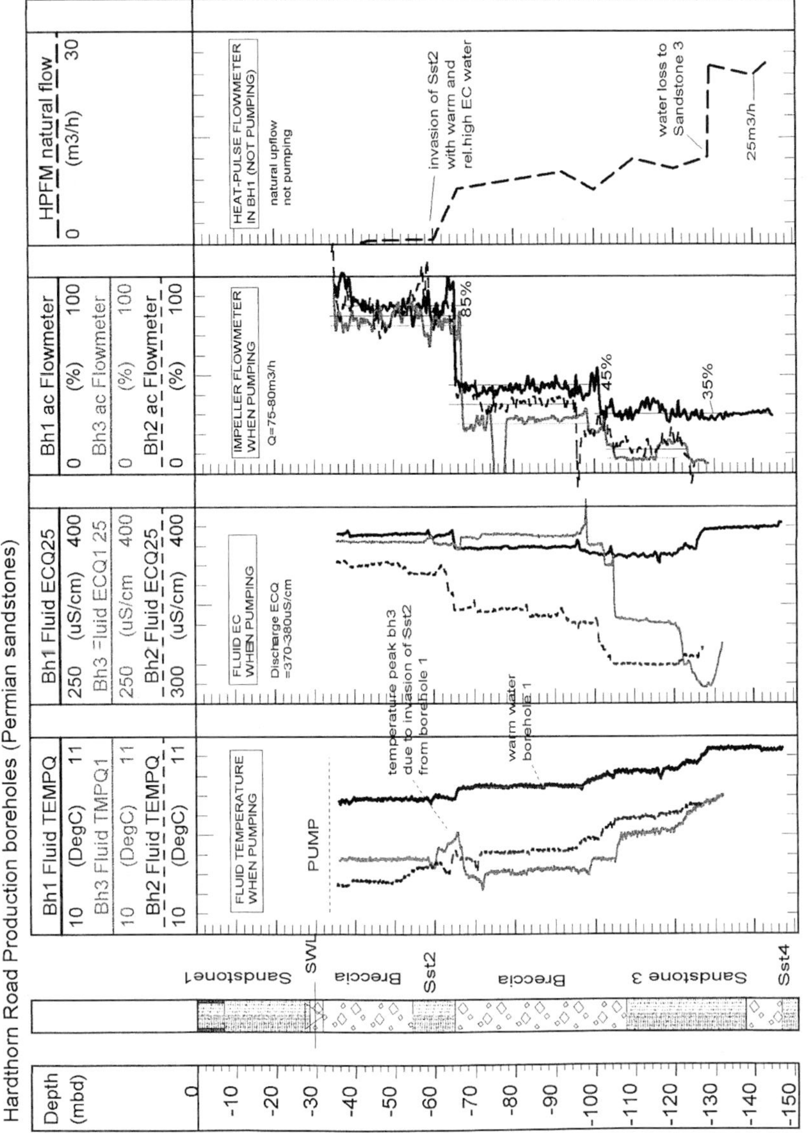

Fig. 7. Comparison of pumped fluid logs, Terregles Production boreholes, Dumfries.

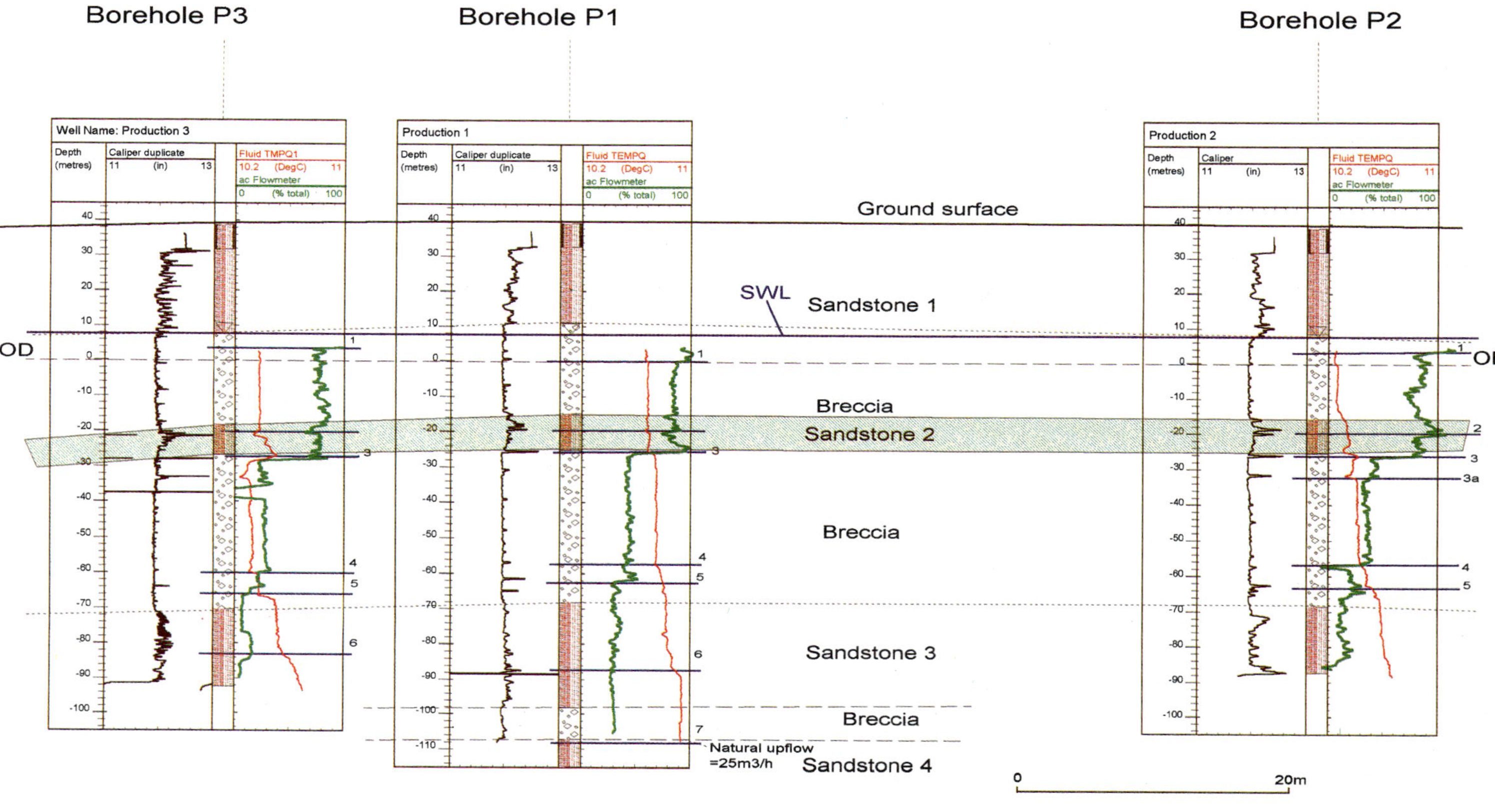

Fig. 8. Hydrogeophysical cross-section, Terregles site, Dumfries.

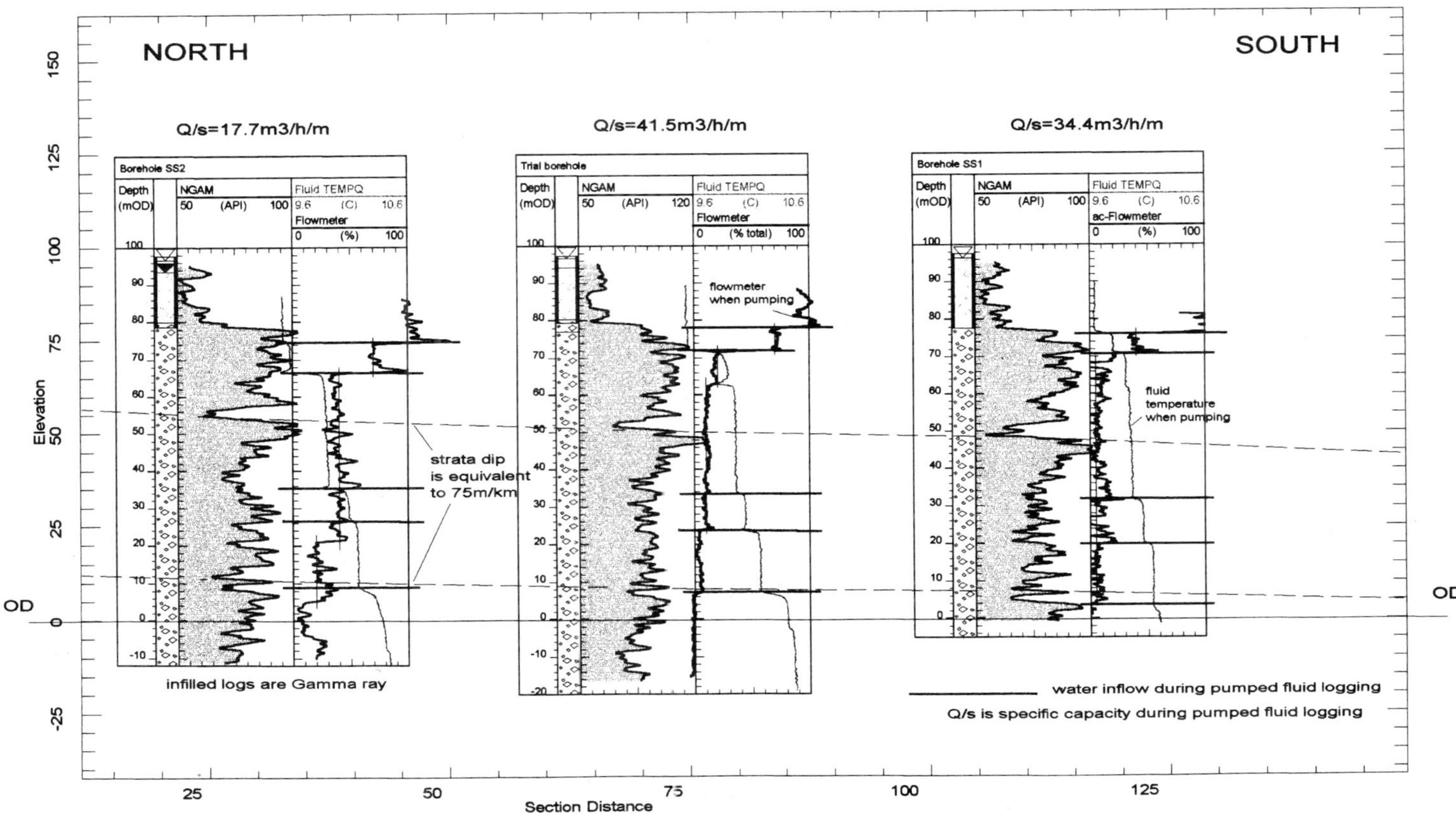

Fig. 9. Scale cross-section showing strata dip and water inflow horizons, Permian breccias, Dyke Farm, Moffat, Scotland.

Artesian flow and recharge dating by CFC's, fractured sandstone aquifer, Isle of Arran

Several boreholes have recently been drilled and tested for public water supply on the Isle of Arran including some in Permian sandstones, a few kilometres inland from the west coast. Five boreholes proved to be artesian and at one site boreholes overflowed at approximately 60 m^3/h. (Fig. 2). Drilling and logging indicated that the sandstones were strongly fractured, cut by igneous intrusions (felsite, dolerite) and that the artesian head was due to recharge to the fractured sandstone aquifer at higher elevation, updip to the north and east.

Formation and fluid logs run in borehole 1C, the deepest drilled, are shown in Fig. 10. The borehole became artesian after 60 m depth (8 m AOD). Fluid temperature logs run in the overflowing borehole revealed eight separate inflows at +29, +13, +5, −1, −13, −22, −56 and −71 m AOD. The fluid temperature of the upward moving mixture decreased at each of these inflows, although there was only a slight change in fluid EC. An impeller flowmeter log run in the natural upflow revealed stepwise changes in flowrate at these positions but also suggested an apparent *loss* of flowrate at the base of the casing. This was also evident on the pumped log. When corrected for cross-section area, the flowrate measured in the casing was still less than that observed in the open hole and it was suspected there was a loss of flow at that point. Further analysis of the flow logging and comparison with other holes nearby suggested it was an artifact of the logging, and most likely due to incomplete centreing of the flowmeter, so that it was sampling a lower velocity nearer the edge of the borehole, unlike the higher and more representative measurements below the casing where the centralizer fits the hole better. The logging results showed that 68% of the total pumped was obtained from below Ordnance Datum, whilst 32% was derived from above. No inflow was recorded between −22 and −56 m AOD.

The chlorofluorocarbons analysis (CFC) of water pumped from boreholes 1, 3 and 5 suggested recharge dates ranging from 1950–1978. The fluid logging clearly demonstrated that the pumped waters represent a mixture of groundwaters circulating in several horizons over a vertical interval of approximately 100 m. A downhole sampler is being developed to obtain samples from the horizons, free of present day atmospheric gases, to measure the CFC and other

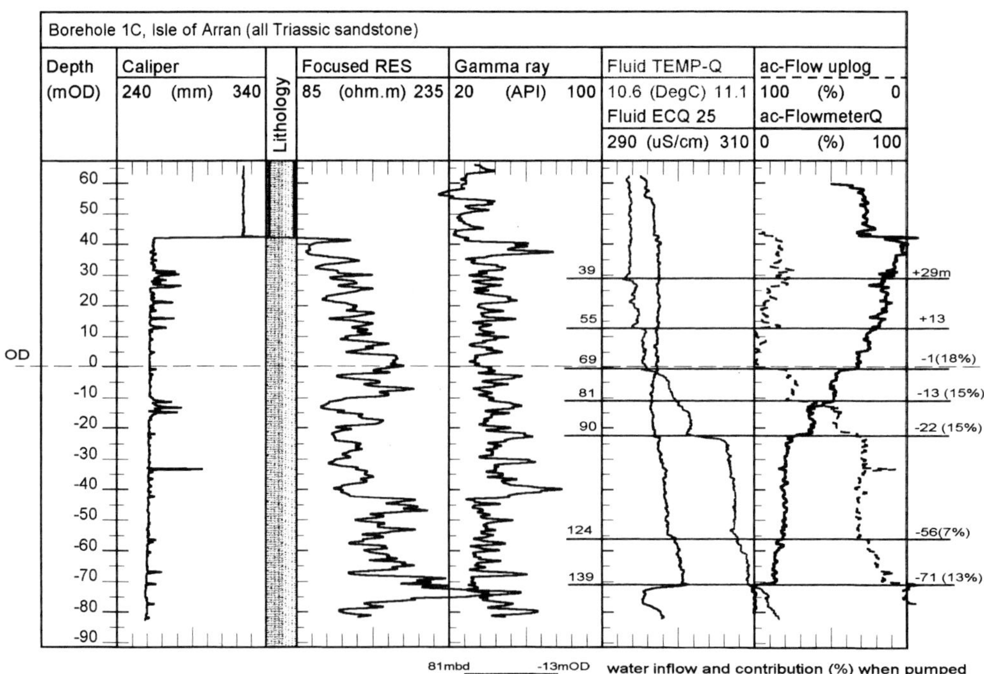

Fig. 10. Water inflows above and below Ordnance Datum, Triassic sandstone, Isle of Arran.

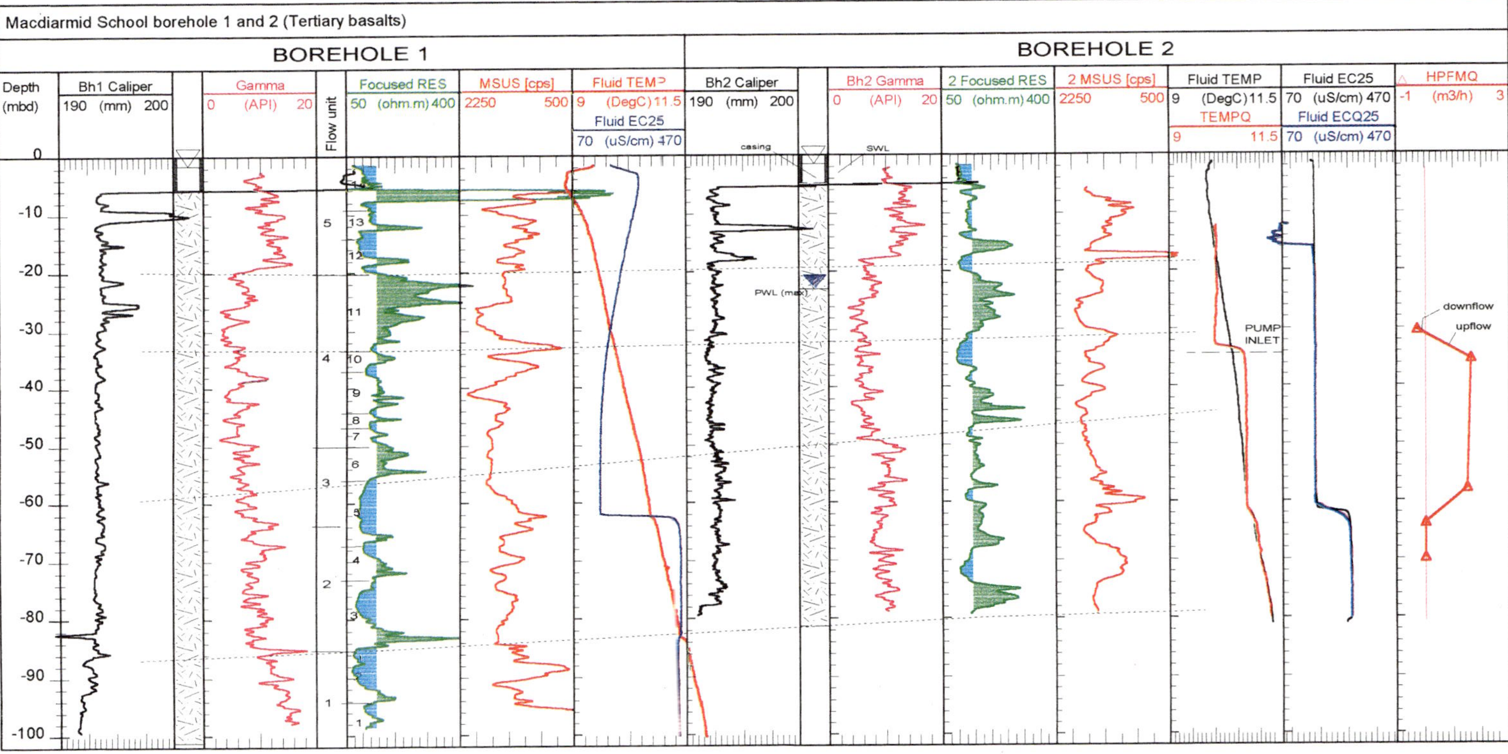

Fig. 11. Formation and fluid logs of Tertiary basalt showing subdivision of lavas and water inflows, Macdiarmid School boreholes 1 and 2, Portree, Isle of Skye.

gas content of the individual circulations. In combination with the flowmeter logging this should provide ages for the individual groundwater circulations making up the flow system, and thus provide recharge dates for the separate groundwaters.

Geological correlation and water inflows, Palaeogene basalts, Isle of Skye

Figure 11 shows formation and fluid logs for two boreholes, 55 m apart, penetrating Palaeogene basalt near Portree on the Isle of Skye. They were drilled in 1998 for groundwater exploration purposes. The formation logs (caliper, gamma ray and magnetic susceptibility) show some similar features in both holes suggesting the lavas dip towards borehole 1 and the gamma ray log (NGAM) indicates that there are probably five distinct flow units present (from 12–30 m in thickness), each having different gamma ray activity, probably due to a different potassium content.

Within the five units defined by the gamma ray log, focused resistivity and magnetic susceptibility measurements resolve individual lava flows, each having a hard massive basalt unit of high resistivity (shown by green infill) and a thicker, softer low resistivity flow top (shown by blue infill). Up to 14 individual hard base, and soft top, lava units are thus recognized by the resistivity log measurements in borehole 1. The magnetic susceptibility measurement, which shows increasing values to the right, reveals that the inferred flow top material has a generally higher susceptibility. This is due to oxidation, weathering and sub-aerial exposure of the flow top surfaces.

Fluid logging indicated water movement at 18, 62, and 84 m depth. Reference to the formation log interpretation reveals that the shallow inflow is associated with the junction between flow units 4 and 5 (lava flows 11 and 12). The deeper inflows occur at junctions between flow units 2 and 3 (within lava flow 5) and flow units 1 and 2 (lava flows 2 and 3). At these junctions, higher susceptibility peaks are also present signifying probable weathering of the surface of the flow. Borehole 2 was pumped at 8.9 m^3/h, with pump suction placed at 33 m depth, and the pumped logs confirmed the inflows at 18 and 62 m depth. Heat-pulse flowmeter measurements showed approximately 20% of the total enters at 18 m and 80% enters at 62 m depth. The pumping water level shown for borehole 2 indicates that the upper inflow was dewatered during the short period of pumping.

The logging showed it is capable of resolving the components of the lava flows and identifying the horizons where water moves. The weathered upper surfaces of the lavas appear to be productive, but the permeability of the particular horizons in the examples shown is low. The logging results and correlations showing the water–bearing horizons related to particular lava flows in the sequence, can be useful for borehole siting and the targeting of groundwater active horizons elsewhere.

Conclusions

Geophysical logging of boreholes drilled for hydrogeological investigations can be used to investigate aquifer layering and the presence of wellbore fluid movement, both of which influence water quality at depth.

Fluid electrical conductivity, fluid temperature and borehole flowmeter measurements, made before and during pumping, provide information on the positions of water inflow in boreholes and their relative contribution to well yield. Fluid temperature and fluid conductivity log profiles are not fixed in space and time but are transient, and their profiles relate to the borehole hydraulics prevailing at the time of measurement.

Wellbore flow is a common feature in boreholes penetrating layered aquifers and is prevalent in hard rock fractured aquifers in hilly terrain typical of many Celtic regions. The wellbore flow may be upwards or downwards, and leads to invasion of low head horizons and mixing of groundwaters from different layers. The effects can be surprising, and in the example at Terregles, water pumped from one borehole contained groundwater from an aquifer not penetrated by the borehole, but transferred to it by upward wellbore flow and invasion of aquifer horizons from another borehole on the same site. Wellbore mixing has implications for water sampling and downhole fluid logging is necessary to identify the wellbore flow and determine the separate inflows contributing to the pumped sample, especially in open hole and long screened boreholes. Sharp stepwise velocity changes on flowmeter profiles reveal concentrated inflow at bedding plane fissures and fractures, and are characteristic of hard rock aquifers, and can be distinguished from smooth intergranular contributions from porous and permeable horizons in granular material.

Scale cross-sections constructed from geophysical logging data, which show the position of water inflows in relation to the lithological layering, help in understanding groundwater occurrence and movement at any particular site in relation to geological structure. They provide insights into the local and regional groundwater flow systems present, and can illustrate the movement of water between boreholes on a site. The compiled information can provide a useful basis for examination of the development of the aquifer, particularly in terms of its Pleistocene history and base level control of flow routes.

The author would like to acknowledge the help and support provided by BGS colleagues during the various logging surveys in the Celtic regions, particularly from Derek Ball, Alan Cripps and John Talbot who assisted with the logging and pumping, and contributed to the interpretations. The paper is published by permission of the Director, British Geological Survey (NERC).

References

ALLEY, W. M. 1993. *Regional Groundwater Quality*. Van Nostrand Reinhold, New York.

BATEMAN, R. M. 1985. *Open-hole log analysis and formation evaluation*. D. Reidel Publishing, Dordrecht.

BOMAN, G. K,, MOLZ, F. J. & BOONE, K. D. 1997. Borehole flowmeter application in fluvial sediments: methodology, results and assessment. *Ground Water*, **35**(3), 443–450.

BOULTON, G. S., CABAN, P. E., VAN GIJSSEL, K., LEIJSNE, A., PUNKARI, M. & VAN WEERT, F. H. A. 1995. The impact of glaciation on the groundwater regime of northwest Europe. *Global and Planetary Change*, **12**, 397–413.

CHAPELLIER, D. 1992. *Well logging in hydrogeology*. Balkema Publishers, Rotterdam.

CHURCH, P. E. & GRAMATO, G. E. 1996. Bias in groundwater data caused by well-bore flow in long-screen wells. *Ground Water*, **34**(2), 262–273.

DESBRANDES, R. 1985. *Encyclopedia of Well Logging*. Institut Français du Pétrole Editions Technip, Paris.

DEWAN, J. T. 1983. *Essentials of modern open-hole log interpretation*. Pennwell Books, Tulsa.

DOVETON, J. H. 2001. Geological application of wireline logs; a keynote perspective. *In*: LOVELL, M. A. & PARKINSON, N. (eds) *Geological Application of Well Logs*. American Association of Petroleum Geology, special volume (in press).

DUDGEON, C. R., GREEN, M. J. & SMEDMOR, W. J. 1975. *Heat-Pulse flowmeter for Boreholes. A new design capable of measuring vertical flows in the range 1–200 mm/sec*. Technical Paper TR4, Water Resources Centre, Medmenham Laboratory, Bucks, UK.

FREEZE, R. A. & CHERRY, J. A. 1979. *Groundwater*. Prentice Hall, New Jersey.

GAUS, I. & Ó DOCHARTAIGH, B. É. 2000. Conceptual modelling of data-scarce aquifers in Scotland: the sandstone aquifers of Fife and Dumfries. *This volume*.

GOSSELL, M. A., NISHIKAWA, T., HANSEN, R. T., IZBICKI, J. A., TABIDIAN, M. A. & BERTINE, K. 1999. Application of flowmeter and depth-dependent water quality data for improved production well construction. *Ground Water*, **35**(5), 729–735.

HUBBERT, M. K. 1940. The theory of groundwater motion. *Journal of Geology*, **48**, 785–944.

KEYS, W. S. & MACCARY, L. M. 1971. *Application of borehole geophysics to water resources investigations*. Techniques of Water Resources Investigations, USGS Book 2 Chapter E1.

LOVELOCK, P. E. R. L. 1969. *Core analysis results from No. 2 borehole at Dumfries factory, ICI*. BGS Technical Report **WD/ST/69/3**.

MOLZ, F. J., BOMAN, G. K., YOUNG, S. C. & WALDROP, W. R. 1994. Borehole flowmeters: field application and data analysis. *Journal of Hydrology*, **163**, 347–371.

——, MORIN, R. H., HESS, A. E., MELVILLE, J. G. & GUVEN, O. 1989. The impeller meter for measuring aquifer permeability variations: evaluations and comparisons with other tests. *Water Resources Research*, **25**, 1677–1683.

—— & YOUNG, S. C. 1993. Development and application of borehole flowmeters for environmental assessment. *Log Analyst*, **3**, 13–23.

REILLY, T. E., FRANKE, O. L. & BENNETT, G. D. 1989. Bias in groundwater samples caused by wellbore flow. *Journal Hydraulics Engineering*, **115**(2), 270–276.

REPSOLD, H. 1989. Well logging in groundwater development. IHP Program, UNESCO, *International contributions to Hydrogeology* Vol. **9** International Association of Hydrogeologists, Heise Hannover.

SCHMIDT, K. D. 1977. Water quality variations from pumping wells. *Groundwater*, **15**(2), 130–137.

TATE, T. K. 1968. *Flow investigation at Dumfries*. BGS Technical Report **WD/ST/68/8**.

TÓTH, JA 1963. A theoretical analysis of groundwater flow in small drainage basins. *Journal of Geophysical Resources*, **68**(16), 4795–4811.

WRIGHT, E. P. & BURGESS, W. G. (eds). 1992. *The Hydrogeology of Crystalline Basement aquifers in Africa*. Geological Society, London, Special Publications, **66**, 183–220.

The water resources of Bardsey, north Wales

M. M. WEBB

Centre for Arid Zone Studies, University of Wales, Bangor, Gwynedd LL57 2UW, UK

Abstract: Bardsey is a small island off the Welsh coast. Water supplies are drawn from spring sources but water shortages are experienced in summer. Sanitation depends on Elsan type toilets. There are limited groundwater resources in the Gwna Mélange, which is of Monian age, with groundwater transport controlled by fracture flow. There is evidence of some groundwater contamination.

Bardsey lies 3 km off the west coast of north Wales, centred at SH120215. The island has a land area of 178 ha, and is some 2.8 km long from north to south and up to 1 km wide from west to east (Fig. 1). There is a north–south change in slope at approximately 40 m AOD. To the east is Mynydd Enlli with slopes in excess of 20° rising to a summit of 167 m AOD. To the west and south, slopes of less than 4° give way to a lowland plain.

This small island is formed of ancient crystalline rocks which are partly till covered. Although the overall water demand is small, groundwater is the main source, both from springs and shallow wells. This paper provides an insight into the issues relating to small island groundwater resource development and management in the Celtic regions.

Bardsey is owned by the Bardsey Island Trust and managed in conjunction with the Countryside Council for Wales. The island is designated as a National Nature Reserve, a Site of Special Scientific Interest, a Special Protection Area and an Environmentally Sensitive Area. Sheep graze over much of the island and there are limited areas of improved pasture and arable cropping. The resident population of less than ten people rises to >50 in summer.

Surface drainage is ephemeral with permanent steams only on the lowland plain. The numerous springs are also mainly seasonal. The island has a temperate maritime climate. The Meteorological Office weather station at the lighthouse recorded a long term (1961–1981) mean annual rainfall of 855 mm, with less than a third of the annual total falling between March and July (Jones 1988).

There is no mains water supply or sewerage. The water supply for the Trust properties is a spring at Site 8 which has a maximum flow rate of 720 l d^{-1} (Table 1), with 8 m^3 tank storage, and is supplemented by rainwater harvesting. Domestic 'grey water' drains through a short section of pipe to the outside of each property and sanitation is in the form of Elsan toilets. Previous work on the island includes a water survey by Pyefinch (1937) and more recent work by Jones & Arnold (1996). The six month long study reported here, was undertaken to facilitate development of the water resources in keeping with the natural environment and the financial constraints of the Bardsey Island Trust.

Bedrock comprises the Gwna Mélange which is part of the Monian System. It is over 2000 m thick and is probably the remnant of a regional mélange unit (Gibbons & McCarroll, 1993). The highly varied clast lithology includes quartzites, limestones, greywacke and feldspathic sandstones, cherty sediments, basic lava, siltstone, and mudstone (Schuster 1977; Jones 1988; Gibbons & McCarroll 1993). Sub-units of the Gwna Mélange are distinguished by the nature and composition of the clasts and matrix. Intrusive material includes granite and Palaeogene dolerite.

Periods of tectonic activity have created folding and faulting, and spatially variable degrees of metamorphosis. There is a major fault to the west of the ridge of Mynydd Enlli with a north-south strike and a dip of 40° to the west. There are numerous smaller faults, and some near vertical faults with westerly strikes.

Till is present over lower lying land.

Data gathering

Water levels of wells and elevations of springs were recorded for the weeks ending 12 July and 27 July 1998 (Table 2). Depth to water table was measured from a known datum with a tape. Data for the week ending 12 July were used to construct a preliminary water level contour map (Fig. 2). Flow rates of the larger springs were measured by timing a known volume (Table 1).

From: ROBINS, N. S. & MISSTEAR, B. D. R. (eds) *Groundwater in the Celtic Regions: Studies in Hard Rock and Quaternary Hydrogeology*. Geological Society, London, Special Publications, **182**, 239–246. 1-86239-077-0/00/

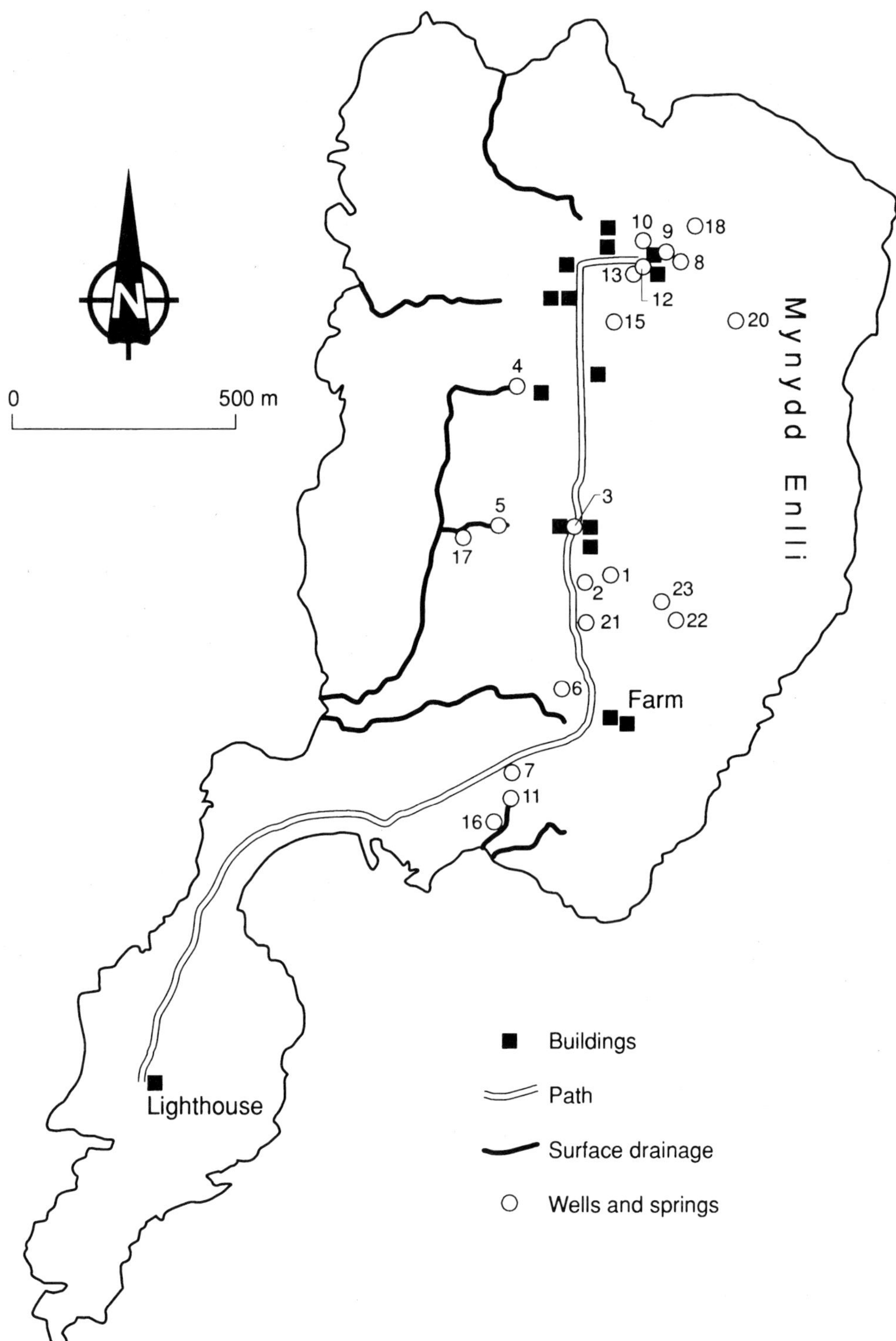

Fig. 1. Bardsey showing location of wells and springs.

Table 1. *Spring flow rates*

Site number	Grid reference	Date	Flow rate (l/min)
1	1201 2160	5/8/98	2.0
1	1201 2160	23/9/98	0.85
8	1217 2219	5/8/98	0.25 to 0.5
8	1217 2219	23/9/98	<0.25
10/12	1213 2220	23/9/98	0.25
11	1182 2112	23/9/98	0.04
18	1218 2224	24/9/98	0.15
20	1225 2209	24/9/98	0.3

At site 18 there are three adjacent springs forming a seepage zone 11.5 m in length.

The Hanna Instruments Palintest Colorimetry Kit was used to measure pH (range 4–11), ammonia (range 0–2.5 mg/l NH_4-N), nitrate (range 0–50 mg l^{-1} NO_3-N), orthophosphate (range 0–5 mg l^{-1} PO_4-P) and iron (range 0–5 mg l^{-1} Fe-total) in the field. Specific electrical conductivity was measured in the laboratory using a Jenway 4010 Conductivity Meter with a 1 cm probe (Table 3). Samples from selected sites were further analysed by the British Geological Survey at Wallingford (Table 4). All samples taken for laboratory analyses were first filtered through a nitrocellulose 0.45 µm membrane filter and stored in polythene containers.

The auger hole slug-test described by Todd (1980, pp. 76–77), which measures horizontal hydraulic conductivity in the immediate vicinity of cylindrical unlined shallow wells, was used at Sites 6, 7, and 17 (Fig. 1). At each well site measurements were made of the diameter and depth of the well and the depth to rest water level.

In the area of the working farm (Fig. 3), water for domestic use is carried from site 6 (Fig. 1), and when the source dries up water is carried over 250 m from site 7. Sites 6 and 7 are shallow wells constructed on natural springs.

To the north and east of the farm the terrain is rugged and the soil layer thin to absent. To the south and west are low-lying arable fields which are partly marshy. Till is present between sites 6 and 7 (Fig. 3) and there are narrow bands in the bedrock lithology within the central section of the survey area which continue in a south-westerly direction towards the coast. All the bedrock lithologies are highly indurated and the limestone bands are metamorphosed. Site 7 and two undeveloped springs at sites 11 and 16 (Fig. 1) merge from between lithological bands, to discharge southward. Site 6 is in marshy ground and appears to be associated with a major north trending fault.

In an attempt to improve the water supply to the farm a surface resistivity survey was carried

Table 2. *Water levels of wells and springs*

Site number	Grid reference	Elevation (m aOD)	Depth to water table (m) 12/7/98	Depth to water table (m) 27/9/98	Change in water level (m)
1	1201 2160	55	0.42	0.55	−0.13*
2	1198 2160	44	0	0.07	−0.07
3	1193 2170	37	3.5	3.5	0
4	1184 2194	23	0.17	0.22	−0.05
5	1181 2167	22	0	0	0
6	1190 2136	24	0.31	0.4	−0.09
7	1181 2117	17	0.3	0.35	−0.05
8	1217 2219	64	0.37*	0.11	+0.25
9	1215 2220	57	0	0	0
10	1213 2220	54	0	0	0
11	1182 2112	11	0	0	0
12	1212 2218	56	0	0	0
13	1211 2217	56	0	0	0
15	1202 2206	55	0.2	0.2	0
16	1182 2106	4	0.2	0.2	0
17	1176 2165	19	0.45	0.5	−0.05
18	1218 2224	69	0	0	0
20	1225 2209	130	0	0	0
22	1217 2148	105	0	Dry	Dry
23	1215 2150	111	0	Dry	Dry
24	1151 2199	15	0	0	0

* Abstraction affecting water level.

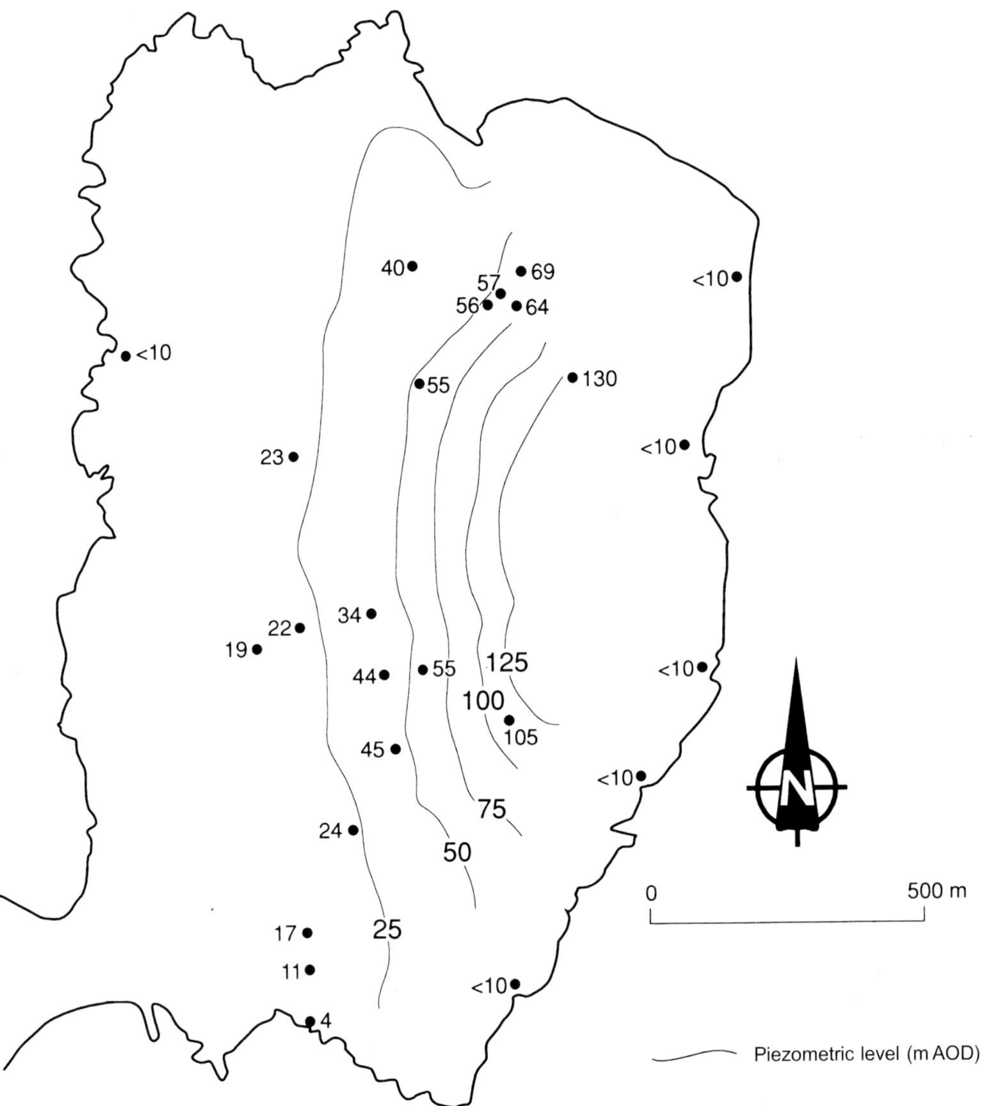

Fig. 2. Water level contours (m AOD).

out. Conductivity measurements in millimhos m^{-1} were made using the Geonics EM31 non-contacting terrain conductivity meter (McNeil 1980). Conductivity was recorded for vertical dipoles (horizontal coils) and horizontal dipoles (vertical coils), and readings were taken at 5 m intervals along field boundaries. Soil depths were measured along the traverses with a soil auger. The examples shown in Fig. 4 indicate likely favourable conditions for a new well in parts of traverse 14.

The groundwater body

Conceptual groundwater system

Using Thornthwaite's approximation to derive potential evapotranspiration (PE), and a short term study on runoff carried out during 1997 by the University of Wales, the outline island-wide water balance can be apportioned ($Mm^3\ a^{-1}$):

$$\text{Rainfall} = \text{PE} + \text{runoff} + \text{infiltration}$$
$$(1.5) \quad (1.1) \quad (0.3) \quad (0.1)$$

Table 3. *Water quality analyses*

Site number	National grid location	Altitude (m AOD)	NH_4-N (mg/l)	Fe_{total} (µg/l)	NO_3-N (mg/l)	PO_4-P (mg/l)	pH	SEC (µS/cm at 25°C)
1	1201 2160	55	0	0	5	0.5	7.5	574
2	1198 2160	44	0.5	0.5	0	0	7	
3	1193 2170	37	1.5	0.5	25	2.5	7.5	806
4	1184 2194	23	0.25	0.5	5	0.5	7	645
5	1181 2167	22	0.25	0	0	0.5	7	781
6	1190 2136	24	0.25	0	40	0	7.5	790
7	1181 2117	17	0	0	0	0	7	944
8	1217 2219	64	0	0.5	0	0	7	489
11	1182 2112	11	0	0	0	0	7	
12	1212 2218	56	0.5	0.5	0	0	7	
14	1193 2189	38	0.25	0	0	0	7.5	
15	1202 2206	55	1.5	4.5	0	0	6.5	307
16	1182 2106	4	0	0	0	0	7	1124
17	1176 2165	19	0.25	0	5	0.5	7.5	694
18	1218 2224	69	0.5	0.5	0	0	7	443
20	1225 2209	130	0.5	0.5	0	0	7	479

Outflows are evident at points around the coast at between 4 and 10 m AOD, suggesting a possible lower boundary to the aquifer system.

The well at Site 17 is situated at 19 m AOD near the main stream channel and is one of the few perennial sources. The tested hydraulic conductivity is relatively high, of the order 10^{-2} m d^{-1} and there may be hydraulic contact with the stream. Site 6 has an intermediate hydraulic conductivity of 10^{-3} m d^{-1}, but this site usually dries up in late summer. Site 7 has the lowest hydraulic conductivity of 10^{-4} m d^{-1}, is perennial and is also one of the few sites on the island not directly associated with faulting.

During the study, the water levels dropped at seven sites not directly influenced by abstraction (Table 2) and flow rates fell at two sites (Table 1). This reflects the dry weather water table recession, and the limited overall groundwater storage capacity.

The water level contour map (Fig. 2) suggests that the elevated Mynydd Enlli area (Fig. 1) is a

Table 4. *Major ion analyses*

Site number	National grid location	Altitude (m aOD)	Na (mg/l)	K (mg/l)	Ca (mg/l)	Mg (mg/l)	B (mg/l)	Li (mg/l)	SO_4 (mg/l)	Cl (mg/l)	NO_3 (mg/l)
1	1201 2160	55	60.1	2.44	35.22	12.73	0.030	<0.001	22.11	110	1.2
5	1181 2167	22	73.3	10.24	54.65	16.36	0.047	0.000	27.68	141	1.2
6	1190 2136	24	73.4	5.73	52.21	20.57	0.041	<0.001	29.53	136	3.5
7	1181 2117	17	78.7	3.02	70.02	35.35	0.047	0.004	34.63	137	0.7
8	1217 2219	64	54.5	1.57	27.71	10.21	0.026	0.000	21.85	94.2	<0.2
16	1182 2106	4	112.9	6.4	71.57	32.1	0.074	0.002	37.39	193	0.8

Site number	National grid location	Altitude (m aOD)	P (mg/l)	Si (mg/l)	Sr (mg/l)	Ba (mg/l)	Mn (mg/l)	Fe (mg/l)	Zn (mg/l)	Cu (mg/l)	Al (mg/l)
1	1201 2160	55	0.03	2.48	0.1226	0.082	0.001	0.00	0.01	0.000	<0.003
5	1181 2167	22	0.18	3.41	0.2402	0.078	0.034	0.034	0.00	0.000	<0.011
6	1190 2136	24	0.02	3.27	0.1701	0.097	0.000	0.000	0.00	<0.002	<0.012
7	1181 2117	17	0.10	3.24	0.247	0.059	0.075	0.075	0.02	<0.001	<0.018
8	1217 2219	64	0.00	1.61	0.1038	0.072	0.005	0.005	0.02	<0.001	<0.004
16	1182 2106	4	0.19	2.59	0.3503	0.063	0.001	0.001	0.02	0.000	<0.011

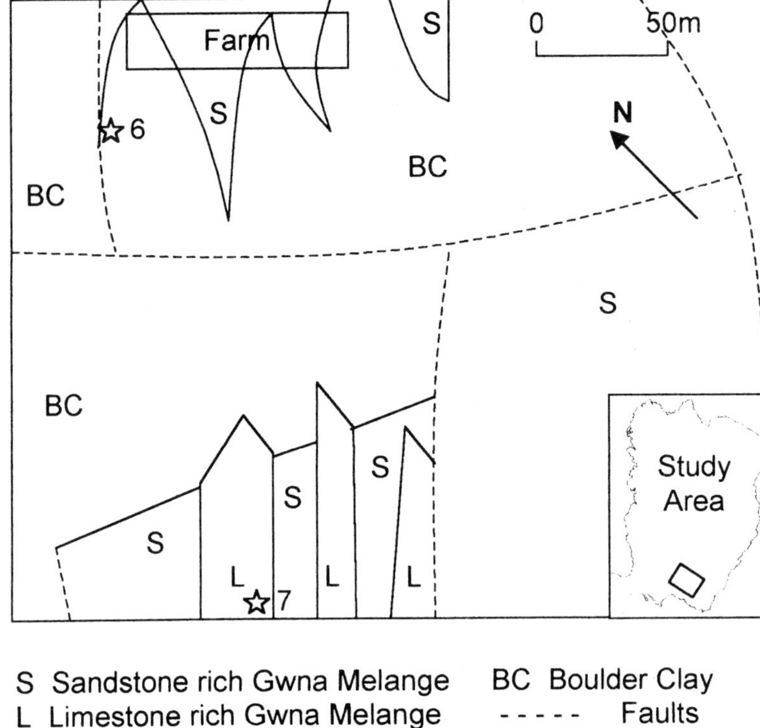

Fig. 3. The farm survey area.

significant recharge zone. The main flowpaths westward are possibly influenced by the main north to south fault, which acts as an impermeable barrier forcing groundwater to discharge as fault springs. Elsewhere, the water table mirrors a subdued form of the surface topography. There is no evidence of any significant easterly groundwater flow.

Groundwater quality

The groundwater is mineralogically potable (DoE, 1993). The mineralization of groundwater increases with residence time and the specific conductivity, which is proportional to the total dissolved solids content of the water, and is inversely related to site elevation and therefore location in the groundwater flowpath (Table 3). The two highest sites, 1 and 8 had lower concentrations of Na, K, Ca, Mg, B SO_4, Cl, Si, Sr and Al than the lower four sites (Table 4).

The groundwater is, however, vulnerable to contamination. This is because it is contained in a shallow fractured aquifer, and there are no water conservation areas or source protection zones. Site 4 for example, is adjacent to, and down gradient from, two small sewage tips and a domestic wastewater drain. Sites 3, 5 and 17 are located on a fault, and are below a sewage tip. Water from Sites 3 and 5 have elevated concentrations of ammonia, nitrate and soluble phosphate. At Site 5 there is an increased level of P-total and a high K/Na ratio. The elevated levels of Fe at site 15 may reflect local storage of scrap iron. Site 16 is a disused trough that collects coastal seepage and it may be contaminated by sea water.

Conclusions

There is a groundwater basin of limited lateral extent and storage. Current water supplies are inadequate to satisfy seasonal demand and there is evidence of localized groundwater contamination, particularly at low elevation.

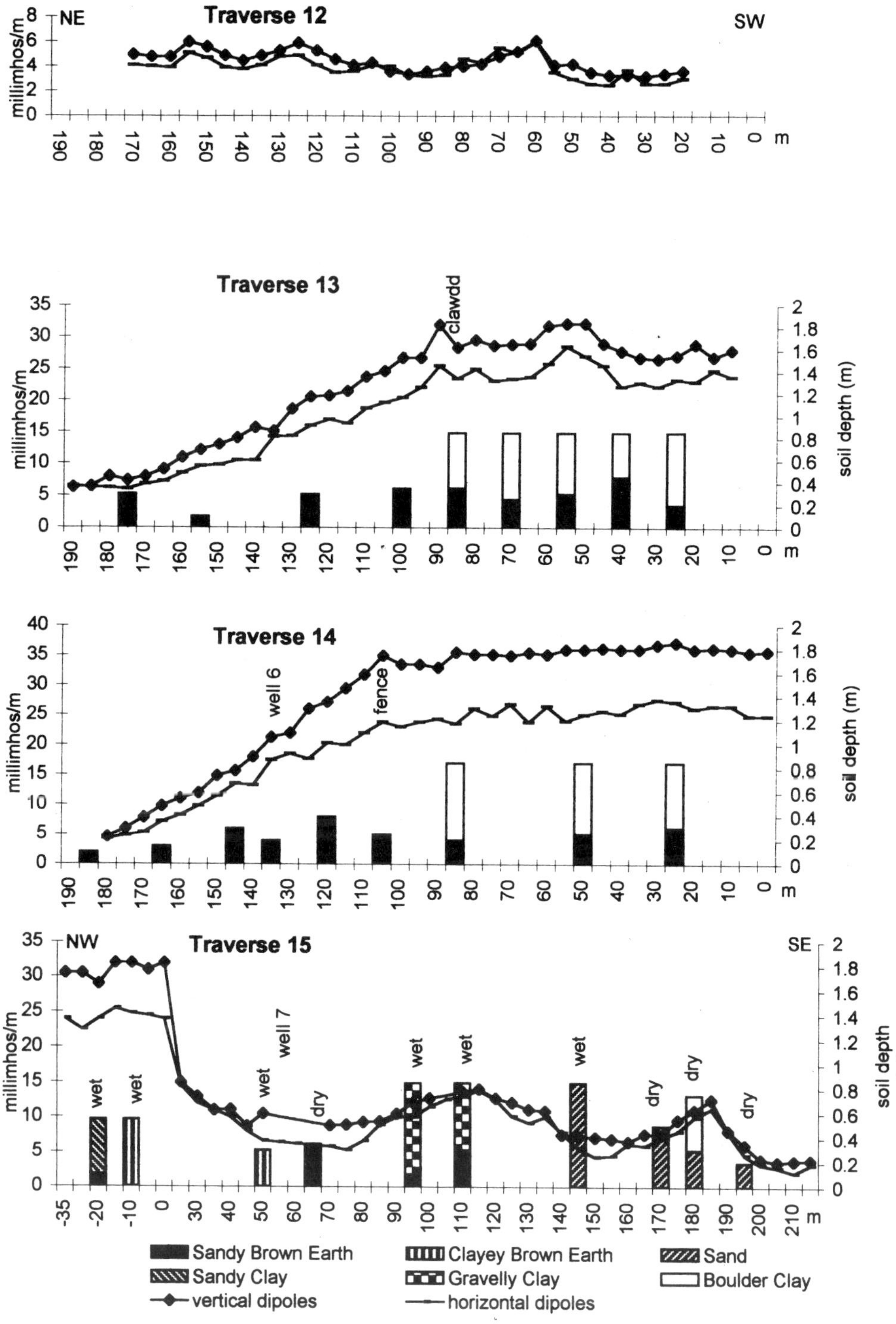

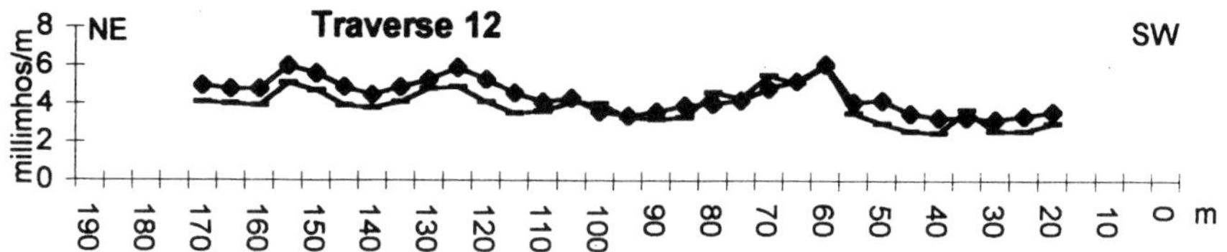

Traverse 13

clawdd

millimhos/m

soil depth (m)

Traverse 14

well 6

fence

millimhos/m

soil depth (m)

NW Traverse 15 SE

wet wet wet well 7 dry wet wet wet dry dry dry

millimhos/m

soil depth (m)

Sandy Brown Earth
Clayey Brown Earth
Sand
Sandy Clay
Gravelly Clay
Boulder Clay
vertical dipoles
horizontal dipoles

Fig. 4. EM31 and soil auger data for traverses 12 to 15 in the farm survey.

The advice, data and equipment provided by the staff of the University of Wales Bangor, Bardsey Bird and Field Observatory, Bardsey Island Trust, the Countryside Council for Wales, Welsh Water, and C. Evans, is gratefully acknowledged. Further thanks are extended to the British Geological Survey for chemical analyses, and to D. Wood for his help with the geology.

References

DoE. 1993. Department of the Environment, Welsh Office, the Scottish Office, Drinking Water Inspectorate. *Manual on Treatment of Private Water Supplies*. Water Research Centre, HMSO, London.

Gibbons, W. & McCarroll, D. 1993. *Geology of the Country Around Aberdaron, Including Bardsey Island*. HMSO, London.

Jones, P. H. 1988. *The Natural History of Bardsey*. The National Museum of Wales, Cardiff.

Jones, R. G. & Arnold, C. J. 1996. *Enlli*. University of Wales Press, Cardiff.

McNeil, J. D. 1980. *Electrical Conductivity of Soils and Rocks*. Geonics Ltd., Ontario, Canada.

Pyefinch, K. A. 1937. The Fresh and Brackish Waters of Bardsey Island (North Wales): a Chemical and Faunistic Survey. *Journal of Animal Ecology*, **6**, 115–137.

Schuster, D. 1977. An Interpretation of Bardsey's Geology. *In: Bardsey Bird and Field Observatory* Report No. **21**, 37–44.

Todd, D. K. 1980. *Groundwater Hydrology*, 2nd edn. John Wiley & Sons, New York.

Role of geographical information systems and groundwater modelling in water resource management for small islands: a case study from St Martin's and St Agnes, Isles of Scilly, UK

DAVID C. WATKINS

Camborne School of Mines, University of Exeter, Redruth, Cornwall TR15 3SE, UK (e-mail: D.C.Watkins@csm.ex.ac.uk)

Abstract: Common to other small islands, water resource management is a critical issue on the Isles of Scilly, UK. This study examined the current practice of water management on the islands of St Martin's and St Agnes in the Isles of Scilly so that a local water management plan could be formulated. The study commenced with a survey of water supplies which included locating wells, rainwater tanks and septic tanks and determining their connections and utility. Wellheads were level surveyed with respect to UK Ordnance Datum, allowing reduced groundwater levels to be monitored. Groundwater is abstracted from shallow dug wells in blown sand or Head deposits and deep wells bored into fractured granite. Hydraulic tests were conducted in seven wells to determine well and aquifer characteristics, although these tests were limited by the low yields of the wells. A geographical information system was used to collate and process information. MODFLOW and MODPATH models were constructed to help interpret the hydrogeology; to determine the impact and sustainability of groundwater abstractions; and to determine the possible extent of source protection zones around wells. The present rate of abstraction is sustainable but, due to large natural variations in the water table level, groundwater resources will always be vulnerable during periods of dry weather. The main threats to existing groundwater supplies are saline intrusion, elevated nutrient levels from agriculture and bacterial contamination from septic tanks.

Freshwater resources are clearly limited on small islands. A lack of surface water sources due to small catchment areas often leaves groundwater abstraction as the only viable freshwater supply, with the inherent risk of saline intrusion. A secondary water resource is provided by utilizing rainwater from roof catchments, and this practice is fairly common on small islands (Falkland 1991). Wastewater must also be disposed of and this is often achieved by the use of septic tanks and final discharge to the sub-surface. This can lead to a conflict in the use of groundwater as a receptor of wastewater whilst also being a supply of freshwater, particularly in the limited confines available to small island communities. Careful water resource management is therefore critical to small islands.

The Isles of Scilly lie in the Atlantic some 40 km west of Cornwall (Fig. 1). Totalling 16 km^2 in extent, they consist of a series of granite outcrops with more than 100 small islands. Five of the islands support a permanent population of around 2000 people but, through tourism, this rises to 5000 people in summer. Some 80% of residents live on the largest island, St Mary's. The other inhabited islands, Tresco, Bryher, St Martin's and St Agnes, each support permanent populations of between 50 and 120 people and summer populations of between 100 and 500 people. The islands of St Martin's and St Agnes are characterized by large seasonal variations in population, small, low-lying land areas and relatively long coastlines (Table 1). No point on St Martin's is more than 370 m from the sea and on St Agnes the furthest location from the coast is 310 m.

Historically, water has always been a critical resource on the Isles of Scilly, particularly for the smaller islands. Water resources are limited on the islands and are under particular stress during the summer months when demand is high. The island of St Mary's derives its water principally from wells abstracting from granite and minor alluvial deposits. It has a central storage and treatment facility and a water distribution network. A water shortage crisis in 1992 resulted in the introduction of a small desalination plant (Garret 1992). In times of high demand, brackish water abstracted from wells near the coast is desalinated by reverse osmosis and blended with water derived from fresh groundwater sources.

From: ROBINS, N. S. & MISSTEAR, B. D. R. (eds) *Groundwater in the Celtic Regions: Studies in Hard Rock and Quaternary Hydrogeology*. Geological Society, London, Special Publications, **182**, 247–268. 1-86239-077-0/00/ \$15.00

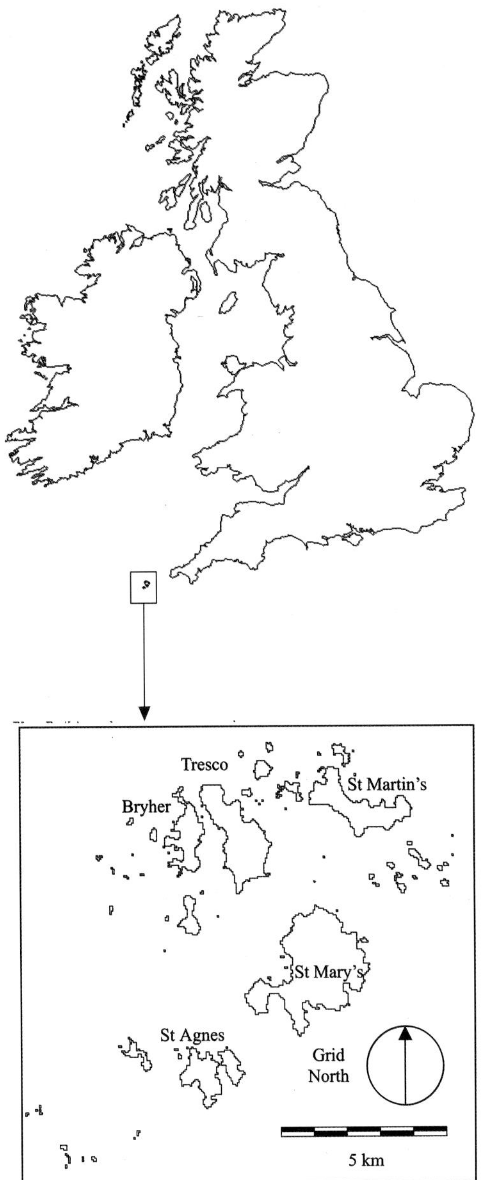

Fig. 1. Location of the Isles of Scilly.

Bryher has a small central water supply system, whilst the other islands, Tresco, St Martin's and St Agnes, are more reliant on individual supplies for each property. These are derived from a combination of groundwater (from wells) and rainwater (collected from roof drainage).

Water resource studies were conducted for the islands of St Martin's and St Agnes during 1997–1998 and 1998–1999, respectively. The principal objectives of the studies were to assess the state and extent of existing resources, identify potential problems and constraints on the development of water resources, and to formulate and propose a future water management plan.

On both islands, a questionnaire was circulated to the permanent residents asking for information on their current water supply and wastewater disposal. All wells and septic tanks were located and each wellhead was level surveyed with respect to Ordnance Datum. A geographical information system (GIS) was used to collate information from the surveys. Hydraulic tests were conducted on selected wells to determine yields and provide an indication of aquifer properties. Distributed numerical models of groundwater flows were constructed to test and improve hydrogeological interpretation and to delineate possible source protection zones around the wells.

Local geology

Barrow (1906) conducted the first full geological survey of the Isles of Scilly. The relevant main units are drift deposits of Holocene wind blown sand, Pleistocene glacial deposits and Head (known locally as Ram) and a bedrock of Carboniferous/Permian granite.

The Isles of Scilly form one of six major outcrops of granite exposed in Devon and Cornwall. It is considered that they are connected at depth and form part of a major set of intrusions, the Cornubian Batholith. The batholith extends some 250 km from Dartmoor in the east to the Isles of Scilly in the west. Other outcrops include Bodmin Moor, St Austell Moor, the Carnmenellis Plateau and Land's End.

Table 1. *Geographical details of St Martin's and St Agnes*

Island	Population		Land area	Coastline	Land < 10 m elevation	Maximum elevation
	Winter	Summer	(km^2)	(km)	(%)	(m AOD)
St Martin's	100	300	2.2	15	25	48
St Agnes	75	250	1.5	18	45	34

The Isles of Scilly pluton is comprised of two separate intrusions, an outer ring of coarse-grained granite pre-dating an inner core of fine-grained granite (Barrow 1906). St Martin's lies entirely in the coarse-grained granite region and St Agnes contains both coarse and fine-grained granites. The coarse-grained granite on Scilly has been dated at 290 Ma (Chen *et al.* 1993).

Deep weathering of the granite, probably initiated in late Tertiary times, was influenced by NW–SE trending structural features (Osman 1928). Fissures in the granite are substantial to depths of 50 to 100 m, and the granite may be classified as a minor aquifer.

During peri-glacial conditions in the Pleistocene the top of the granite was exposed to extreme weathering. The Head consists of frost shattered angular granitic fragments in a sandy matrix. Solifluction features are widespread throughout the Head, but also in many instances quartz veins can be traced through the granite and continue through the Head indicating *in situ* weathering.

The glacial deposits can be divided into glacial diamictite and glacial outwash gravel, derived from the incursion of the Irish Sea ice (Mitchell & Orme 1967). The diamictites only crop out at a few exposures on the north of Bryher, Tresco and St Martin's. They consist of red, brown and grey sandstones and flints in a clayey matrix. Scattered gravels, marking the outwash deposits, are also only present in some northern extremities of these islands. Some of the interglacial peat deposits on Scilly were radiocarbon dated, using the organic material, by Scourse (1991) with ages between 18 000 and 25 000 years.

Blown sand, up to a few metres in thickness, dominates much of the low-lying surface of the islands. The sands are coarse and rounded, white in colour and composed mainly of quartz and feldspars.

Development of water resources on St Martin's and St Agnes

Up until about 1960, water was drawn by bucket and rope or by hand pump from shallow hand dug wells founded in the blown sand deposits or the Head. There are ten dug wells on St Martin's, of which four are still in use, and eight on St Agnes, five of which are currently used. In the 1960s a few deeper wells were bored into the granite on each island. These were fitted with hand pumps or surface pumps with petrol driven generators. Abstraction of deep groundwater was therefore limited by the suction lift of the pumps.

Between 1985 and 1987, mains electricity was first introduced to the islands. This led to an increase in the standard of service provided to the holiday industry and a corresponding increase in water demand. Mains electricity also led to the introduction of the submersible pump and its use in deeper wells. Since the mid-1980s many wells were constructed. These were often located in a rather haphazard fashion and with no regard to water resource issues such as the sustainability of the quantity and quality of supplies. There are presently 32 bored wells on St Martin's, of which five are abandoned, and 19 on St Agnes, four of which are abandoned. The reasons for abandonment include low yields and poor water quality due to either saline intrusion or high bacteria levels.

Wells on St Martin's and St Agnes have relatively low yields and the pumps rarely have flow control valves. It is common practice to install automatic electrical cut-out and cut-in circuits on the submersible pumps. When the pump is running, the drawdown increases toward the level of the pump and the well bore storage is emptied. A low-level probe is installed just above the submersible pump and when the drawdown reaches this probe the electrical supply is cut off. The water in the well bore recovers to the position of a high level cut-in probe and the electrical supply is restored so that pumping continues. The well yield is therefore achieved by repeated cycles of pumping and recovery. A typical well construction is illustrated in Fig. 2. Many of the wells and wellheads, however, are far more basic than that illustrated in Fig. 2. Dug wells abstract groundwater exclusively from the drift deposits whilst bored wells are drilled where the water table is present in the granite.

Rainwater collection from roofs has always been an integral part of the water resources on these islands. Rain intercepted by the roof of a dwelling is fed via guttering into a storage tank. From there it is pumped into a header tank in the roof space and gravity fed into the household pipework. In a few cases ultra-violet filters are used to treat the water but generally any form of treatment is rare. The capacity of rainwater tanks on St Martin's and St Agnes ranges from 0.15 m^3 to 80 m^3, with a typical tank holding around 10 m^3 of rainwater.

Commonly, households rely on rainwater storage for all domestic water supplies during the winter months and top up their storage tanks with groundwater abstracted from wells during the summer. Apart from the tourist accommodation facilities, whose populations are purely seasonal and mainly groundwater reliant, approximately equal volumes of rainwater and groundwater are utilized throughout the year. A few properties are totally reliant

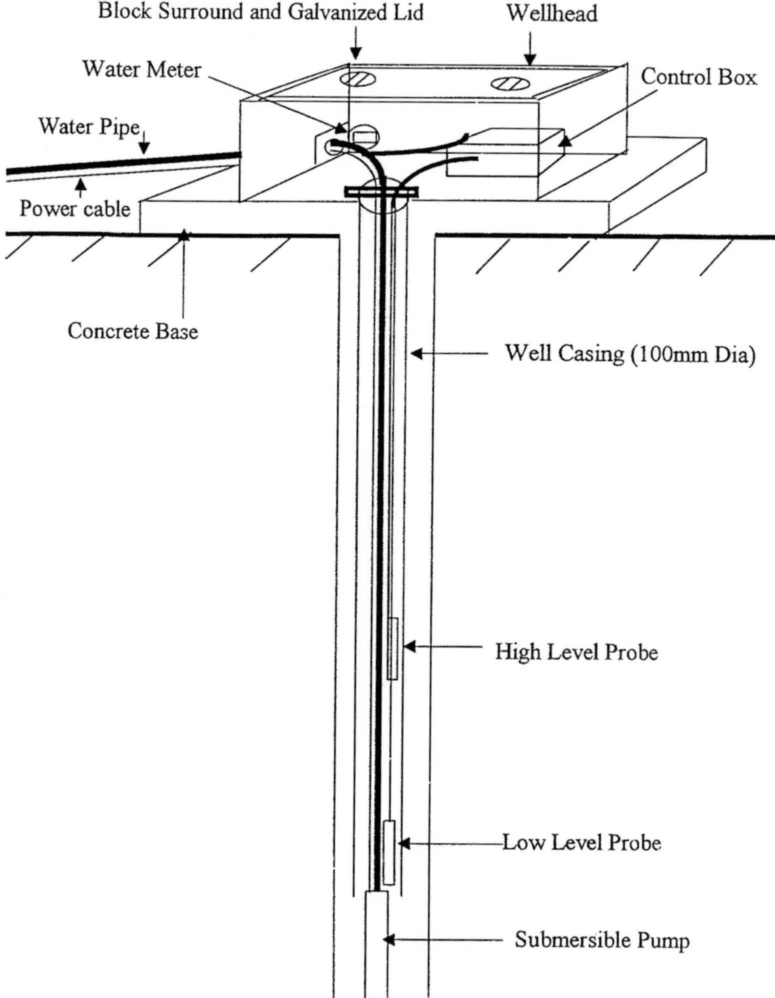

Fig. 2. Typical bored well construction. Not to scale.

on rainwater and a few rely totally on groundwater.

In all domestic residences, wastewater is disposed of to septic tanks with subsequent discharge of effluent to soakaways. This creates a potential conflict in the use of groundwater as a supply of drinking water and as a receptor of wastewater, especially in the limited confines available to these small island communities. On each island, the public conveniences located at the main quaysides are flushed with seawater pumped into header tanks at high tide.

Some properties share wells and some small distribution networks exist based on one or more wells. In some instances each property pumps from a shared well on particular days of the week, by informal agreement. Septic tanks can also serve more than one property.

Previous studies

Previous water supply studies have concentrated primarily on the island of St Mary's. A succession of consultant's reports has guided the development of water resources on this island during the past sixty years (Balfour & Son 1939; Hickling 1939; Lapworth & Partners 1967; Binnie & Partners 1971; SWWA 1975, 1976, 1984; Young 1991). The University of Birmingham and University College London have collected hydrogeological data through a series of MSc projects during the 1970s and 1980s.

Bryher and St Agnes were the specific subjects of hydrogeological studies by University College London through MSc projects by D. Shaw and D. Morgan in 1983. At around the same time the water resources of St Agnes were assessed and potential problems identified on behalf of the Isles of Scilly Council (Ashworth 1983).

In 1996, water supply problems became evident on St Martin's. A group of residents concerned about the sustainability of their groundwater supplies contacted the National Groundwater Centre of the Environment Agency of England and Wales for advice. Though the Environment Agency has no jurisdiction for water supply on the Isles of Scilly, the request instigated a desk study on the water resources of St Martin's (Fletcher 1996). The report identified the following issues for consideration: sustainable development of resources; potential for saline intrusion; collection, storage and collation of data; groundwater modelling.

Data collection

On both islands, general information derived from local knowledge was obtained through the use of a questionnaire. This provided background information prior to fieldwork during which detailed information was sought. Wells were located and water levels monitored. Hydraulic tests were conducted on some wells but the tests were limited by the hydraulic characteristics of the wells.

The questionnaire

The questionnaire that was circulated to each property on the islands was in five parts. Part 1 provided details of the property and the number of winter and summer residents. Part 2 enquired into the amount of water used by each property, what it was used for (domestic, agricultural or other uses) and where it comes from. Part 3 was on water supply and contained questions on rainwater storage and wells. The disposal of wastewater was the subject of Part 4. Finally, Part 5 allowed the respondent to add any comments that may be of interest to the study. Non-returned questionnaires were followed up during the fieldwork phase so that 100% coverage was achieved.

Table 2. *Numbers of water supply and disposal systems*

Island	Number of wells	Rainwater tanks	Septic tanks
St Martin's	42	26	42
St Agnes	27	25	38

Evaluation of the returned questionnaires enabled all wells, septic tanks and rainwater storage tanks to be located and their details to be recorded (Table 2 & Fig. 3). It allowed the small shared networks and the supply to each household to be identified and the routes for wastewater to be determined.

Monitoring

Once each of the wellheads was located and reduced to Ordnance Datum by levelling with respect to local bench marks, a monitoring programme was initiated. Water level dip meters were left with a nominated resident and a communal dip meter lodged at the Post Office on each island. Groundwater levels were measured in selected wells on a weekly basis by the well owners. The results show that water level variations in response to rainfall are large. The rest level of some wells fluctuated up to 10 m between winter and summer and a 5 m drop in phreatic level was noted in the granite of St Martin's in response to a dry February in 1998 (Fig. 4).

Figure 5 shows the phreatic level contours recorded on one day in September 1997 and September 1998 for St Martin's and St Agnes, respectively. Care was taken to ensure that all of the wells had fully recovered from pumping so that the measurements related to rest water levels. However, it was noticed that tidal influences affected the levels in some of the wells nearest to the coast.

Well tests

A programme of hydraulic testing was implemented on selected wells. The wells chosen were those that had a ready means of storing pumped water; it was felt to be imprudent to be seen to be wasting water as part of a water resources assessment and at a time when water was in short supply.

No observation wells were present within a reasonable distance of the pumped wells, so only single well tests could be conducted. A particular problem experienced with the pumping tests was the fact that, due to the lack of flow control valves, the wells cannot be pumped continuously without the low-level probe acting as a cut-out.

(a)

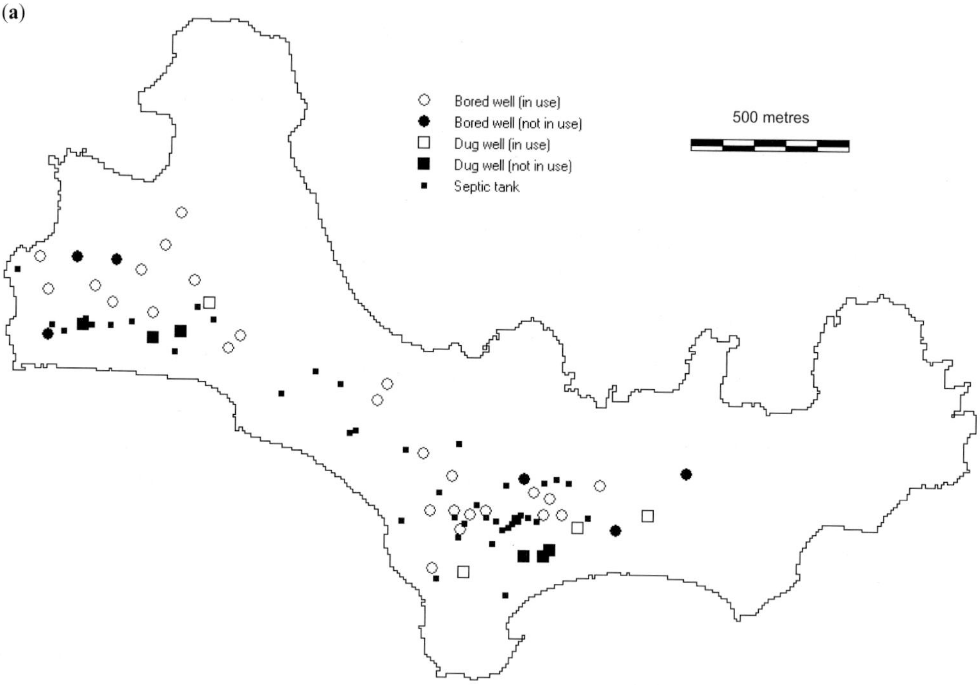

(b)

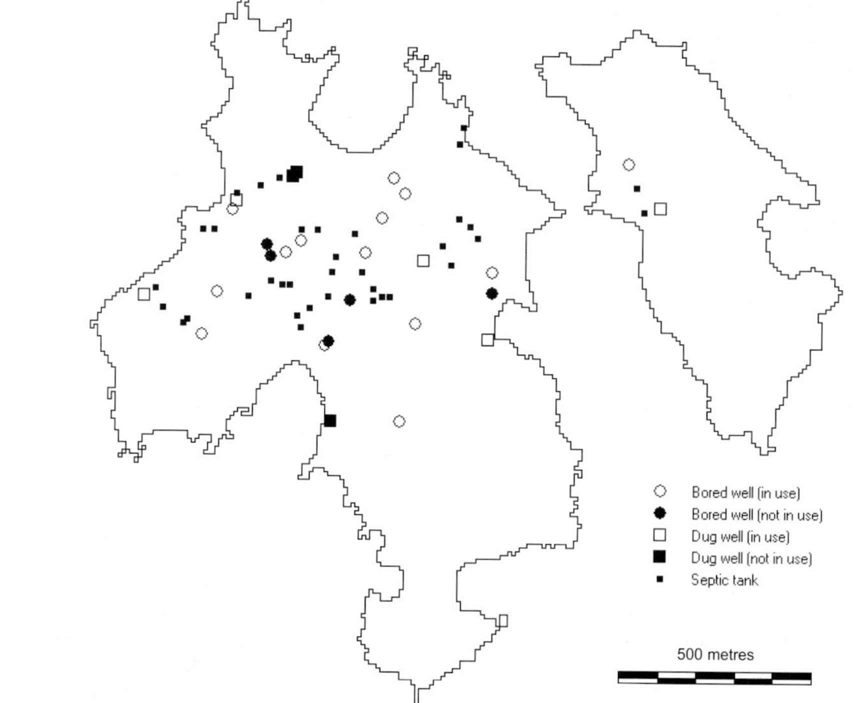

Fig. 3. Locations of dug wells, bored wells and septic tanks on (**a**) St Martin's and (**b**) St Agnes.

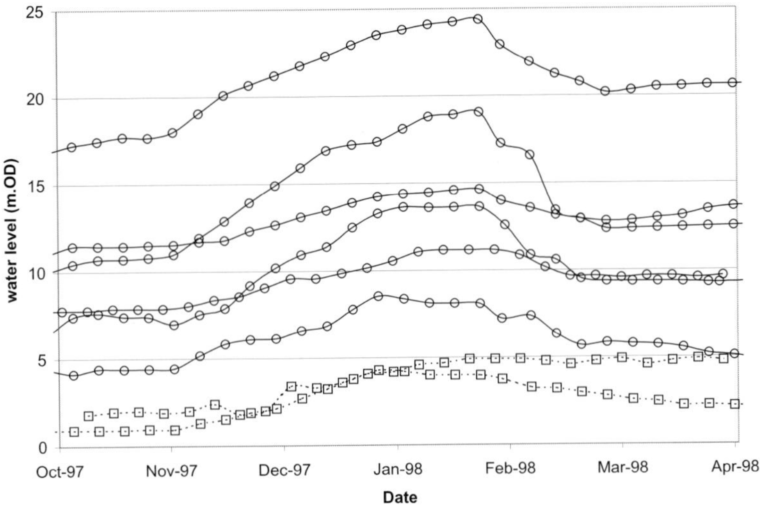

Fig. 4. Well monitoring on St Martins during the winter of 1997–98. The solid lines and circles are bored wells in granite and the broken lines and squares are dug wells in the superficial deposits.

The wells were pumped to this maximum drawdown and recovery was monitored. For the analysis, the recovery was treated as a rising head test using the Hvorslev analysis and a geometrical shape factor to obtain an estimate of the hydraulic conductivity. Theis (log-log) and Cooper–Jacob (semi-log) analyses were also conducted on the time-drawdown data, and applied to the well radius, but were felt to be unreliable due to the short discharge times achieved and lack of observation wells. Due to the difficulty in maintaining a constant discharge, the maximum sustainable yield was estimated by calculating the volumetric inflow to each well during the first 1 m of recovery. This flow rate equates to the steady discharge if the water level in the well was maintained near the low-level probe situated just above the pump.

Figure 6 shows the drawdown and recovery for one of the wells on St Martin's. The well was pumped at a constant rate of 10 litres per minute (14 m^3/day). Figure 7 shows the recovery and relative drawdown for one of the wells on St Agnes. In this case the water column was removed in just three minutes, pumping at a rate of 10 litres per minute. The results are summarized in Table 3. The apparent transmissivity was estimated as the product of hydraulic conductivity and the height of the water column in the well when at rest; the aquifer is assumed to extend to the base of the well. The transmissivity estimates agreed fairly well (within an order of magnitude) with the few estimates available from log-log and semi-log analyses, despite the tenuous nature of the estimates. The granite of St Agnes appears to be significantly less permeable than that of St Martin's, though there is insufficient evidence to say whether this is due to differences between the coarse- and fine-grained granites.

Water quality

Water samples collected from most of the wells and previous analyses, held privately, were collated. The groundwater can be described as aggressive, of low alkalinity and undersaturated with respect to many mineral phases (Banks *et al.* 1998). The major ion chemistry is dominated by sodium chloride derived from sea spray and reflects the exposed aspect of the islands. Additionally, agriculture and sewage disposal have left their impact in terms of elevated nitrate and potassium levels and the local presence of faecal coliform bacteria.

Geographical Information System

The results of the questionnaires were compiled on a database and linked to the geographical locations of wells and septic tanks through a geographical information system (GIS) (Fig. 3) using the Idrisi software (Eastman 1997). Attribute files within the GIS were used to provide a database of information. Table 4 illustrates the structure of

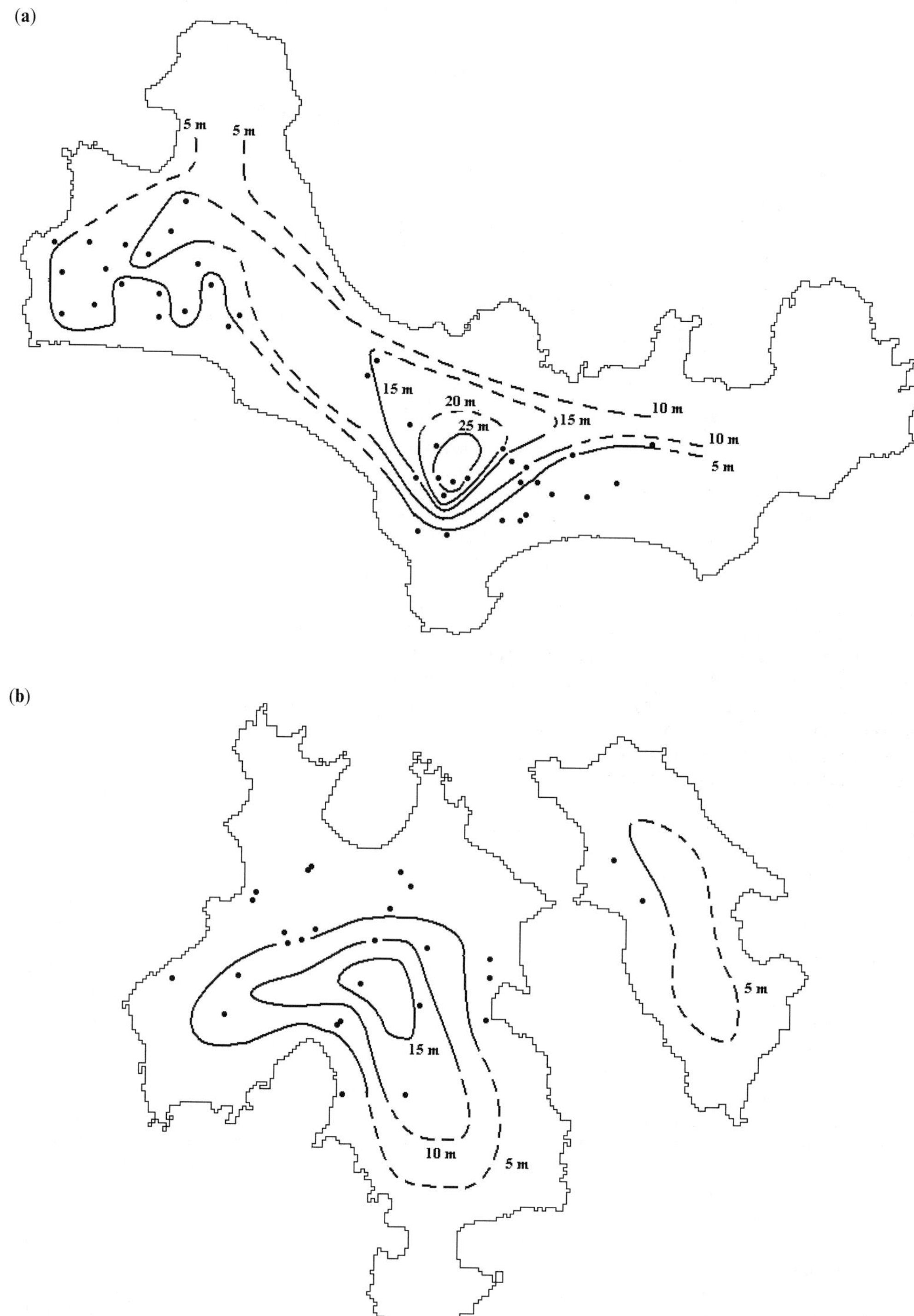

Fig. 5. Inferred groundwater level contours based on measured well levels when at rest. (**a**) St Martin's – 19 September 1997, (**b**) St Agnes – 15 September 1998.

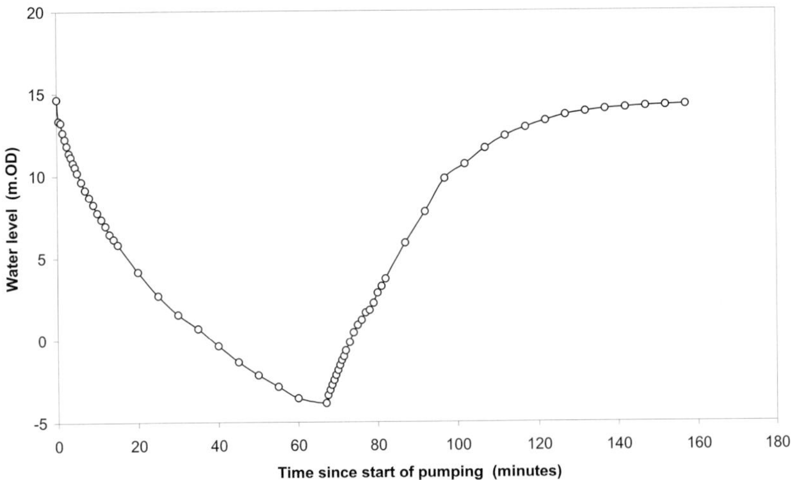

Fig. 6. Time-drawdown response for well #7 on St Martin's. The well was pumped until the low-level probe cut in, when the drawdown reached 18.5 m after 67 minutes. The well was then allowed to recover fully.

the GIS database and shows the linkages used for cross-referencing between the fields.

The GIS was also used to hold a digital elevation model (DEM) of the islands, using spot heights on a 10 m grid spacing (O.S. 1992*a*, *b*). St Agnes consists of two islands: the main island of St Agnes and The Gugh, lying to its east. St Agnes and The Gugh are separated at high tide but connected by a substantial sand bar at low tide. For most instances they can be treated as one island.

The raster DEM (Fig. 8) was pared to encompass only the areas of interest, defined by the main islands under study. Minor islands and rocky islets were then deleted from the DEM. The DEM was then translated to a Boolean image containing values of one in cells containing land and zero for cells over sea. This image provides a mask that was used to filter regions outside of the areas of interest from any calculations made using the GIS. Any raster image can be filtered through multiplication by the Boolean mask image. The DEM was also used to generate polygon vector files representing the coastlines corresponding to the mean high water and mean sea level (Fig. 8).

The well locations were overlain on a blank raster image and a distance operation was run to calculate distances from wells. This raster image was then translated into a Boolean image by setting a value of one to all values less than or equal to 50 m and a zero to all those greater than 50 m. The resulting image was then converted to a polygon vector file representing 50 m source protection zones (Fig. 9) and violations of these zones by septic tanks were recorded.

Numerical modelling

Flow modelling

A distributed numerical groundwater flow model was constructed for each island. These were based on horizontal flow in unconfined aquifers receiving areal recharge from rainfall infiltration. The finite difference code MODFLOW (McDonald & Harbaugh 1984) was used for the flow modelling.

The GIS database utilized a 10 m regular grid spacing for raster format data. To provide consistency with the GIS and allow easy transfer and integration of data, a regular, compatible, 10 m grid and block-centred finite difference scheme was used for the numerical model. This required a domain of 331 columns and 223 rows for St Martin's, and 236 columns and 206 rows for St Agnes. The GIS mask image was used to assign inactive cells to regions covered by the sea.

A constant head boundary was applied to the coastline for each island, defined by the mean high water level as a polygon vector file determined from the GIS. The mean high water was felt to be more representative of groundwater levels near the coast than the mean sea level, as a seepage face tends to be established along beaches and cliffs. The base of the aquifer was set at a uniform level of 20 m below sea level.

(a)

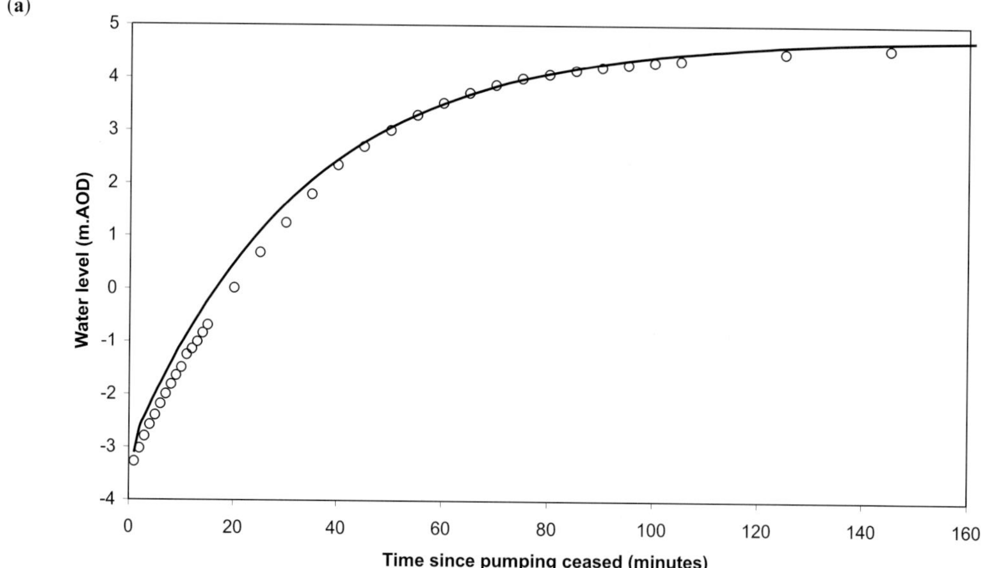

(b)

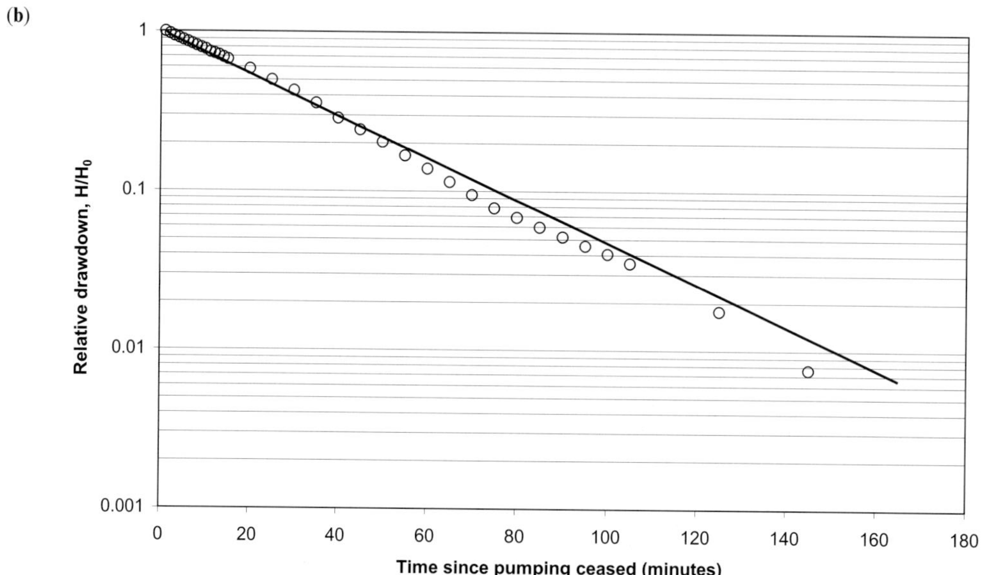

Fig. 7. Recovery data for well # 19 on St Agnes. The circles refer to the field data and the solid line is the response calculated for a hydraulic conductivity, k, of 0.009 m per day, the value obtained from the rising head test analysis. (**a**) Time-drawdown response during recovery, (**b**) Relative drawdown for rising head test analysis.

For simplicity, the models were only run under steady state conditions. Average annual rainfall is around 800 mm a^{-1}. A uniform recharge rate of 0.001 m/day (365 mm a^{-1}) was chosen, as this value is consistent with recharge values used in previous studies on Scilly (e.g. Burgess *et al.* 1976). The conceptual flow system modelled was therefore one of uniform areal recharge leading to groundwater flow outward from the centre of the islands, increasing in magnitude as recharge is added to the flow which eventually discharges to the sea along the coast.

First, a uniform aquifer with the hydraulic properties representative of the granite alone

Table 3. *Hydraulic characteristics obtained from pumping tests on selected wells*

Island and well ref. no.	Ground level (m.OD)	Level of base of well (m.OD)	Rest water level (m.OD)	Well diameter (m)	Hydraulic conductivity (m/d)	Apparent transmissivity (m^2/d)	Specific capacity (m^2/d)	Maximum sustainable yield (m^3/d)
St Martin's								
#7	+35.9	−9.1	+14.7	0.13	0.045	0.68	0.66	11.5
#19	+36.0	−4.0	+17.8	0.13	0.038	0.76	0.43	5.0
#25	+19.6	−10.4	+3.7	0.13	0.47	6.5	8.1	25.0
#26	+31.0	−11.6	+24.8	0.13	0.078	2.7	1.1	3.1
St Agnes								
#4	+18.0	−25.0	+6.1	0.10	0.013	0.42	0.40	11.6
#5	+17.6	−14.6	+6.0	0.10	0.022	0.25	0.28	2.7
#19	+15.3	−39.7	+4.9	0.12	0.009	0.42	0.39	2.8

Table 4. *Some of the GIS database files. Bold italics indicate linkages used to cross-reference the data*

GIS file:	DEM	Wells	Well data	Properties	Waste water	Septic tank	SPZ
File type:	Raster	Point vector	Attribute	Attribute	Attribute	Point vector	Polygon vector
Data held:	Spot heights at 10 m grid spacing	***Well ID***	***Well ID***	***Property ID***	***Septic tank ID***	***Septic tank ID***	***Well ID***
		co-ordinates	***Properties served***	Name and address	***Properties served***	co-ordinates	50 m radius
			Pump (surface or submersed)	No. of residents (summer and winter)	Capacity of septic tank		400 day travel time
			Type (dug well or bored)	Details of rainwater collection system	Age of septic tank		Catchment zone
			Depth and diameter	Details of wastewater disposal system	How often emptied		
			Date drilled	***Septic tank ID***	When last emptied		
			Reduced level				

was used. The granite was allocated a hydraulic conductivity of 0.04 m/day for St Martin's and 0.02 m/day for St Agnes, broadly reflecting the results of the well tests (see Table 3). The resulting groundwater head contours (Fig. 10), in both cases of St Martin's and St Agnes, were predicted to be above ground level in many places, without field evidence of springs, and did not accurately reflect the groundwater flow patterns determined from the well monitoring (see Fig. 5). It was considered that the impact of the higher permeability blown sand and head deposits was being neglected and that these, lying above the granite, were acting as highly transmissive drains regulating groundwater levels on the lower parts of the islands (Fig. 11).

It would be possible to model this arrangement by utilizing a two-layer model with the lower permeability granite comprising the lower layer and the higher permeability deposits the upper layer, which would be dry at high elevations. However, it was felt that this extra complexity was not warranted and the islands were modelled as single-layer heterogeneous systems. It was assumed that all land below 10 m above sea level (see Table 1) was comprised of high permeability blown sand or Head and that all land above 10 m elevation was lower permeability granite. By using this assumption, the zones with different hydraulic properties were automatically generated from the DEM held on the GIS. No data were available on the permeability of the drift deposits but the purpose of the modelling was to refine the hydrogeological understanding of the system rather than to expect an accurate simulation of flows. The sand/Head was arbitrarily assigned a hydraulic conductivity of 4 m per day for both islands. This time, a good correlation was found between modelled and measured groundwater levels, in terms of representing the regional flow patterns. The results are shown in Fig. 12, which can be compared with the field data shown in Fig. 5.

For the third run of each model, the wells were included and every well (including all abandoned wells) was set at a discharge rate of 5 m^3 per day. Whilst some wells are capable of pumping more than this (Table 3), it was assumed in the models that pumping is continuous and thus represented a gross overestimate of actual abstractions, which generally only take place in summer.

Comparing pre-pumping groundwater heads with post-pumping heads allows the drawdowns

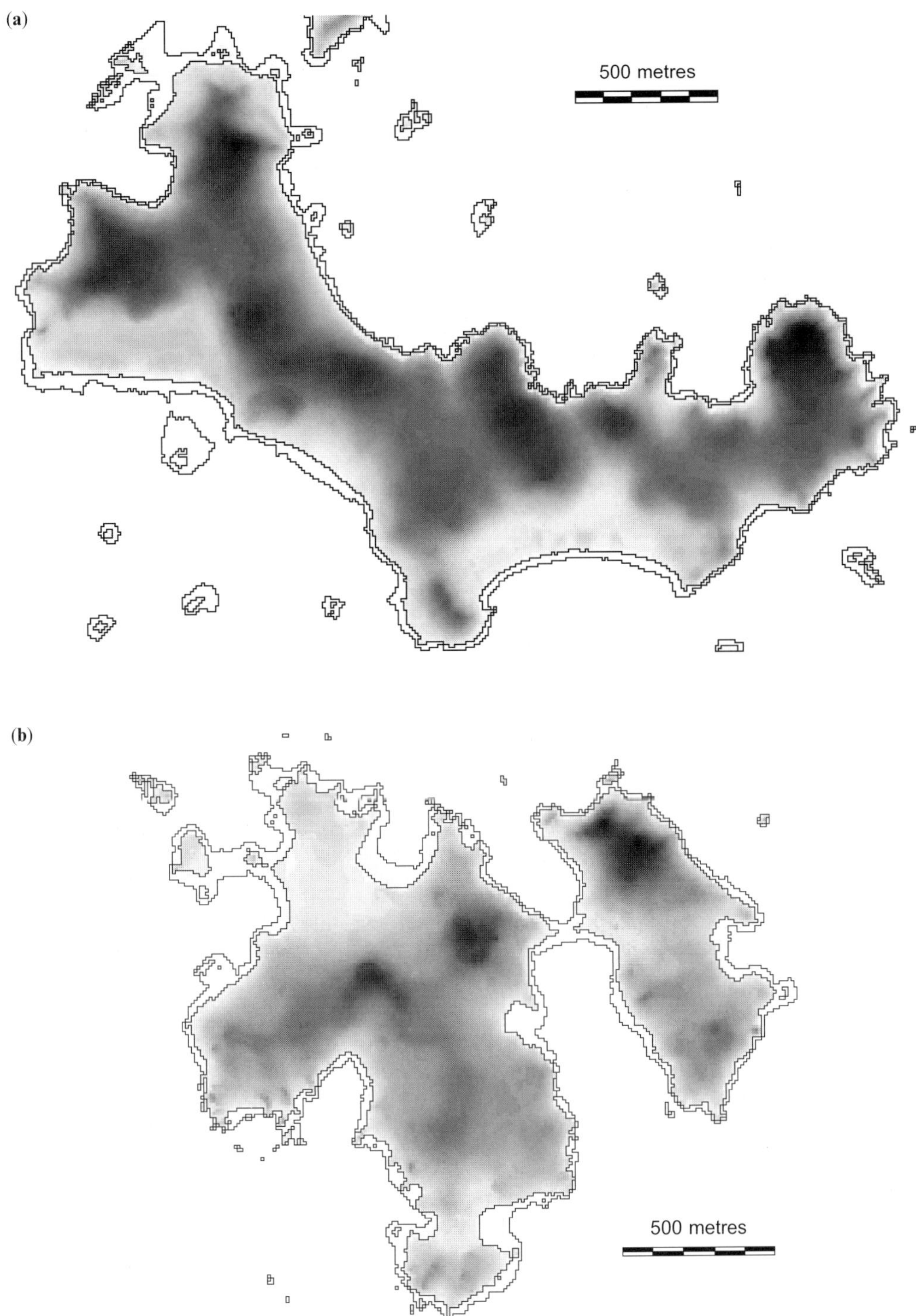

Fig. 8. Digital elevation models of St Martin's and St Agnes. The darker shading corresponds with the higher elevations. The coastlines shown correspond to the Mean High Water and the Mean Sea Level. (**a**) St Martin's, (**b**) St Agnes.

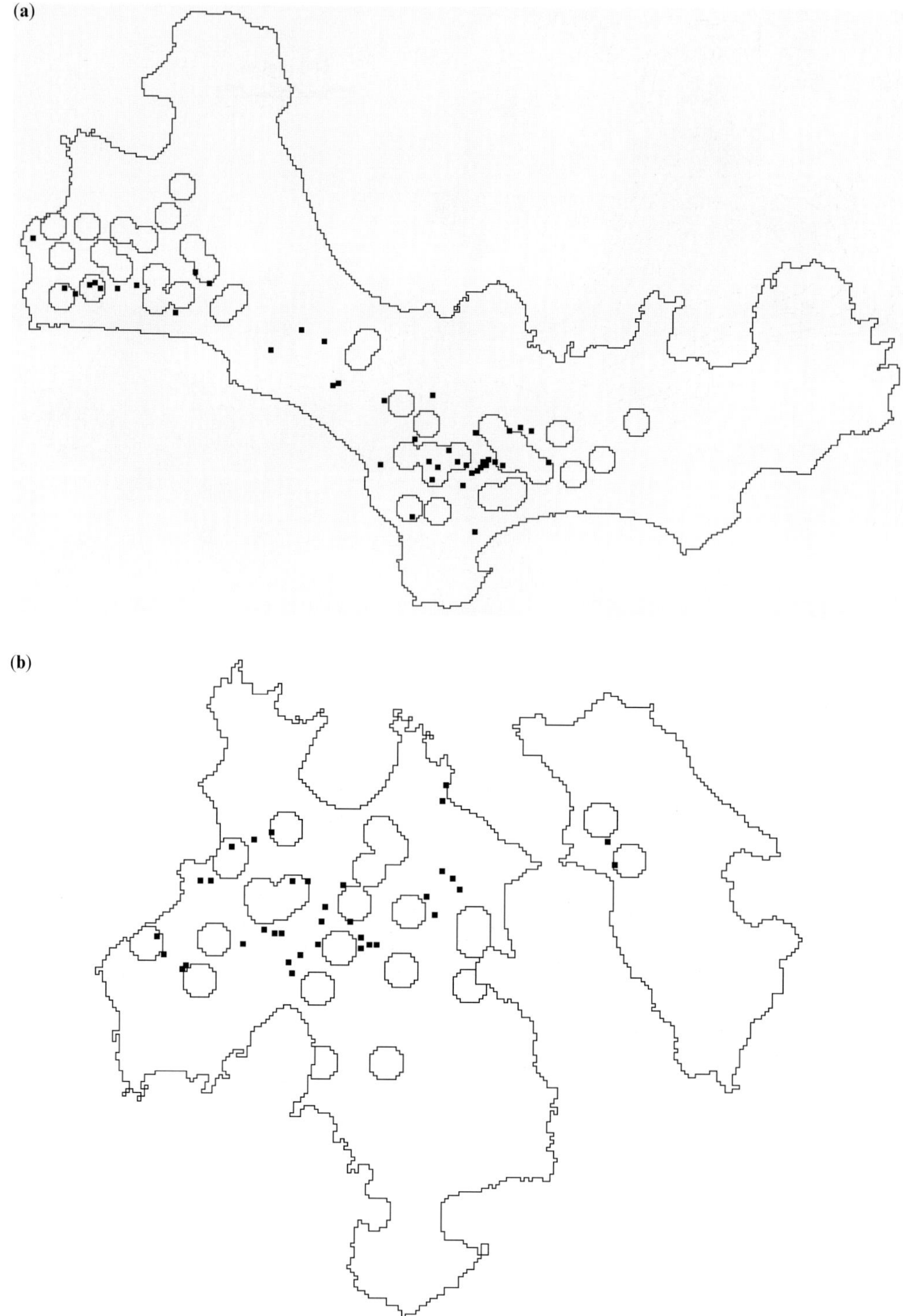

Fig. 9. 50 m source protection zones around wells and the locations of septic tanks. Results derived from the GIS. (**a**) St Martin's, (**b**) St Agnes.

(a)

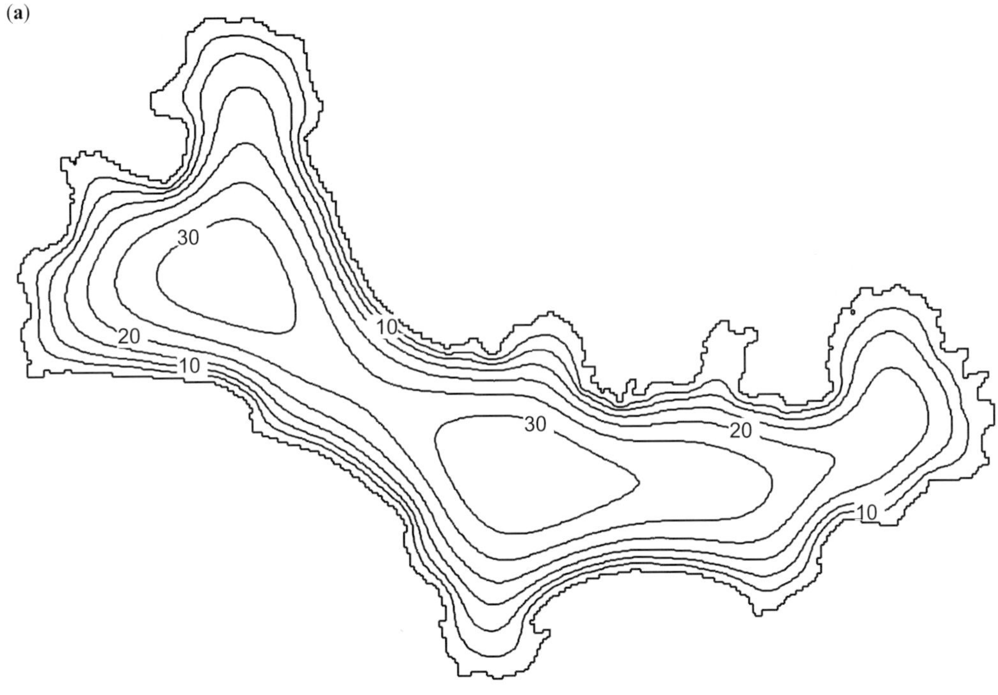

(b)

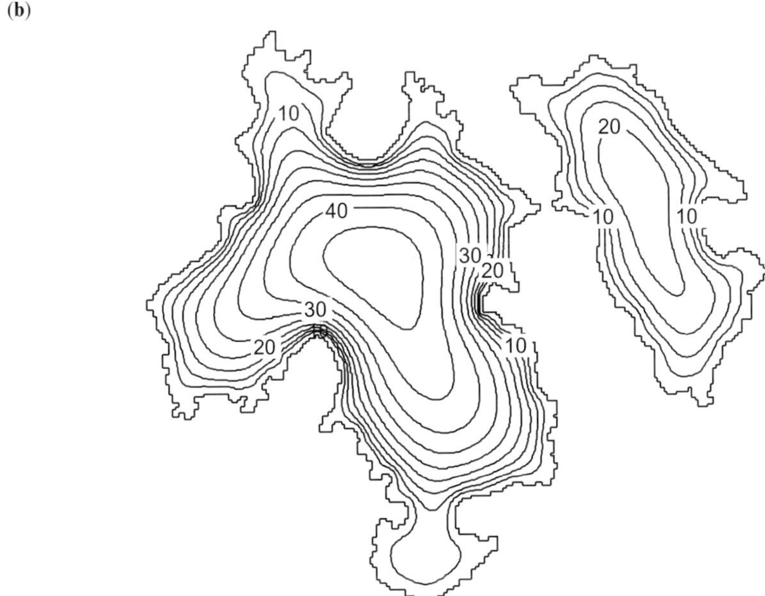

Fig. 10. Phreatic head contours resulting from a numerical simulation of groundwater flows in a homogeneous aquifer. Contours are shown in 5 m intervals. (**a**) St Martin's, (**b**) St Agnes.

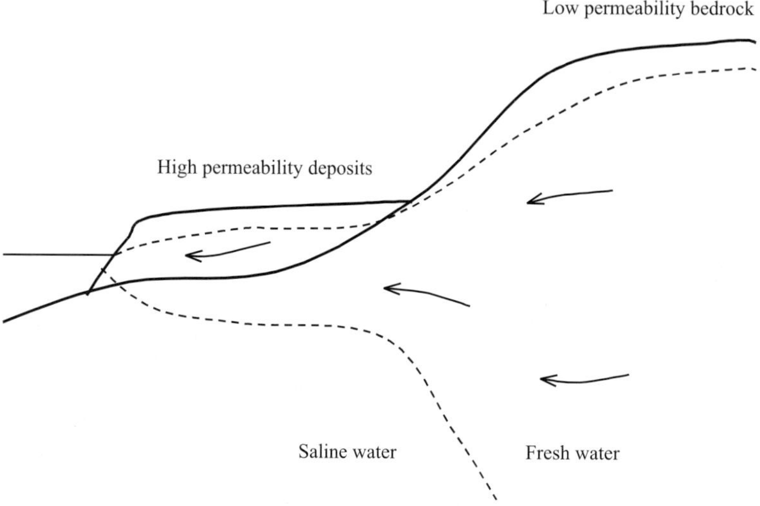

Fig. 11. Cross-section through a conceptual model of high permeability blown sand or Head deposits overlying a low permeability granite aquifer. Not to scale.

to be determined (Fig. 13). These may depress the groundwater levels by 2 to 3 m over substantial areas but in view of the natural variations in response to recharge (Fig. 4) such groundwater abstractions are felt to be sustainable. The quantities of water abstracted in the models, 210 m^3 per day for St Martin's and 135 m^3 per day for St Agnes, represent slightly less than 10% of the throughflow for each island, based on the product of the modelled recharge and land areas. Actual groundwater abstraction is estimated to be around 20 m^3 per day for St Martin's and 15 m^3 per day for St Agnes during the summer months, approximately 1% of throughflow and around 10% of the modelled steady abstraction.

The groundwater flow models provide a framework for analysis and interpretation. They may be refined and improved as more hydrogeological data become available and may be used to test assumptions and the relative impact of present and future actions.

Pathline modelling

To determine the areas of influence of the well abstractions, a particle tracking routine was used. This was achieved using the code MODPATH (Pollock 1989) which is designed to interact with the output from the groundwater flow model MODFLOW. In reality, groundwater flows through the granite in discrete fissures and actual pathways are likely to be tortuous and complex. The flow model applies a continuum approach to the aquifer and so modelled flow paths can only be regarded as broad estimates of possible flow paths. The particle tracking routine was used to define inner, outer and catchment source protection zones around each well based on 50 day, 400 day and total capture zones (NRA 1992).

The granite was assigned a porosity of 1% and the sand/Head a porosity of 25% to allow groundwater velocities to be calculated. Both values were chosen arbitrarily in the absence of supporting data.

For each well, virtual particles were placed around the grid cell, containing the well, on the four vertical faces and the water table surface. The code was then run as a backward-tracking routine to trace the paths of the particles and delineate the full catchment zone for each well. The results of the particle tracking were also used to delineate the capture zones with a time limit of 50 days and 400 days to test the limits of the inner and outer source protection zones, respectively.

The particle paths were first formatted as line vectors on the GIS and then converted to raster files by identifying cells within which the line vectors are present. The raster image was then converted to a polygon vector file encompassing the zones (Fig. 14).

In every case the 50 m criterion (Fig. 9) was found to be much larger than the 50 day travel time distance. The 400 day travel time distance in many cases was also found to be less than the 50 m criterion.

(a)

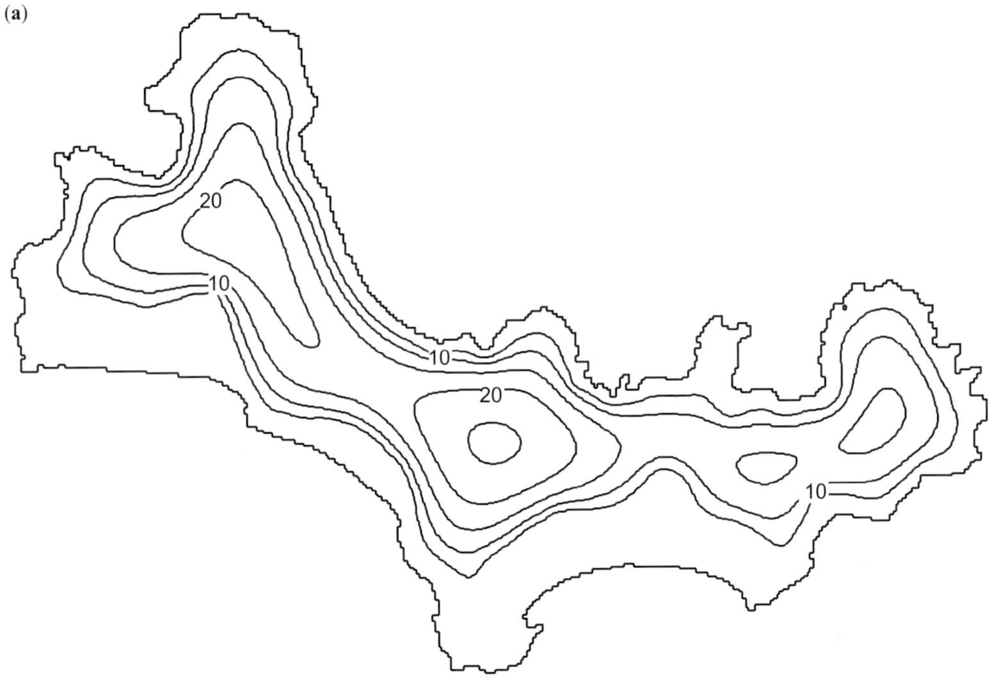

(b)

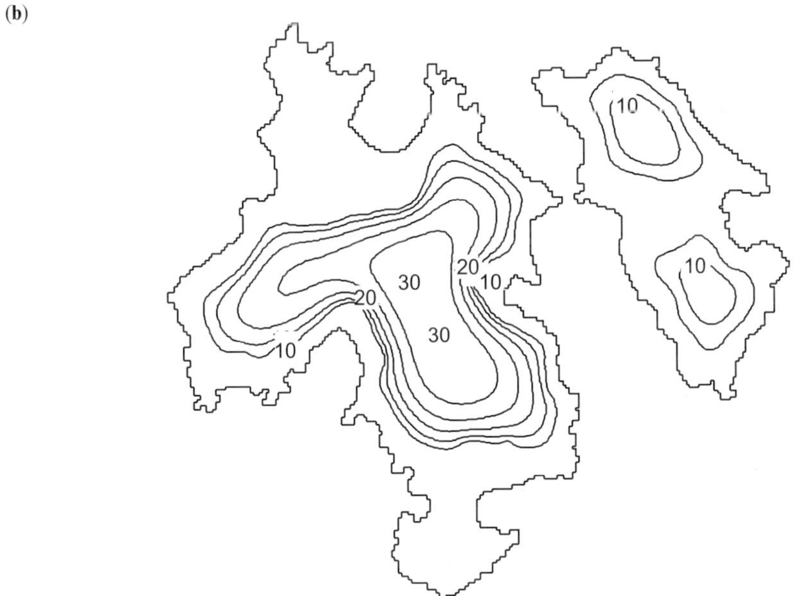

Fig. 12. Phreatic head contours resulting from a numerical simulation of groundwater flows in a heterogeneous aquifer. Contours are shown in 5 m intervals. (**a**) St Martin's, (**b**) St Agnes.

(**a**)

(**b**)

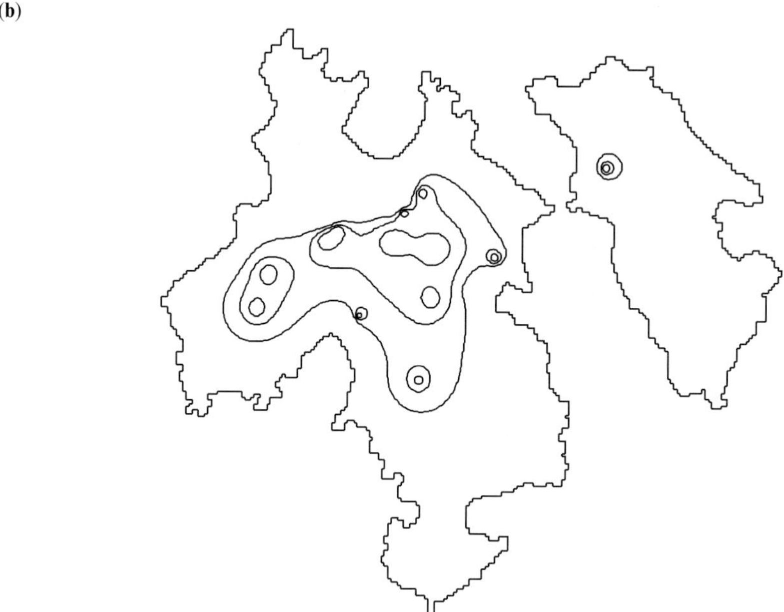

Fig. 13. Contours showing drawdowns of 1 m, 2 m and 3 m. The drawdowns shown reflect the steady-state conditions when all wells are pumping 5 m^3 per day each. Drawdowns are small in wells located in the more permeable Head and blown sand deposits. (**a**) St Martin's, (**b**) St Agnes.

Fig. 14. Well catchment zones as derived from the results of numerical particle tracking and the positions of septic tanks. (**a**) St Martin's, (**b**) St Agnes.

One further particle tracking simulation was run for each island. This time a single virtual particle was placed in each septic tank. The routine was run in a forward-tracking direction to find each particle's destination and, in some cases, this was found to be a well. Some wells, which had already been taken out of action due to the presence of faecal coliform bacteria, were highlighted in this manner.

Saline intrusion

Being small islands, saline intrusion of seawater into the aquifer is a potential problem. This can result from wells being positioned too close to the coast, with their bases too deep or being pumped at too high a rate.

It is possible to numerically model the effects of saline intrusion but this would require a fully three-dimensional, variable density fluid, advection–dispersion model, the complexity of which was considered to be outside the scope of the study. Instead, a simple analytical formula was used to estimate the maximum yield that can be safely abstracted from each well, based on the well depth, rest water level and aquifer hydraulic conductivity. When a well is pumped the pressure on the saline interface is reduced. This creates a rise in the saline interface (upconing) which mirrors the drop in groundwater head in the aquifer. According to Bear & Dagan (1964), the level of upconing can be described by the following equations:

$$B = \frac{GQ}{2\pi kL}$$

where B, is the height of upconing of the saline interface induced by pumping; Q, is the rate of abstraction; k, is the aquifer hydraulic conductivity; L, is the distance between the saline interface, without pumping, and the base of the well (Fig. 15); and

$$G = \frac{\rho_f}{\rho_s - \rho_f}$$

where ρ_f, is the density of freshwater and ρ_s, is the density of seawater. Taking the density of freshwater to be 1000 kg/m^3 and the density of seawater to be 1025 kg/m^3, provides a value for G of 40. The position of the saline interface was taken to be at an elevation below sea-level equal to 40 times the height of the rest water above sea-level, in accordance with above relationship between densities and the Gyben–Herzberg principle. To provide a factor of safety to account for the fact that the saline interface is not a sharp boundary, the height of upconing was set to half the distance from the rest saline interface to the base of the well, $B = 0.5L$. The discharge required to create this condition can therefore be expressed as:

$$Q = \frac{\pi k}{40} L^2 \quad \text{or as} \quad Q = \frac{\pi k}{40}(40Z_1 - Z_2)^2$$

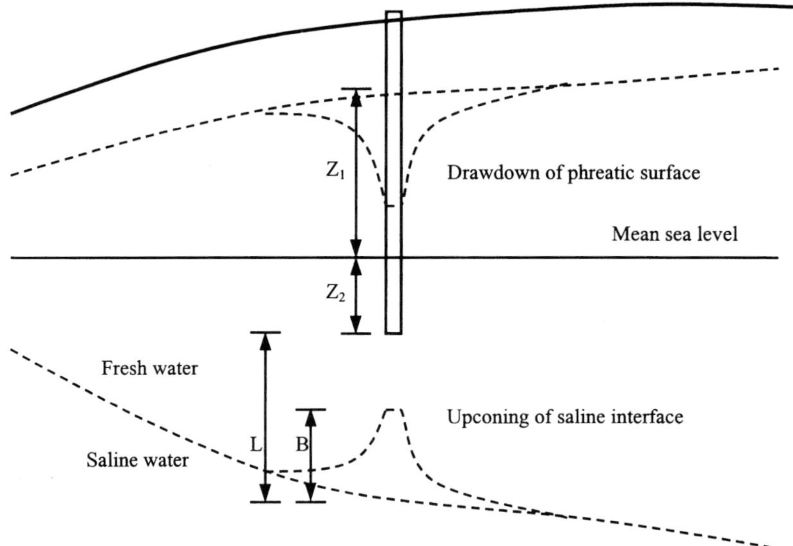

Fig. 15. Cross-section through an aquifer showing upconing of the saline interface in response to pumping of a well. Not to scale.

Table 5. *Summary of results of calculations to determine the vulnerabilities of wells to saline intrusion*

Island	Number of wells in each vulnerability category			
	High	Medium	Low	Insufficient data
St Martin's	9	3	22	8
St Agnes	8	8	3	8

where Z_1, is the rest level of groundwater in the well and Z_2, is the level of the base of the well, both expressed with respect to mean sea level (Fig. 15).

This procedure was used to classify the wells in terms of high, medium or low vulnerability to contamination by saline intrusion. This was done on an arbitrary basis whereby if the maximum safe abstraction rate, Q was less than 20 m^3/day the vulnerability was deemed to be high, between 20 and 100 m^3/day was taken to be a medium vulnerability and more than 100 m^3/day to be a low vulnerability. For a number of wells there were insufficient data to perform the calculation. This was due either to the lack of a dip reading on the well when at rest, or to the lack of a plumb depth measurement because of obstructions by equipment in the well. The results of these calculations are summarized in Table 5.

Water management plan

The purpose of the water resource studies was to provide information that would allow a water management plan (WMP) to be formulated for each island. The WMP described the current state and extent of existing resources, listed a series of actions that were required to maintain and protect the resource and set out a code of practice relating to the management of wells, rainwater collection systems and septic tanks.

A schedule of wells was prepared listing their status, condition and use. The condition of the wellheads was highly variable and ranged from very poor to excellent. Wellheads should be secure and provide protection against the direct ingress of surface water. The well casings should stand proud of the ground surface and the annulus should be tightly sealed. Corroded pipes should be replaced. New wells should have flow meters and sampling taps fitted. The wellheads should be kept clean and tidy.

Rainwater collection should not be used with roofs that contain bituminous materials in their construction. Rainwater tanks should contain a secure lid and an overflow pipe. The delivery pipe should contain a coarse inlet filter and a mechanism for redirecting the first flush of water after a dry spell. There should be provision for access for regular inspection and cleaning. It was recommended that water collected from roofs should pass through an ultra-violet filter before consumption, to counteract the threat of contamination by bird droppings.

The position of new septic tanks should take due regard of the location of wells, and should not be located within 50 m of a well or directly up hydraulic gradient from a well. The effluent should be regularly inspected and the tanks emptied on a regular basis. The spreading of sewage sludge should not take place within 50 m of a well or within the well catchment zone.

It was felt that present abstraction rates are sustainable. The abstraction rates used in the numerical models were overestimated by a factor of about ten with no particular detrimental effects. However, due to the low yields of the wells, an increase in abstraction would require the drilling of new wells rather than greater use of the existing wells. Any new wells would need to be positioned with due regard to the locations of existing septic tanks and areas in which saline intrusion is liable to occur. Whilst further abstraction is possible, groundwater resources will always be vulnerable during dry summers. The fall of the water table could cause the level in wells to drop to a position whereby only limited drawdown can occur, reducing the yield of the wells. This may occur irrespective of whether the groundwater is being exploited for water supplies.

Conclusions

This study provides a case study of the data requirements for the assessment of water resources on small islands. It allowed the water resources on the islands of St Agnes and St Martin's to be assessed and documented, and enabled a water management plan to be formulated.

The numerical flow modelling provided a valuable insight into the hydrogeology of the islands and illustrated that groundwater resources do not presently limit the extent of further development of water resources on the islands. However, the groundwater resources will always be stressed during a particularly dry summer, irrespective of the abstraction rates, if the water table drops significantly in wells limiting the drawdown that can be achieved. The locations of any new wells may be constrained by the locations of existing septic tanks. The pathline modelling provided information on possible source protection zones around wells and highlighted the risks posed by septic tank effluent. The GIS proved to be a useful method of integrating data from the various components of the studies and also as an analytical tool for source protection zone and vulnerability mapping. In particular, the GIS and the distributed numerical models worked together in a complementary fashion.

Integration of the GIS and the numerical models proved to be a suitable and effective technique for collating and organizing the data. The methodology used may be applicable to the water resource management of other small island communities.

Thanks are due to the residents of St Martin's and St Agnes for their hospitality and their help and enthusiasm toward the project; without their input the studies could not have been successful. Ian Hewer, Jason Coomber (for St Martin's), Victoria Cousins and Leslie McCormack (for St Agnes) are thanked for their hard work on their projects, whilst undergraduate students at Camborne School of Mines. The role of the Duchy of Cornwall in providing funds for the work is very gratefully acknowledged.

References

ASHWORTH, J. M. 1983. *Hydrogeology of St Agnes*. Report to the Council of the Isles of Scilly.

BALFOUR & SON. 1939. *Water Supply of the Isles of Scilly, St Mary's Island*. Report to the Council of the Isles of Scilly.

BANKS, D., REIMANN, C. & SKARPHAGEN, H. 1998. The comparative hydrochemistry of two granitic island aquifers: The Isles of Scilly, UK, and the Hvaler Islands, Norway. *Science of the Total Environment*, **209**(2–3), 169–183.

BARROW, G. 1906. The Geology of the Isles of Scilly. *Memoirs of the Geological Survey*, Explanation of sheets 357 and 360, HMSO, London.

BEAR, J. & DAGAN, G. 1964. Moving Interface in Coastal Aquifers. *Proceedings of the American Society of Civil Engineers*, **99**, 193–215.

BINNIE & PARTNERS. 1971. *Report on the Water Resources of St Mary's*. Report to the Council of the Isles of Scilly.

BURGESS, W. G., CLOWES, U. R., LLOYD, J. W. & MARSH, J. M. 1976. Aspects of the Hydrogeology of St Mary's, Scilly Isles. *Proceedings of the Ussher Society*, **3**, 382–389.

CHEN, Y., CLARK, A. H., FARRAR, E., WASTENEYS, H. A. H. P., HODGSON, M. J. & BROMLEY, A. V. 1993. Diachronous and independent histories of plutonism and mineralization in the Cornubian Batholith, southwest England. *Journal of the Geological Society*, **150**, 1183–1191.

EASTMAN, R. J. 1997. *Idrisi for Windows*. Clark Laboratories for Cartographic Technology, Clark University, Ma., USA.

FALKLAND, A. 1991. *Hydrology and Water Resources of Small Islands: A Practical Guide*. Studies and reports in hydrology, 49. UNESCO, Paris.

FLETCHER, S. 1996. *Water Consumption on St Martin's, Isles of Scilly*. National Groundwater Centre, Solihull.

GARRET, P. 1992. Taking the Salt Out of Seawater. *Water Bulletin*, **525**, 14–15.

HICKLING, G. 1939. *Water Supply of St Mary's, Isles of Scilly*. Report to the Council of the Isles of Scilly.

LAPWORTH & PARTNERS. 1967. *Water Resources of St Mary's, Isles of Scilly*. Report to the Council of the Isles of Scilly.

MCDONALD, M. G. & HARBAUGH, A. W. 1984. *A Modular Three-Dimensional Finite-Difference Ground Water Flow Model*. US Geological Survey Open-File Report **83–875**.

MITCHELL, G. F. & ORME, A. R. 1967. The Pleistocene Deposits of the Isles of Scilly. *Quarterly Journal of the Geological Society of London*, **123**, 59–92.

NRA. 1992. *Policy and Practice for the Protection of Groundwater*. National Rivers Authority, Bristol.

O.S. 1992*a*. *Landform Profile digital topographic data for 1:10 000 sheet SV90NE*. Ordnance Survey, Southampton.

——1992*b*. *Landform Profile digital topographic data for 1:10 000 sheet SV91NE*. Ordnance Survey, Southampton.

OSMAN, C. W. 1928. The granites of the Scilly Isles and their relation to the Dartmoor Granite. *Quarterly Journal of the Geological Society of London*, **84**, 258–290.

POLLOCK, D. W. 1989. *Documentation of Computer Programs to Compute and Display Pathlines Using Results from the, US Geological Survey Modular Three-Dimensional Finite-Difference Ground-Water Flow Model.*, US Geological Survey Open-File Report **89–381**.

SCOURSE, J. D. 1991. Late Pleistocene Stratigraphy and Paleobotany of the Isles of Scilly. *Philosophical Transactions of the Royal Society of London*, **B334**, 405–448.

SWWA. 1975. *Report on Water Supplies on St Mary's*. South West Water Authority, Exeter.

——1976. *Report on the Groundwater Resources of St Mary's, Isles of Scilly and Recommendations on Their Management and Further Development*. South West Water Authority, Exeter.

——1984. *Water Supplies in the Isles of Scilly*. South West Water Authority, Exeter.

YOUNG, C. P. 1991. *Final Technical Report on Waste Disposal Options, Isles of Scilly*. Water Research Centre Report **CO 2874**.

Index